Highway Design 2018

Transportation Research Record: Journal of the Transportation Research Board Volume 2672, Issue 39

ISBN: 978-1-5108-8247-8

Printed from e-media with permission by:

Curran Associates, Inc.
57 Morehouse Lane
Red Hook, NY 12571

Some format issues inherent in the e-media version may also appear in this print version.

Published by Sage Publications USA

ISBN (Print) 978-1-5108-8247-8
ISBN 2018 Printed Set (All Issues) 978-1-5108-7735-1

Printed by Curran Associates, Inc. (2019)

Additional copies of this publication are available from:

Curran Associates, Inc.
57 Morehouse Lane
Red Hook, NY 12571 USA
Phone: 845-758-0400
Fax: 845-758-2633
Email: curran@proceedings.com
Web: www.proceedings.com

Volume 2672 Number 39 December 2018

TRR
JOURNAL OF THE TRANSPORTATION RESEARCH BOARD

Contents

Highway Design 2018

Article

2018 K. B. WOODS AWARD: Outstanding Paper in Design and Construction

Articles

TRB
TRANSPORTATION RESEARCH BOARD

CORE SPONSORS

Transportation Departments of the 50 States and the District of Columbia

Federal Government

U.S. Department of Transportation
- Federal Aviation Administration (FAA)
- Federal Highway Administration
- Federal Motor Carrier Safety Administration (FMCSA)
- Federal Railroad Administration (FRA)
- Federal Transit Administration
- National Highway Traffic Safety Administration (NHTSA)
- Office of the Assistant Secretary of Transportation for Research and Technology (OST-R)

U.S. Air Force Civil Engineer Center (AFCEC)
U.S. Coast Guard (USCG)
U.S. Department of Energy
U.S. Department of the Interior (DOI)
U.S. Environmental Protection Agency (EPA)

Nongovernmental Organizations

American Public Transportation Association (APTA)
Association of American Railroads (AAR)
California Air Resources Board

The National Academies of
SCIENCES • ENGINEERING • MEDICINE

The **National Academy of Sciences** was established in 1863 by an Act of Congress, signed by President Lincoln, as a private, nongovernmental institution to advise the nation on issues related to science and technology. Members are elected by their peers for outstanding contributions to research. Dr. Marcia McNutt is president.

The **National Academy of Engineering** was established in 1964 under the charter of the National Academy of Sciences to bring the practices of engineering to advising the nation. Members are elected by their peers for extraordinary contributions to engineering. Dr. C. D. Mote, Jr., is president.

The **National Academy of Medicine** (formerly the Institute of Medicine) was established in 1970 under the charter of the National Academy of Sciences to advise the nation on medical and health issues. Members are elected by their peers for distinguished contributions to medicine and health. Dr. Victor J. Dzau is president.

The three Academies work together as the National Academies of Sciences, Engineering, and Medicine to provide independent, objective analysis and advice to the nation and conduct other activities to solve complex problems and inform public policy decisions. The Academies also encourage education and research, recognize outstanding contributions to knowledge, and increase public understanding in matters of science, engineering, and medicine.

Learn more about the National Academies of Sciences, Engineering, and Medicine at **www.national-academies.org.**

The **Transportation Research Board** is one of seven major programs of the National Academies of Sciences, Engineering, and Medicine. The mission of the Transportation Research Board is to increase the benefits that transportation contributes to society by providing leadership in transportation innovation and progress through research and information exchange, conducted within a setting that is objective, interdisciplinary, and multimodal. The Board's varied activities annually engage about 7,000 engineers, scientists, and other transportation researchers and practitioners from the public and private sectors and academia, all of whom contribute their expertise in the public interest. The program is supported by state transportation departments, federal agencies including the component administrations of the U.S. Department of Transportation, and other organizations and individuals interested in the development of transportation.

Learn more about the Transportation Research Board at **www.TRB.org**.

Article

TRR
JOURNAL OF THE TRANSPORTATION RESEARCH BOARD

Transportation Research Record
2018, Vol. 2672(39) 1–10

DOI: 10.1177/0361198118759005
journals.sagepub.com/home/trr
SAGE

Identification and Calculation of Horizontal Curves for Low-Volume Roadways Using Smartphone Sensors

Jonathan S. Wood[1] and Shaohu Zhang[2]

Abstract

Horizontal curves are a contributing factor to the number of observed roadway crashes. Identifying locations and geometric characteristics of horizontal curves plays a crucial role in crash prediction and prevention. However, most states in the USA face a challenge in maintaining detailed and high-quality roadway inventory databases for low-volume rural roads due to the labor-intensive and time-consuming nature of collecting and maintaining the data. This paper proposes a low-cost mobile road inventory system for two-lane horizontal curves based on off-the-shelf smartphones. The proposed system is capable of accurately detecting horizontal curves by exploiting a K-means machine learning technique. Butterworth low-pass filtering is applied to reduce sensor noise. Extended Kalman filtering is adopted to improve the GPS accuracy. Chord method-based radius computation and superelevation estimation are introduced to achieve accurate and robust results despite the low-frequency GPS and noisy sensor signals obtained from smartphones. This study implements this method using an Android-based smartphone and tests 21 horizontal curves in South Dakota. The results demonstrate that the proposed system achieves high curve identification accuracy as well as high accuracy for calculating curve radius and superelevation.

Two-lane roadways comprise the predominant road type in most countries. There are more than three million miles of two-lane highways in the USA, 90% of which carry traffic volumes of fewer than 2000 vehicles per day (*1*). According to the Fatality Analysis Reporting System (FARS), in 2014 66.4% (29,796 out of 44,858) fatal crashes occurred on two-lane roadways. Also, 20.1% (7656 out of 38,046) of single- and two-vehicle fatal crashes occurred along horizontal curves (*2*). Knowledge of locations and geometric characteristics of the roadway curves on low-volume roads play a crucial role in crash prediction and prevention. Although some states collect and store roadway curve data, including curve-related information (i.e., degree of curvature) on US and state highways, it is typically incomplete curve information (e.g., lacking information such as superelevation) (*3*). There is high demand for this information on rural roads for use in engineering design, safety management, and determining appropriate advisory speeds. This is a particularly important issue given that the Federal Highway Administration (FHWA) has mandated that all highway transportation agencies (state, local, and so forth) survey all roadway horizontal curves by December 31, 2019 (*4*).

Mobile asset data collection vehicles equipped with sensors such as LiDAR (light detection and ranging), GPS (Global Positioning Signal), and inertial measurement units (IMU) can provide transportation agencies with location information and roadway design elements, including horizontal curve properties. However, users need to be familiarized with detailed and tedious device procedures as well as post-data-collection processes. In addition, the cost for a survey vehicle with this equipment can be prohibitive for many agencies and researchers. Smartphones that integrate GPS, IMU, and advanced computing technologies have become more accessible owing to affordable price, product innovation, and people's desire to be connected. As more people own these devices, there are greater opportunities for data acquisition in intelligent transportation systems, which turns them into low-cost, real-time mobile sensor platforms.

This study provides a novel approach to automatically identify and measure horizontal curves using smartphone sensors. Smartphones have low-cost sensors chips. These low-cost sensors allow smartphones to have GPS and accelerometer technology without causing a significant increase in the cost of purchasing smartphones. However, the low cost is associated with a reduction in accuracy compared to more expensive sensors. Thus, when using these sensors, it is critical to utilize methods that improve the sensors' accuracy. This requires a system that (1) accounts for and reduces

[1]Department of Civil and Environmental Engineering, South Dakota State University, Brookings, SD
[2]Department of Computer Science, North Carolina State University, Raleigh, NC

Corresponding Author:
Address correspondence to Jonathan S. Wood: woohon@gmail.com

sensor (e.g., gyro and accelerometer) measurement errors and outliers; (2) reduces the location error caused by the weak GPS signals or GPS outage; and (3) provides a cost-effective real-time mobile system that detects horizontal curves and calculates their parameters accurately and reliably (e.g., radius and superelevation).

Literature Review

Many agencies and researchers have shown an interest in the extraction and identification of horizontal curves. Common curve identification techniques employed involve geographic information systems (GIS) based tools, satellite imagery processing, and mobile asset data collection. GIS-based tools (*3,5,6*) can extract and identify horizontal curves from GIS roadway maps in large-scale networks. However, the GIS-based techniques require high-quality GIS data to reduce identification errors from transverse measurement errors and low vertex resolution of GIS roadway centerlines. In addition, it is not possible to evaluate the superelevation of horizontal curves based on standard GIS data and tools because of lack of data or data resolution that is not fine enough. Image processing using high-resolution satellite imagery (*7–10*) can retrieve geometric characteristics of some typical curves; the drawbacks are that the accuracy greatly relies on image resolution and that it requires processing of many high-resolution images. Thus, these methods incur high computational costs. Also, none of these methods can extract information on the superelevation for horizontal curves.

A survey vehicle equipped with a GPS receiver, inertial system, and other sensors (including LiDAR) is another way to collect data and construct roadway asset inventories. The collected raw GPS data are postprocessed to extract horizontal curve components. Harkey et al. (*11*) discussed how the azimuth data obtained from the instrumented vehicle can be a useful tool for identifying curves and tangents on the roadway. However, the data collection and the postprocessing methods need to be improved to measure the radius and length of horizontal curves accurately. Carlson et al. (*12*) identified and tested 10 techniques (i.e., basic ball bank indicator [BBI], advanced BBI, chord length, compass, field survey, GPS, lateral acceleration, plan sheet, speed advisory plate, and vehicle yaw rate) to obtain horizontal curve radii. Based on the results, the in-vehicle GPS method was recommended to record roadway alignment for field personnel. To evaluate the accuracy of horizontal alignment data from multiple commercial roadway inventories, Findley et al. (*13*) applied five different techniques (i.e., chord method, GIS method, vendor data, survey data, and design data) to obtain curve parameters. The results were compared with the vendors' data, and the authors argued that agencies should consider the limitations of each technique to collect appropriate data to meet their needs. Laser scanning technology (*14*) can provide highly accurate survey data, which can be utilized to extract horizontal and vertical alignment of a curve. However, its cost is higher than that of other approaches – it was reported that the cost per mile to obtain the required highway inventory dataset for photo/video log, satellite/aerial imagery, and GPS data logger were \$72, \$107, and \$700, respectively, while the mobile LiDAR was \$915 (*15*). In addition, users must be familiarized with operating instructions. Finally, the postprocessing of raw data from these systems is time-consuming and labor-intensive.

Most smartphones are equipped with a series of sensors, including an accelerometer, gyroscope, barometer, magnetic sensor, and GPS. This richness in sensors enables support for roadway inventory in a low-cost manner. Recently, studies have been published using smartphone sensing in vehicles. GPS-related applications are applied to position capturing (*16,17*) or to vehicle tracking (*18*). Accelerometers have been used in studies to measure potholes (*19,20*) or pavement roughness (*21,22*). Won et al. (*23*) applied the barometer sensor of a smartphone to extract unique elevation signatures of driving routes. Zhang et al. (*24*) presented a mobile system to detect horizontal curves by synthesizing smartphone sensor data to generate curve models by applying a support vector machines learning technique.

Methodology

The basic concept of the proposed system is to use the rotation rate around the *z*-axis of the gyroscope to identify when a vehicle negotiates a horizontal curve. The corrected GPS location can be applied to identify the PC (point of curvature) and PT (point of tangent) of a curve, which can then be used to calculate the curve radius. Figure 1 illustrates the coordinate relationship between the vehicle and smartphone.

The system has four main system modules: data collection, data correction, curve identification, and curve calculation, as shown in Figure 2. The system takes real-time sensor readings from the smartphone – including the GPS, accelerometer, and gyroscope readings – in the data collection module. The data correction module reduces the noise of the raw input data. The Butterworth low-pass filter is applied to smooth the acceleration rate (m/s^2) and rotation rate (rad/s). GPS and speed data are corrected by applying the extended Kalman filter (EKF) algorithm. The curved road segments can be easily detected based on the smoothed rotation rate around the *z*-axis. Once the filtered curved segment data is identified, the curve identification module implements a K-means machine learning algorithm to evaluate the segment data and then determine if it is the horizontal curve. The obtained sensor data from the curves are imported to the curve calculation module. The filtered GPS data are used to identify the PC and PT of a curve, then to calculate the radius. Superelevation is calculated using the obtained radius, speed, and acceleration rate in the lateral direction.

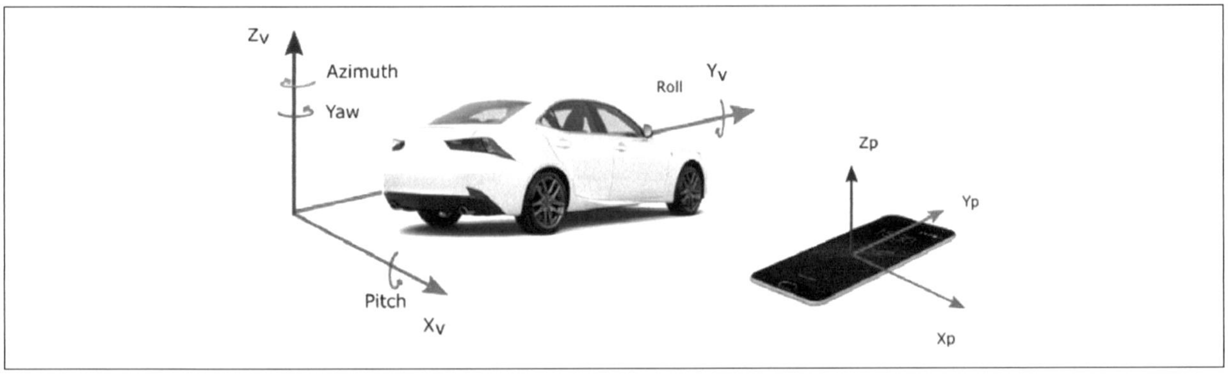

Figure 1. The coordinate systems of a smartphone and a vehicle.

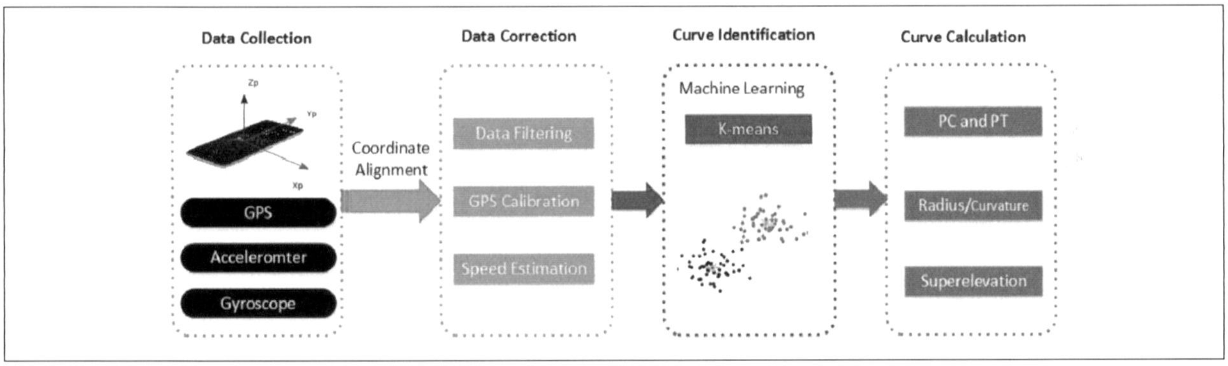

Figure 2. System architecture.

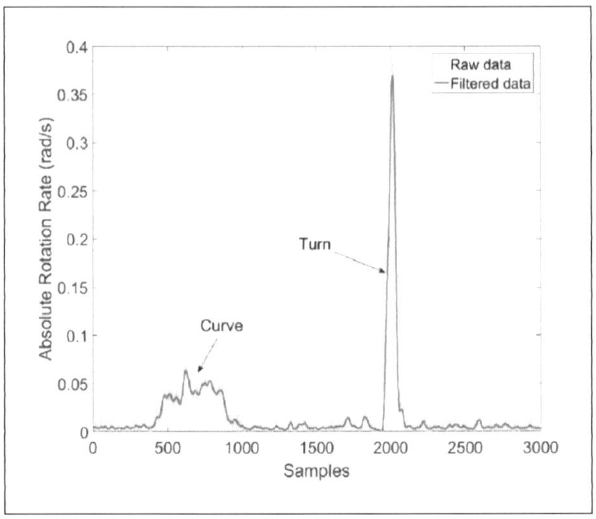

Figure 3. Variation of the angular speed around the z-axis.

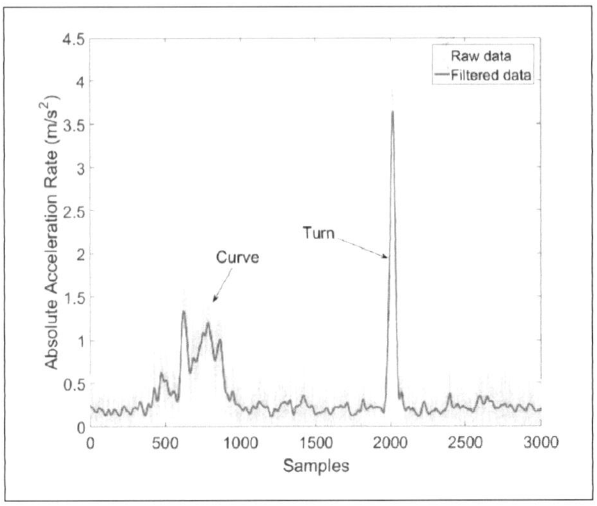

Figure 4. Variation of the acceleration rate on the *x*-axis.

Butterworth Low-Pass Filter

It is crucial to minimize measurement noise and outliers to achieve accurate curve detection and speed estimation. The Butterworth low-pass filter is well known for its effectiveness to reduce noise of high-frequency measurements *(25)*. This technique is adopted to remove the noise of raw accelerometer and gyroscope data collected using a smartphone. Figures 3 and 4 plot measurements of rotation rate around the *z*-axis and acceleration rate on the *x*-axis for a turning and a roadway

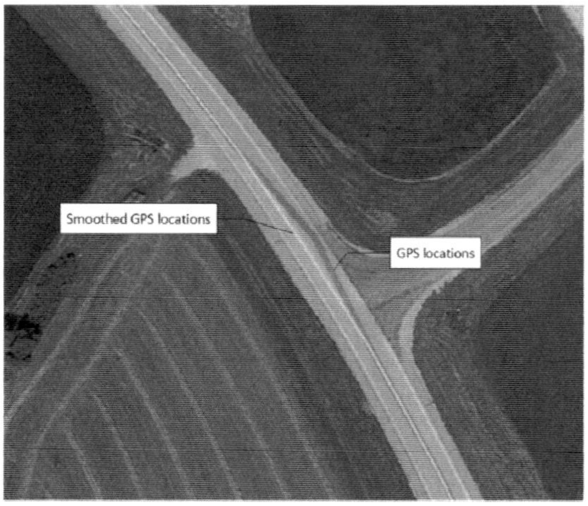

Figure 5. Raw GPS locations (red) versus smoothed GPS locations (green).

curve, respectively. It is shown that the noise is effectively removed by applying the Butterworth low-pass filter.

Extended Kalman Filtering

This system requires accurate measurement of the vehicle's GPS location and speed to derive the horizontal superelevation. The smartphone can acquire the speed reading from the embedded GPS module when the GPS signal is strong. The acceleration rate and rotation rate from IMU readings can be used to calibrate the vehicle's speed and locations when the GPS signal is weak or there is a signal outage.

The EKF is the nonlinear version of the Kalman filter, which linearizes an estimate of the current mean and covariance. The EKF is considered the de facto standard in the theory of nonlinear state estimation, navigation systems, and GPS. Assuming the process has a state vector x_k, and letting $f()$ denote the nonlinear function, the recursion equations for EKF are given as follows *(26)*:

$$x_k = f\left(x_{k-1}, u_{k-1}, w_{k-1}\right) \tag{1}$$

$$z_k = h\left(x_k, v_k\right) \tag{2}$$

where u_{k-1} is the control vector; w_{k-1} and v_k represent the process and measurement noises, which are both assumed to be zero mean multivariate Gaussian noises with covariance Q_k and R_k, respectively. The nonlinear function $h()$ relates to the state x_k to the measurement z_k.The state vector x is given in Equation 3:

$$x = \begin{Bmatrix} lat, lon, att, v_x, v_y, v_z, w_x, \\ w_y, w_z, acc_x, acc_y, acc_z \end{Bmatrix} \tag{3}$$

where v is the speed, w is the rotation rate, and acc is the acceleration rate. When the GPS signal is weak or there is an outage, the speed can be evaluated by the acceleration rate, and the GPS can be estimated by the rotation rate and speed. Figure 5 demonstrates that the GPS accuracy is improved by using the EKF.

Identification of Horizontal Curves

To distinguish between horizontal curves and lane change or turning maneuvers, this study adopted a K-means machine learning technique. K-means is one of the common "clustering" unsupervised machine learning techniques. It partitions n observations into k clusters in which each observation belongs to the cluster with the nearest mean value. Suppose a dataset D contains n objects in Euclidean space, which can be partitioned into k clusters *(27)*. A centroid-based partitioning technique uses the centroid of a cluster C_i to represent the ith cluster. The centroid can be determined by the mean of the clusters *(27)*.

The Euclidean distance $dis\left(p, c_i\right)$ measures the distance between the object point p and the centroid c_i. The sum of squared error between all objects in C_i is used to determine the centroid of cluster C_i, which can be defined as *(27)*:

$$SSE = \sum_{i=1}^{k} \sum_{p \in C_i} dis\left(p, c_i\right)^2 \tag{4}$$

where p represents any given observation, and c_i is the centroid of cluster C_i.

With K-means clustering, a threshold is selected to remove the data from the tangent road sections. It was observed that the standard deviation and mean of the rotation rate around the z-axis have a set of clustering features. Consequently, this study was able to use K-means clusters analyzing the rotation rate around the z-axis and determines whether that movement is a turn or a curve. A K-means algorithm, with K = 2, is applied to partition the training dataset into two clusters which represent curves and turns, respectively. Specifically, a centroid-based partitioning technique (*28*) was used to obtain the centroid of a cluster C_i to represent clusters. As shown in Figure 6, 50 curves and 50 turns were used to train the clustering centroids for the standard deviation and the mean of the rotation rate around the z-axis. To identify horizontal curve displacement, a threshold (0.01 rad/s) of the rotation rate was set to obtain the points in array A which represent the curved section of the vehicle's position (e.g., lane change, turns, and curve). The array A, representing a curved road, is considered if the Euclidean distance with the centroid of curve clustering c_i is the smallest one, based on Equation 5:

$$d = dis\left(A, c_i\right) \tag{5}$$

where d is the Euclidean distance between A and the ith centroid of the mean and standard deviation of the rotation

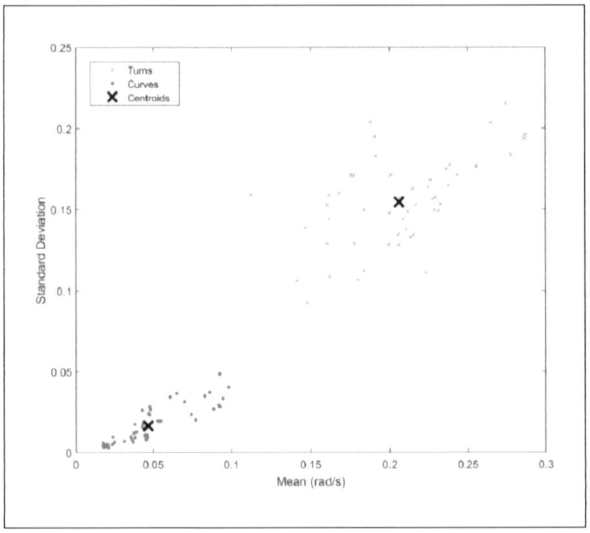

Figure 6. Cluster assignments and centroids of the rotation rate around the z-axis.

rate around the z-axis, and A is an array of the filtered rotation rate around the z-axis.

Calculation of Horizontal Curves

Chord Offset Method. A chord offset method is adopted to calculate curve radius. To improve the accuracy, three location measurements were recorded within the boundaries of PC and PT for each horizontal curve. In Figure 7, R represents the radius of the circle in feet; C is the chord length in feet; and H is the middle ordinate in feet. The radius in H and C can be derived by using:

$$\frac{C^2}{4} + (R-H)^2 = R^2 \tag{6}$$

$$R = \frac{H}{2} + \frac{C^2}{8H} \pm \frac{\text{Lane width}}{2} \tag{7}$$

$$D = \frac{5729.38}{R} \tag{8}$$

The GPS frequency rate is 1 Hz in the majority of modern smartphones. Owing to the slow rate, it cannot properly capture the locations of the PC and PT for roadway horizontal curves using GPS alone. The gyroscope has a much higher output rate, which can be used to identify the locations of PC and PT. More specifically, the count of samples in rotation rate over a specified threshold when the vehicle is traversing a curved road segment is used. This calculated proportion can be used to find the location between a point on the tangent and the next point on the curved road segment.

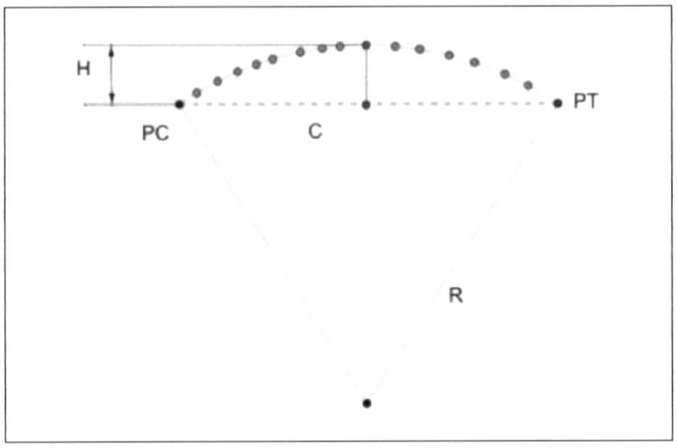

Figure 7. Chord offset method for curve radius measurement.

The chord method can be used to estimate the radius of the curve. This method can be sensitive to where middle ordinate measurements are taken. To improve the accuracy of curve radius estimates, the three neighboring GPS points around the PC and PT (within the extents of the curve) can be used to obtain nine chord lengths (C) and nine middle ordinate (H) measurements, respectively. Equation 9 was used to calculate the chord lengths based on the observed data. Then the calculated R_i was averaged using the weighted average method shown in Equation 10:

$$C_i = dis(PC_j, PT_k) \tag{9}$$

$$R_{avg} = \frac{\sum_{i=1}^{9} R_i H_i}{\sum_{i=1}^{9} H_i} \tag{10}$$

Superelevation Estimation. The lateral acceleration and vehicle speed measurements from the smartphone can be used to calibrate the superelevation. According to the AASHTO *Green Book* (*29*):

$$\frac{0.01e + f}{1 - 0.01ef} = \frac{v^2}{15R} \approx 0.01e + f \tag{11}$$

$$R = \frac{v^2}{15(0.01e + f)} \tag{12}$$

$$e = 100\left(\frac{v^2}{15R} - f\right) \tag{13}$$

where e is the superelevation in percent, f is the side-friction factor (which is equivalent to X_{acc}/g, where X_{acc} is the acceleration rate on the side of the vehicle measured by the smartphone), and v is the vehicle speed in miles per hour. Thus, using the calculated radius, vehicle speed, and lateral acceleration from the smartphone sensors (to obtain an estimate of f), superelevation is estimated.

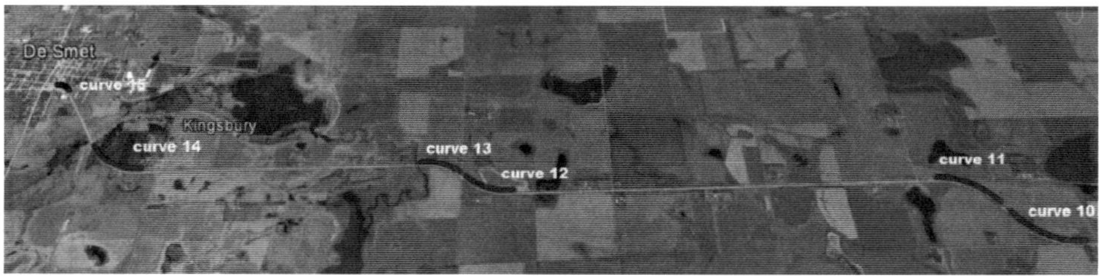

Figure 8. Test section with six identified horizontal curves.

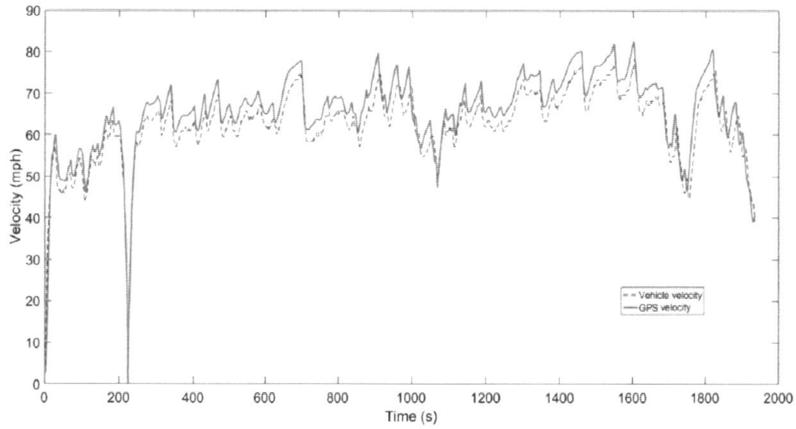

Figure 9. Comparison of vehicle velocity and calibrated velocity.

Field Test and Evaluation

A Samsung Galaxy S7 Edge with the Android 5 operating system was used to evaluate the proposed methodology. The GPS was collected at 1 Hz. Gyroscope and accelerometer were at 20 Hz. A total of 100 miles of two-lane highway located on Brookings County and Kingsbury County, South Dakota were selected as the test area. This study considered the curve grade greater than 1 degree of curvature (i.e., curve radius smaller than 5,730 ft). A total of 21 horizontal curves covering different radii were included. For each of these curves, roadway inventory data from the South Dakota Department of Transportation (SDDOT) were used to compare with the estimates. These locations are shown in Figure 8.

Speed Assessment

Vehicle speed plays an important role in this system. To evaluate the speed accuracy, an On-Board Diagnostics (OBD-II) sensor was used to record the velocity from vehicle the speedometer as a reference. Figure 9 illustrates the profile of the calibrated (i.e., GPS) velocity V_t and the reference (i.e., speedometer) velocity V_r from OBD-II sensor.

As shown in Figure 9, the absolute difference between the two speed values is 0–4.5 mph with a mean value of 3.24 mph. The mean absolute percentage error (MAPE) in Equation 14 was used to evaluate the speed deviation:

$$\text{MAPE} = \frac{1}{N}\sum_{n=1}^{N}\left|\frac{V_r - V_t}{V_r}\right| \tag{14}$$

where N is the number of samples.

The MAPE results indicated that there was approximately 6% difference between the GPS and vehicle speeds. This finding is consistent with SenSpeed *(30)*, which utilizes smartphone sensors to sense driving conditions and employs the accelerometer sensor in the smartphone to achieve an average speed error of 2.12 km/h (1.32 mph). Given that the vehicle speedometer is not perfectly accurate, this suggests that the speed precision obtained from the smartphone is acceptable.

Table 1. Radius Measurement without Lane Width Adjustment

No.	Highway and mileage reference marker	Design		Survey 1		Survey 2		Survey 3		Mean	
		G (%)	R (ft)	R	Error	R	Error	R	Error	Average	Error
1	US13@125	5.5	1042	1180	13.24%	1001	3.91%	1040	0.19%	1073	3.05%
2	US13@120	5	1146	1184	3.34%	1166	1.76%	1203	4.94%	1184	3.35%
3	US13@113.28	1.5	3820	3601	5.73%	3676	3.75%	3582	6.21%	3620	5.23%
4	US14@435	2.75	2083	2072	0.55%	2050	1.61%	2083	0.03%	2068	0.73%
5	US14@438	2.5	2292	2296	0.18%	2306	0.61%	2291	0.02%	2298	0.26%
6	US14@402	3	1910	1900	0.51%	1943	1.72%	1866	2.28%	1903	0.36%
7	US14@400.75	11.25	509	455	10.67%	541	6.19%	544	6.87%	513	0.80%
8	US14@397	2.5	2292	2263	1.26%	2268	1.03%	2280	0.49%	2271	0.93%
9	US14@397	2.5	2292	2281	0.48%	2391	4.34%	2287	0.20%	2320	1.22%
10	US14@386	2.75	2083	2083	0.02%	2096	0.60%	2105	1.06%	2095	0.55%
11	US14@385	2.75	2083	2078	0.27%	2046	1.79%	2005	3.75%	2043	1.94%
12	US14@382	2.75	2083	2073	0.50%	2036	2.25%	1956	6.13%	2022	2.96%
13	US14@381.69	2.75	2083	1963	5.80%	1902	8.71%	1972	5.36%	1945	6.62%
14	US14@379.08	2.75	2083	2108	1.20%	2155	3.45%	2058	1.22%	2107	1.14%
15	US14@378.71	10	573	557	2.86%	594	3.65%	550	3.93%	567	1.05%
16	US14B@421.41	3	1910	1895	0.77%	1961	2.66%	1876	1.77%	1911	0.04%
17	US14E@417.08	3	1910	1904	0.33%	1947	1.95%	1909	0.07%	1920	0.52%
18	US30@364.02	4	1432	1495	4.40%	1438	0.40%	1493	4.25%	1476	3.02%
19	US30@365	5	1146	1188	3.71%	1247	8.80%	1214	5.90%	1216	6.14%
20	US30@370	2	2865	2928	2.22%	2961	3.36%	2549	11.03%	2813	1.82%
21	US30@370	2	2865	2663	7.03%	2589	9.62%	2950	2.97%	2734	4.56%
Average difference				3.10%		3.44%		3.27%		2.20%	

Table 2. Radius Measurement with Lane Width Adjustment

No.	Highway and mileage reference marker	Design		Survey 1		Survey 2		Survey 3		Mean	
		G (%)	R (ft)	R	Error	R	Error	R	Error	Average	Error
1	US13@125	5.5	1042	1173	12.57%	1008	3.24%	1033	0.86%	1071	2.82%
2	US13@120	5	1146	1177	2.73%	1159	1.15%	1196	4.33%	1177	2.74%
3	US13@113.28	1.5	3820	3594	5.92%	3683	3.57%	3589	6.03%	3622	5.17%
4	US14@435	2.75	2083	2078	0.26%	2056	1.32%	2089	0.26%	2074	0.44%
5	US14@438	2.5	2292	2290	0.08%	2300	0.35%	2297	0.24%	2296	0.17%
6	US14@402	3	1910	1906	0.20%	1937	1.41%	1872	1.96%	1905	0.25%
7	US14@400.75	11.25	509	461	9.49%	535	5.01%	538	5.69%	511	0.41%
8	US14@397	2.5	2292	2256	1.56%	2275	0.73%	2273	0.80%	2268	1.03%
9	US14@397	2.5	2292	2288	0.17%	2384	4.03%	2294	0.11%	2322	1.32%
10	US14@386	2.75	2083	2091	0.34%	2088	0.24%	2098	0.70%	2092	0.43%
11	US14@385	2.75	2083	2085	0.09%	2054	1.43%	2013	3.39%	2051	1.58%
12	US14@382	2.75	2083	2080	0.14%	2029	2.61%	1963	5.77%	2024	2.84%
13	US14@381.69	2.75	2083	1970	5.44%	1910	8.35%	1979	5.00%	1953	6.26%
14	US14@379.08	2.75	2083	2116	1.56%	2148	3.09%	2065	0.86%	2110	1.26%
15	US14@378.71	10	573	564	1.55%	586	2.34%	558	2.62%	569	0.61%
16	US14B@421.41	3	1910	1901	0.45%	1955	2.34%	1882	1.46%	1913	0.14%
17	US14E@417.08	3	1910	1898	0.64%	1953	2.26%	1915	0.25%	1922	0.62%
18	US30@364.02	4	1432	1489	3.99%	1432	0.02%	1487	3.83%	1470	2.60%
19	US30@365	5	1146	1195	4.28%	1240	8.23%	1220	6.47%	1218	6.33%
20	US30@370	2	2865	2935	2.45%	2954	3.13%	2555	10.81%	2815	1.74%
21	US30@370	2	2865	2670	6.80%	2596	9.39%	2943	2.75%	2736	4.48%
Average difference				2.89%		3.06%		3.06%		2.06%	

Table 3. Superelevation Evaluation

		Survey 1			Survey 2			Survey 3			Mean	
No.	Design elevation (%)	R	Elevation 1 (%)	Error	R	Elevation 2 (%)	Error	R	Elevation 3 (%)	Error	Average	Error
1	5.8	1173	4.0	1.8	1008	7.3	−1.5	1033	5.7	0.1	5.7	0.1
2	5.8	1177	6.8	−1.0	1159	6.2	−0.4	1196	4.1	1.7	5.7	0.1
3	4.6	3594	3.8	0.8	3683	3.0	1.6	3589	5.5	−0.9	4.1	0.5
4	5.6	2078	3.6	2.0	2056	3.2	2.4	2089	7.1	−1.5	4.6	1.0
5	6	2290	6.8	−0.8	2300	7.1	−1.1	2297	3.9	2.1	5.9	0.1
6	5.8	1906	3.8	2.0	1937	7.0	−1.2	1872	3.2	2.6	4.7	1.1
7	4	461	4.5	−0.5	535	5.0	−1.0	538	5.8	−1.8	5.1	−1.1
8	5.8	2256	5.8	0.0	2275	2.9	2.9	2273	6.2	−0.4	4.9	0.9
9	5.8	2288	3.9	1.9	2384	7.4	−1.6	2294	3.7	2.1	5.0	0.8
10	6	2091	3.3	2.7	2088	8.5	−2.5	2098	2.7	3.3	4.8	1.2
11	6	2085	6.8	−0.8	2054	2.4	3.6	2013	7.9	−1.9	5.7	0.3
12	6	2080	3.8	2.2	2029	8.5	−2.5	1963	4.4	1.6	5.6	0.4
13	6	1970	7.8	−1.8	1910	3.5	2.5	1979	8.0	−2.0	6.4	−0.4
14	5.8	2116	3.6	2.2	2148	8.0	−2.2	2065	3.3	2.5	5.0	0.8
15	2.6	564	4.8	−2.2	586	2.0	0.6	558	5.3	−2.7	4.0	−1.4
16	5.6	1901	5.0	0.6	1955	6.2	−0.6	1882	5.1	0.5	5.4	0.2
17	4	1898	4.2	−0.2	1953	3.2	0.8	1915	4.2	−0.2	3.9	0.1
18	5.8	1489	7.0	−1.2	1432	5.3	0.5	1487	7.3	−1.5	6.5	−0.7
19	6	1195	6.6	−0.6	1240	7.5	−1.5	1220	6.9	−0.9	7.0	−1.0
20	4	2935	1.8	2.2	2954	2.6	1.4	2555	5.2	−1.2	3.2	0.8
21	4	2670	3.7	0.3	2596	3.7	0.3	2943	2.0	2.0	3.1	0.9
Average difference			1.32			1.56			1.59		0.66	

Curve Identification and Calculation

Three runs of each survey were measured to evaluate the stability of the proposed approach. All 21 horizontal curves were identified. Table 1 shows the radius measurements without adjusting for the lane width. The results indicated that the average radius difference of three runs is 3.10%, 3.44%, and 3.27%, respectively. The mean radius difference is 2.2%. Table 2 shows the radius with the adjustment of the lane width. The average radius difference of three runs is 2.89%, 3.06%, and 3.06%, respectively. The average error is 2.06%.

The results indicated that drivers might reduce speed when the vehicle is approaching a curve, which generates increased sensor noise, while the sensor data have less noise when leaving the curve because of the reduced speed and smoother driving. This is consistent with previous research findings (*31–33*). For typical superelevation transitions, 90% of the transition occurs before the PC and after the PT. Thus, the majority of the lengths of horizontal curves have full superelevation. Figure 10 shows one case of a superelevation profile measurement with a smartphone. The design superelevation is 4%, while the measured superelevation is 4.2%. Analysis of data from all of the curves indicated that estimating the average of superelevation using data between the 15th and 90th percentile of the length of the curve results in values that are consistent with the design superelevation from the SDDOT roadway inventory.

Table 3 shows the superelevation measurements for the 21 curves. In three runs, the average difference between design superelevation and measured superelevation is 1.32%, 1.56%, and 1.59%, respectively. The overall superelevation difference (i.e., using the average of the three values) is 0.66%. It should be noted that the comparison is between the smartphone-measured superelevation and design superelevation (which may be slightly different from the as-built superelevation). However, as-built superelevation information was not available. Thus, an inclinometer was used to measure the superelevation values for each curve. Natural variation means the measured values changed depending on the specific location that the inclinometer was placed. Using multiple inclinometer measurements at each curve, the superelevation was within 1% of the design superelevation for each of the measurements. Thus, the results in Table 3 are within a reasonable range, particularly when using the mean value of multiple runs.

Conclusions and Discussion

This study presented the design, implementation, and evaluation of a mobile system for low-cost collection of a horizontal curve inventory using a standard smartphone. Although the GPS frequency from smartphones is only 1 Hz, the field test demonstrated that the proposed approach can achieve desirable radius measurement accuracy for sharp curves. The

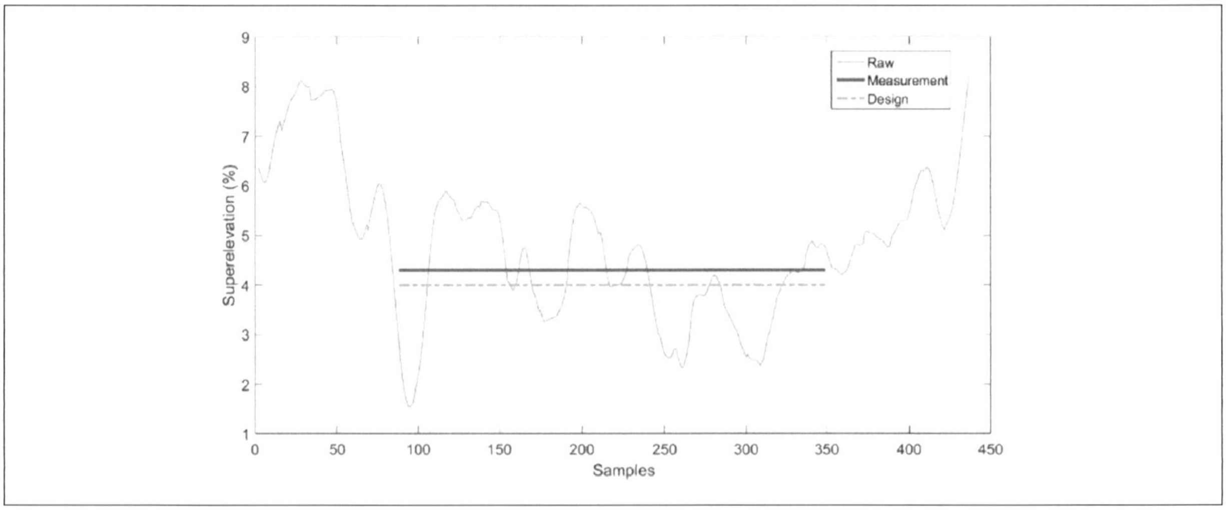

Figure 10. Comparison of design superelevation and measured superelevation.

average error is around 3%. Since the highest accuracy of GPS is 2–5 m, adjusting for differences in lane width does not significantly affect the accuracy of radii estimation. However, multiple runs can achieve higher accuracy. The accuracy of superelevation estimates relies on the accuracy of curve radius, vehicle speed, and acceleration from the smartphone. Improving their accuracy can mean increased accuracy for superelevation estimation.

Off-the-shelf smartphone technology is a potential alternative tool for horizontal curve surveys on account of its low cost and ease of application. However, the EKF approach for correcting GPS is sustainable only over a few seconds when there is a GPS outage. Longer GPS outage will result in larger location error. In addition, this research focused on applying this technology to measuring simple, sharp horizontal curves. Future work should include compound, reverse, and spiral curves using smartphone sensors. The GPS frequency from current smartphones is only 1 Hz. Increasing the frequency of the GPS measurements could potentially improve the radius estimation. GPS/IMU sensor fusion should be considered to increase the frequency of GPS in future work.

References

1. Zeeger, C. *NCHRP Report 362: Roadway Widths for Low-Traffic Volume Roads*. HRB, National Research Council,, Washington, D.C., 1994.
2. National Highway Traffic Safety Administration. *Traffic Safety Facts 2014*. National Highway Traffic Safety Administration, Washington, D.C., 2015, p. 87.
3. Li, Z., M. V. Chitturi, A. R. Bill, D. Zheng, and D. A. Noyce. Automated extraction of horizontal curve information for low-volume roads. *Transportation Research Record: Journal of the Transportation Research Board*, 2015. 2472: 172–184.
4. U.S. Department of Transportation. *Manual on Uniform Traffic Control Devices for Streets and Highways*. U.S. Department of Transportation, Washington, D.C., 2009.
5. Xu, H., and D. Wei. Improved Identification and Calculation of Horizontal Curves with Geographic Information System Road Layers. *Transportation Research Record: Journal of the Transportation Research Board*, 2016. 2595: 50–58.
6. Findley, D. J., C. V. Zegeer, C. A. Sundstrom, J. E. Hummer, W. Rasdorf, and T. J. Fowler. Finding and Measuring Horizontal Curves in a Large Highway Network: a GIS Approach. *Public Works Management & Policy*, Vol. 17, No. 2, 2012, pp. 189–211.
7. Yang, Y., and C. Zhu. Extracting Road Centrelines from High-resolution Satellite Images Using Active Window Line Segment Matching and Improved SSDA. *International Journal of Remote Sensing*, Vol. 31, No. 9, 2010, pp. 2457–2469.
8. Shi, W., Z. Miao, and J. Debayle. An Integrated Method for Urban Main-Road Centerline Extraction from Optical Remotely Sensed Imagery. *IEEE Transactions on Geoscience and Remote Sensing*, Vol. 52, No. 6, 2014, pp. 3359–3372.
9. Miao, Z., B. Wang, W. Shi, and H. Wu. A Method for Accurate Road Centerline Extraction from a Classified Image. *IEEE Journal of Selected Topics in Applied Earth Observations and Remote Sensing*, Vol. 7, No. 12, 2014, pp. 4762–4771.
10. Miao, Z., W. Shi, H. Zhang, and X. Wang. Road Centerline Extraction from High-resolution Imagery Based on Shape Features and Multivariate Adaptive Regression Splines. *IEEE Geoscience and Remote Sensing Letters*, Vol. 10, No. 3, 2013, pp. 583–587.
11. Harkey, D., C. Yi, and J. Feaganes. Evaluation and Validation of Automated In-Vehicle Data Collection System for Developing Roadway Alignments. *Transportation Research Record: Journal of the Transportation Research Board*, 2004. 1897: 164–172.
12. Carlson, P. J., M. B. Kit Black, and E. R. Rose. Comparison of Radius-Estimating Techniques for Horizontal Curves. *Transportation Research Record: Journal of the Transportation Research Board*, 2005. 1918: 76–83.

13. Findley, D. J., J. E. Hummer, W. Rasdorf, and B. T. Laton. Collecting Horizontal Curve Data: Mobile Asset Vehicles and Other Techniques. *Journal of Infrastructure Systems*, Vol. 19, No. 1, 2012, pp. 74–84.
14. Kim, J., J.-C. Lee, I.-J. Kang, S.-Y. Cha, H. Choi, and T.-G. Lee. Extraction of Geometric Information on Highway Using Terrestrial Laser Scanning Technology. *International Archives of the Photogrammetry, Remote Sensing and Spatial Information Sciences*, Vol. 37, 2008, pp. 539–544.
15. Jalayer, M., H. Zhou, J. Gong, S. Hu, and M. Grinter. A Comprehensive Assessment of Highway Inventory Data Collection Methods. *Journal of the Transportation Research Forum*, Vol. 53, 2014, pp. 73–92.
16. Qin, X., S. Zhang, and W. Wang. Avoiding Roadway Departure Crashes with an In-Vehicle Head-Up Display. Presented at the Transportation Research Board 94th Annual Meeting, 2015.
17. Aichinger, C., P. Nitsche, R. Stütz, and M. Harnisch. Using Low-cost Smartphone Sensor Data for Locating Crash Risk Spots in a Road Network. *Transportation Research Procedia*, Vol. 14, 2016, pp. 2015–2024.
18. Lee, S., G. Tewolde, and J. Kwon. Design and Implementation of Vehicle Tracking System Using GPS/GSM/GPRS Technology and Smartphone Application. *IEEE World Forum on Internet of Things (WF-IoT)*, Seoul, South Korea, 2014, pp. 353–358.
19. Ghose, A., P. Biswas, C. Bhaumik, M. Sharma, A. Pal, and A. Jha. Road Condition Monitoring and Alert Application: Using In-Vehicle Smartphone as Internet-Connected Sensor. *2012 IEEE International Conference on Pervasive Computing and Communications Workshops (PERCOM Workshops)*, Lugano, Switzerland, 2012, pp. 489–491.
20. Raj, S., A. Jain, and R. Misra. Smartphone Sensing for Large Data Set Collection of Potholes. *Proc., 11th Annual International Conference on Mobile Systems, Applications, and Services*, Taipei, Taiwan, 2013, pp. 517–518.
21. Islam, S., W. Buttlar, R. Aldunate, and W. Vavrik. Measurement of Pavement Roughness Using Android-based Smartphone Application. *Transportation Research Record: Journal of the Transportation Research Board*, 2014. 2457: 30–38.
22. Buttlar, W. G., and M. S. Islam. *Integration of Smart-Phone-Based Pavement Roughness Data Collection Tool with Asset Management System*. NEXTRANS Project No. 098IY04. University of Illinois at Urbana-Champaign, 2014.
23. Won, M., Z. Shaohu, A. Chekuri, and S. H. Son. Enabling Energy-Efficient Driving Route Detection Using Built-in Smartphone Barometer Sensor. *2016 IEEE 19th International Conference on Intelligent Transportation Systems (ITSC)*, Rio de Janeiro, Brazil, 2016, pp. 2378–2385.
24. Zhang, S., M. Won, and S. H. Son. *Low-Cost Realtime Horizontal Curve Detection Using Inertial Sensors of a Smartphone*. Presented at the 84th IEEE Vehicular Technology Conference, Montréal, Canada, 2016.
25. Guo, H., M. Yu, J. Liu, and J. Ning. Butterworth Low-Pass Filter for Processing Inertial Navigation System Raw Data. *Journal of Surveying Engineering*, Vol. 130, No. 4, 2004, pp. 175–178.
26. Haykin, S. S. *Kalman Filtering and Neural Networks*. Wiley, New York, 2001.
27. Steinley, D. K-means Clustering: A Half-Century Synthesis. *British Journal of Mathematical and Statistical Psychology*, Vol. 59, No. 1, 2006, pp. 1–34.
28. Han, J., J. Pei, and M. Kamber. *Data Mining: Concepts and Techniques*. Elsevier, Amsterdam, 2011.
29. American Association of State Highway and Transportation Officials. *A Policy on Geometric Design of Highways and Streets*, 6th ed. AASHTO, Washington, D.C., 2011.
30. Yu, J., H. Zhu, H. Han, Y. J. Chen, J. Yang, Y. Zhu, Z. Chen, G. Xue, and M. Li. Senspeed: Sensing Driving Conditions to Estimate Vehicle Speed in Urban Environments. *IEEE Transactions on Mobile Computing*, Vol. 15, No. 1, 2016, pp. 202–216.
31. Figueroa Medina, A., and A. Tarko. Speed Factors on Two-lane Rural Highways in Free-Flow Conditions. *Transportation Research Record: Journal of the Transportation Research Board*, 2005. 1912: 39–46.
32. Hu, W., and E. T. Donnell. Models of Acceleration and Deceleration Rates on a Complex Two-Lane Rural Highway: Results from a Nighttime Driving Experiment. *Transportation Research Part F: Traffic Psychology and Behaviour*, Vol. 13, No. 6, 2010, pp. 397–408.
33. Wood, J., and E. Donnell. Stopping Sight Distance and Horizontal Sight Line Offsets at Horizontal Curves. *Transportation Research Record: Journal of the Transportation Research Board*, 2014. 2436: 43–50.

The Standing Committee on Geometric Design (AFB10) peer-reviewed this paper (18-04114).

2018 K. B. WOODS AWARD: Outstanding Paper in Design and Construction

TRR
JOURNAL OF THE TRANSPORTATION RESEARCH BOARD

Transportation Research Record
2018, Vol. 2672(39) 11–19

DOI: 10.1177/0361198118783162
journals.sagepub.com/home/trr
SAGE

Safety Evaluation of Geometric Design Criteria: Intersection Sight Distance at Unsignalized Intersections

Scott Himes[1], Richard J. Porter[1], and Kimberly Eccles[1]

Abstract

A Policy on Geometric Design of Highways and Streets, 6th Edition, (publisher: AASHTO), provides design criteria for minimum sight distances, including intersection sight distance (ISD). An understanding of the relationship between ISD and safety at stop-controlled intersections is needed, with potential applications of this knowledge to both performance-based design and substantive road safety management practices. To establish this relationship, crash, traffic, and geometric data were collected from 832 two-lane minor unsignalized intersection approaches in Ohio, and Washington. The data were analyzed using a cross-sectional study design to quantify the relationship between safety and ISD. The analyses indicated that the expected number of target crashes are associated with available ISD. Target crash frequencies increase as available ISD decreases. Results suggest that ISD is associated with expected crash frequency in a non-linear fashion. The sensitivity of the expected number of target crashes to changes in ISD is highest when ISD is shorter, and decreases as ISD increases (i.e., the safety benefit of increasing ISD from 300 to 600 ft is substantially larger than the safety benefit of increasing ISD from 1,000 to 1,300 ft). The results also suggest that the impacts of ISD on crash frequencies vary as a function of the major road two-way annual average daily traffic and the major road speed limit. The sensitivity of the expected number of crashes to changes in ISD increases as both traffic volume and speed limit increase. Crash modification functions for each of the target crash types were estimated using the regression models.

A Policy on Geometric Design of Highways and Streets, 6th Edition (the Green Book) (*1*), published by the American Association of State Highway and Transportation Officials (AASHTO), provides design criteria for minimum sight distances, including intersection sight distance (ISD). The ISD criteria in the Green Book vary according to minor road traffic control, design speed of the major road, and turning movement from the minor road. Minimum ISD values are based on driver gap-acceptance behavior. Assumptions are made on physical conditions (e.g., object height and driver eye height), vehicle performance capabilities, and driver behavior.

The relationship between ISD and safety is not fully understood. Several studies (*2–8*) have attempted to quantify the relationship. Early studies of ISD found removing sight obstructions reduced crashes by as much as 30% to 67% (*2*, *3*). More recent studies have found that increased ISD results in more moderate crash reductions (*4–6*) and perhaps even adverse effects (*7*). Harwood et al. examined multiple studies and through the work of an expert panel determined that improving limited sight distance in each quadrant of the intersection can reduce total crashes by 5% to 17% (*4*). Himes et al. found that greater ISD is associated with fewer target crashes and target fatal and injury crashes through the development of crash modification functions (CMFunctions) (*8*). For their research, and for this paper, target crashes included only those involving a vehicle on the major road colliding with vehicles turning or departing from the minor road approach associated with the measured ISD. Additionally, the safety impact of ISD was found to be associated with major and minor road traffic volumes.

An understanding of the relationship between ISD and safety at stop-controlled intersections is needed, with potential applications of this knowledge to both performance-based design and substantive road safety management practices. The recent work by Himes et al. made significant strides toward addressing this need, but the methodology used at the time limited their ability to fully explore and capture three key features of the ISD-safety relationship that are relevant to designers (*8*):

1) ISD is very likely associated with the expected number of target crashes in a non-linear fashion. The

[1]VHB, Venture I, Raleigh, NC

Corresponding Author:
Address correspondence to Scott Himes: shimes@vhb.com

Table 1. Sites by State Stratified by Major Road Type and Area Type

	NC		OH		WA		Total	
Major route facility type	Rural	Urban	Rural	Urban	Rural	Urban	Rural	Urban
2-lane major	65	43	112	53	72	63	249	159
4-lane major undivided	40	46	23	33	7	47	70	126
4-lane major divided	51	41	41	19	48	28	140	88
Grand total	156	130	176	105	127	138	459	373

Note: NC = North Carolina; OH = Ohio; WA = Washington.

sensitivity of the expected number of target crashes to changes in ISD is expected to be highest when ISD is shorter, and decrease as ISD increases (e.g., the safety benefit of increasing ISD from 300 to 600 ft is expected to be substantially larger than the safety benefit of increasing ISD from 1,000 to 1,300 ft). If this is true, an "inverse ISD" specification is likely to be appropriate.

2) The impact of ISD on target crashes and target fatal and injury crashes is expected to vary with speed (e.g., the safety benefit of increasing ISD from 300 to 600 ft is expected to be substantially larger on high-speed roads compared with low-speed roads). A variable that captures speed and ISD interactions is needed to capture this effect. With operating speeds not available, speed limit could serve as a surrogate for operating speed.
3) The impact of ISD on target crashes and target fatal and injury crashes is also expected to vary with major and minor road traffic volumes. Variables that capture volume and ISD interactions are needed to capture these effects.

By adapting the modeling approach described in Le and Porter, this paper extends the work by Himes et al. to address these features of the ISD-safety relationship, offering new insights to inform performance-based design decisions (*8*, *9*). The remainder of this paper provides a brief overview of the data, the analytical method used, the resulting models for target crashes and target fatal and injury crashes, and an application methodology for practical use of the results of this research.

Data Summary

The research by Himes et al. describes the methodology developed to collect intersection approach-level data on the minor road approaches of stop-control intersections (*8*). In particular, available ISD was measured at each site from the decision point on the minor approach that is located approximately 14.5 ft from the edge of the nearest through lane and 4 ft from the minor road centerline (as defined in the Green Book). The project team used 1,320 ft as the maximum available ISD. For values greater than 1,320 ft, a value of 1,321 ft was recorded. A digital level tool was used to measure the vertical grade (in percentage of slope) on the major road approaches at 100-, 250-, and 500 ft from the intersection on both approaches. Only intersections with minor road approaches with measured grades less than or equal to 3% positive grade were included.

The project team collected data at 832 intersection approaches in North Carolina (NC), Ohio (OH), and Washington (WA). All three states are part of the Federal Highway Administration (FHWA) Highway Safety Information System (HSIS). HSIS contains quality crash, traffic, and roadway data for state-maintained roads in these three states. These states were selected for the reason that they could individually and collectively offer completeness, quality, and availability of data; regional and terrain diversity; and ease of access for data collection. With an overarching goal of relating ISD to safety, the sites were selected as representative of typical unsignalized intersections representing a range of volumes, facility types, and geometries. Sites were selected independent of historic crash frequency (i.e., they were not selected based on high-crash occurrence). For the 832 intersection approaches, the average target crash frequency was 0.22 crashes per approach per year. The average target fatal and injury crash frequency was 0.10 crashes per approach per year. Therefore, the results are generalizable and should provide conservative estimates of CMFs for intersections with a history of ISD-related crashes.

Three types of major roads were considered in this study (i.e., two lane, four lane divided, and four lane undivided), each of which intersects with a two-lane minor road. Table 1 shows the distribution of major road types by area type in the three states.

Initially, 3 years of crash data were obtained for each site (2008–2010). Since the average crash frequency for 3 years was quite low, the project team extended the analysis period by 1 year, to include 2011. Additionally, the 2008 HSIS data for NC has quality issues, so 2008 was excluded and years 2009–2012 were used instead. For both WA and OH, 2012 HSIS data were not available; therefore 2008–2011 HSIS data were utilized. Table 2 provides the summary statistics, including the minimum, maximum, and average value for continuous variables. Observations are included at the approach-direction level, as applied to a minor road approach (i.e., left or right direction from one minor road approach).

Table 2. Data Summary for Continuous Variables

Continuous variable	Min	Max	Mean	SD
Major AADT	272	35,671	11,395	7,963
Minor AADT	42	15,637	1,317	1,444
Available ISD (ft)	99	1,321	1,001	358
Median width (ft)	0	96	10.3	19.7
Posted speed limit (mph)	20	70	48.2	9.6
Shoulder width (ft)	0	14	2.7	2.8
Grade at 500 ft (%)	−8.7	7.9	−0.1	2.2
Intersection angle (degrees)	8	90	80.1	13.0
Total target crashes (crashes/year)	0	19	0.44	1.18
Target fatal and injury crashes (crashes/year)	0	12	0.20	0.66

Note: AADT = annual average daily traffic; ISD = intersection sight distance; SD = standard deviation.

Analytical Method

A cross-sectional study design with count regression modeling was used to quantify the relationship between safety and ISD. A cross-sectional study design is a type of observational study used to analyze a representative sample of observations at a specific point in time. The safety effect is estimated by taking the ratio of the estimated expected crash frequency for two groups, one with the feature of interest (or with a feature of interest under a proposed condition) and the other without the feature of interest (or with a feature of interest under an existing condition). In this case, the feature of interest is the ISD under a proposed condition versus an existing condition. For this method to work, the two groups should be similar in all regards except for the feature of interest. In practice, this is difficult to accomplish and multivariable regression models were used to estimate the safety effects of one feature while controlling for other characteristics that vary among sites.

In this case, expected crash frequency was the dependent variable of interest and explanatory variables listed in Table 2 were considered for the right-hand side of the model, including traffic volumes and other roadway and operational characteristics. Regression coefficients were estimated during the modeling process for each of the explanatory variables. The coefficients represent the expected change in the dependent variable (expected crash frequency) due to a unit change in the explanatory variable, all else being equal.

The current state-of-the-practice for developing statistical road safety models of this type is to assume a log-linear relationship between expected crash frequency and site characteristics. Generalized linear modeling techniques were applied to develop the models, and a log-linear relationship was specified using a negative binomial error structure. The negative binomial error structure also has advantages over the Poisson distribution in that it allows for over-dispersion that is often present in crash data (i.e., the variance is greater than the mean). It is typical in statistical road safety modeling for this "additional dispersion" to be specified as a dispersion parameter multiplied by the expected number of crashes squared (NB-2 model). Therefore, the models of the expected numbers of target crashes for this analysis were structured as shown in Equation 1:

$$\lambda_i = \text{EXP}\left(X_i\beta + \varepsilon_i\right) \tag{1}$$

where

λ_i = expected number of target crashes for the ith observation, with an observation in this analysis being one minor road and major road approach combination,

X_i = a matrix of explanatory variables associated with λ_i, including available ISD,

β = a vector of parameters to be estimated that quantify the relationships between the explanatory variables and λ_i,

ε_i = a disturbance term, where EXP(ε_i) is gamma-distributed with a mean equal to 1 and variance equal to α_i.

In the NB-2 model, the variance in the expected number of crashes (quantifying the unexplained diversity in the same observations) is then written as $\lambda_i + \alpha_i\lambda_i^2$.

There are several potential sources of bias in the development of statistical road safety models. These are described in the following list with an explanation of why they were dismissed or how they were addressed.

- Selection of appropriate functional form. Functional form relates to the overall form of the model and the form of each independent variable. The current state-of-the-practice was used for the overall form of the model (i.e., log-linear relationship), and exploratory data analysis techniques were used to identify an appropriate form for each predictor.
- Accounting for state-to-state differences. Data form three states were used to develop the models, and indicator variables were considered to identify the respective state for each site.
- Correlation among independent variables. Correlation refers to the degree of association among variables. A high degree of correlation among the predictor

variables makes it difficult to determine a reliable estimate of the effects of specific predictor variables. The correlation matrix was examined to determine the extent of correlation among the independent variables, and used to prioritize variables for inclusion.
- Overfitting of prediction models. Overfitting is related to the law of diminishing returns. At some point, it is not worth adding any more independent variables to the model because they do not significantly improve the model fit. Overfitting also increases the opportunity to introduce correlation in the model, and the opportunity for small samples when indicator variables are considered. Several combinations of predictor variables were considered, and relative goodness-of-fit measures were employed to penalize models with more estimated parameters.
- Low sample mean and sample size. Low sample mean was mitigated by expansion of the number of sites where data were collected. Sample size was not an issue for target crashes but was a potential concern for other crash types (e.g., right-turn crashes). This was mitigated by only analyzing target crashes and target fatal and injury crashes.
- Endogenous independent variables. Endogeneity arises when one or more of the independent variables depend on the dependent variable. For example, left-turn lanes may be installed due to the frequency of left-turn crashes at an intersection. A cross-sectional model that predicts left-turn crash frequency based on the presence of left-turn lanes and other factors may conclude that left-turn lanes increase left-turn crashes. Endogeneity was dismissed as a concern as study approaches were selected independent of crash history, and available ISD is an inherent intersection characteristic and not a treatment or countermeasure.
- Omitted variable bias. It is difficult to completely account for the potential effects of omitted variable bias in an observational cross-sectional study. In this case, omitted variable bias was addressed to the extent possible by carefully considering the roadway and traffic characteristics that should be included in the models before conducting field and desktop data collection.

The appropriate model was determined after exploratory data analysis, according to the procedure outlined in Hauer (*10*). The main effects of ISD as well as ISD, speed, and traffic volume interactions were the primary variables of interest in the matrix of explanatory variables, X_i. However, a number of other traffic and geometric variables were included to decrease unexplained variation in expected crash frequency and to try to minimize omitted variable bias. Omitted variable bias would involve over- or underestimating the safety effects of ISD due to other variables that influence crash frequency and are correlated with ISD, but are excluded from the model. However, the breadth of the model specifications had to be balanced with available data. Statistical significance guided model specification, but sound underlying theory and parameter stability throughout model testing informed specifications more than achieving a *p*-value of 0.05 or less for all parameter estimates.

Results

Table 3 presents the final models for target crashes and target fatal and injury crashes. Models for target fatal and incapacitating injury crashes, target angle crashes, and target daytime crashes were also explored; this uncovered similar trends between the frequency of these crash types and ISD. In some cases, the statistical significance of estimated model parameters decreased due to the smaller numbers of these more refined target crash type definitions.

The following provides definitions of the three main variables capturing the safety effects of available ISD and interpretations of the estimated parameters.

Speed Limit/ISD interaction (spdlmt/ISD): This variable represents an interaction between major road speed limit and available ISD. It is the key model variable capturing the safety effects of available ISD. Its value is the major road speed limit divided by the available ISD for a subject minor road/major road approach combination. The functional form of the variable, combined with the positive estimated model parameter, indicate the following:

- The expected number of target and target fatal and injury crashes increases as available ISD decreases.
- ISD is associated with expected target and target fatal and injury crash frequency in a non-linear fashion. The sensitivity of the expected number of target and target fatal and injury crashes to changes in ISD is highest when ISD is shorter, and decreases as ISD increases (i.e., the safety benefit of increasing ISD from 300 to 600 ft is substantially larger than the safety benefit of increasing ISD from 1,000 to 1,300 ft).
- The sensitivity of the expected number of target and target fatal and injury crashes to changes in ISD increases as the speed limit on the major road increases (e.g., the safety benefit of increasing ISD from 300 to 600 ft is expected to be substantially larger on high-speed roads compared with low-speed roads).

Low major road AADT/ISD interaction (LmajAADT/ISD): This variable represents an interaction between major road two-way annual average daily traffic (AADT) and ISD. Its value is zero if the major road AADT is greater than 5,000. Its value is one divided by available ISD for the applicable minor road approach direction (i.e., the inverse of available ISD) if the major road AADT is less than or equal to 5,000. The negative parameter represents the finding that the effect of ISD on the expected number of target crashes is smaller on lower-volume roads than on higher-volume roads.

Table 3. Model Estimation Results for Expected Number of Target Crashes.[a]

	Target crashes		Target FI crashes	
Variable	Coefficient	SE	Coefficient	SE
Constant	−8.147	1.382	−8.234	1.431
β1 (log-majAADT)	0.244	0.120	0.115	0.117
β2 (log-minAADT)	0.536	0.072	0.498	0.090
β3 (LmajAADT/availSD)	−243.0	185.3	NA	NA
β4 (MmajAADT/availSD)	−177.8	86.53	NA	NA
β5 (LMmajAADT/avaidlSD)	NA	NA	−155.5	108.6
β6 (LT)	0.334	0.120	0.507	0.152
β7 (fourleg)	0.845	0.126	0.953	0.162
β8 (median)	−0.016	0.166	0.215	0.210
β9 (spdlmt)	−0.021	0.008	−0.009	0.010
β10 (spdlmt/ISD)	7.194	2.450	6.335	2.915
β11 (grd500)	−0.061	0.026	−0.054	0.035
β12 (lnT)	1 (exposure)			
Dispersion (α)	2.66	0.278	3.136	0.155

Note: majAADT = major road annual average daily traffic; minAADT = minor road annual average daily traffic; LmajAADT = low major road annual average daily traffic; availSD = available intersection sight distance; MmajAADT = mid major road annual average daily traffic; LMmajAADT = low to mid major road annual average daily traffic; LT = left direction; spdlmt = speed limit; grd500 = grade at 500 feet.
[a]N = 1653; Log-likelihood = −1296.89; Pseudo R^2 = 0.055.

Medium major road AADT/ISD interaction (MmajAADT/ISD): This variable represents a second interaction between major road two-way AADT and ISD. Its value is zero if the major road AADT is less than or equal to 5,000 or greater than 15,000. Its value is one divided by available ISD for the applicable minor road approach direction (i.e., the inverse of available ISD) if the major road AADT is greater than 5,000 and less than or equal to 15,000. The negative parameter represents the finding that the effect of ISD on the expected number of target crashes is smaller on medium-volume roads than on higher-volume roads. However, the parameter is less negative than for the low-volume interaction, representing the finding that the effect of ISD on the expected number of target crashes is larger on medium-volume roads than on lower-volume roads.

Interpretations for other geometric variables can be made by taking the exponents of the coefficients. The interpretations, based on Table 3, include the following:

- Major road AADT. A unit increase in log major road AADT results in a 28% increase in target crashes and a 12% increase in target fatal and injury crashes. Note that a categorical variable for major road AADT is also included in an interaction with available ISD.
- Minor road AADT. A unit increase in log minor road AADT results in a 71% increase in target crashes and a 65% increase in target fatal and injury crashes. Note that a categorical variable for minor road AADT is also included in an interaction with available ISD.
- Left-direction indicator. The base condition is the right direction from an approach. The left direction is associated with 40% more target crashes and 66% more target fatal and injury crashes than the right direction.
- Four-leg intersection indicator. The base condition is a three-leg intersection. Four-leg unsignalized intersections experience 133% more target crashes and 159% more target fatal and injury crashes than three-leg intersections.
- Presence of a median indicator. The base condition is no median division. The presence of a median division is associated with 2% fewer target crashes and 24% more target fatal and injury crashes.
- Major road posted speed limit. A unit increase in posted speed limit is associated with a 2% reduction in target crashes and a 1% reduction in target fatal and injury crashes. Note that major road posted speed limit is also included in an interaction with available ISD.
- Grade on the major road approach at 500 ft from the intersection. A unit increase in percent grade at 500 ft is associated with a 6% reduction in target crashes and a 5% reduction in target fatal and injury crashes.

The general trends and directions of safety effects were consistent between the target crash model and the target fatal and injury crash model reported in Table 3 with the following exceptions:

- The smaller sample sizes corresponding to target fatal and injury crashes did not allow the fatal and injury model to distinguish between "low" and "medium" volume regions in the major road volume/ISD interaction. Therefore, a different interaction was created that combines low and medium volumes

(LMmajAADT/availSD). Its value is zero if the major road AADT is greater than 15,000. Its value is one divided by available ISD for the minor road major road approach combination (i.e., the inverse of available ISD) if the major road AADT is less than or equal to 55,000. The negative parameter still represents the finding that the effect of ISD on the expected number of target fatal and injury crashes is smaller on lower- and medium-volume roads than on higher-volume roads.

- The sign of the median variable, included to capture possible safety effects resulting from left-turning drivers being able to take refuge in the median, changed but the estimated parameter remained statistically insignificant.

The models presented in Table 3 were used to create CMFunctions, which can be used to calculate the effect of changing available ISD for an approach direction. The CMF calculation for total target crashes uses this CMFunction and is presented in Equation 2.

$$CMF_T = \frac{CMF_{T_1}}{CMF_{T_2}} \tag{2}$$

where

CMF_T = CMF for target crashes for an approach direction,

CMF_{Ti} = Target crash CMF for condition of interest i (where i = 1 for proposed condition and i = 2 for existing condition).

Equation 3 provides the CMFunction used to determine the target crash CMF for the proposed or existing condition.

$$CMF_{T_i} = \frac{\exp\left(\begin{array}{l} -0.021 \times PSL + \frac{7.194 \times PSL}{ISD_i} + \frac{-243.009 \times LowAADT_{maj}}{ISD_i} \\ + \frac{-177.826 \times MidAADT_{maj}}{ISD_i} \end{array}\right)}{\exp\left(\begin{array}{l} -0.021 \times PSL + \frac{7.194 \times PSL}{ISD_{base}} + \frac{-243.009 \times LowAADT_{maj}}{ISD_{base}} \\ \frac{-177.826 \times MidAADT_{maj}}{ISD_{base}} \end{array}\right)} \tag{3}$$

where

PSL = Posted speed (in mph),

$LowAADT_{maj}$ = 1 if major road AADT ≤5,000; otherwise 0,

$MidAADT_{maj}$ = 1 if 5,000 < major road AADT ≤ 15,000; otherwise 0,

ISD_i = proposed or existing available intersection sight distance for the condition of interest i (where i = 1 for proposed condition and i = 2 for existing condition) (in ft),

ISD_{base} = base intersection sight distance for an approach direction (in ft). For practical applications, this value is assumed to be 1,320 ft.

The CMFunction for target fatal and injury crashes uses this CMFunction and is presented in Equation 4.

$$CMF_{TFI} = \frac{CMF_{TFI_1}}{CMF_{TFI_2}} \tag{4}$$

where

CMF_{TFI} = CMF for target fatal and injury crashes for an approach direction,

CMF_{TFI_i} = target fatal and injury crash CMF for condition of interest i (where i = 1 for proposed condition and i = 2 for existing condition).

Equation 5 provides the CMFunction used to determine the target fatal and injury crash CMF for the proposed or existing condition.

$$CMF_{TFI_i} = \frac{\exp\left(\begin{array}{l} -0.009 \times PSL + \frac{6.335 \times PSL}{ISD_i} \\ + \frac{-155.504 \times LowMidAADT_{maj}}{ISD_i} \end{array}\right)}{\exp\left(\begin{array}{l} -0.009 \times PSL + \frac{6.335 \times PSL}{ISD_{base}} \\ + \frac{-155.504 \times LowMidAADT_{maj}}{ISD_{base}} \end{array}\right)} \tag{5}$$

where $Low\ MidAADT_{maj}$ = 1 if major road AADT ≤15,000; otherwise 0.

Application

The CMFunctions are relatively complex and were translated into 12 practical charts for practitioners in design, planning, operations, and traffic safety concerned with improving intersection safety. The charts provide safety guidance for three separate bins of major road AADT for target crashes and two separate bins of major road AADT for target fatal and injury crashes. The project team identified the following three major road AADT ranges for target crashes:

- Low major road AADT. The major road AADT is less than or equal to 5,000.
- Mid major road AADT. The major road AADT is greater than 5,000 but less than or equal to 15,000.
- High major road AADT. The major road AADT is greater than 15,000.

The project team divided the major road AADT into two categories for target fatal and injury crashes. The low to mid major road AADT category consists of major road AADT less than or equal to 15,000. The high major road AADT category consists of major road AADT greater than 15,000.

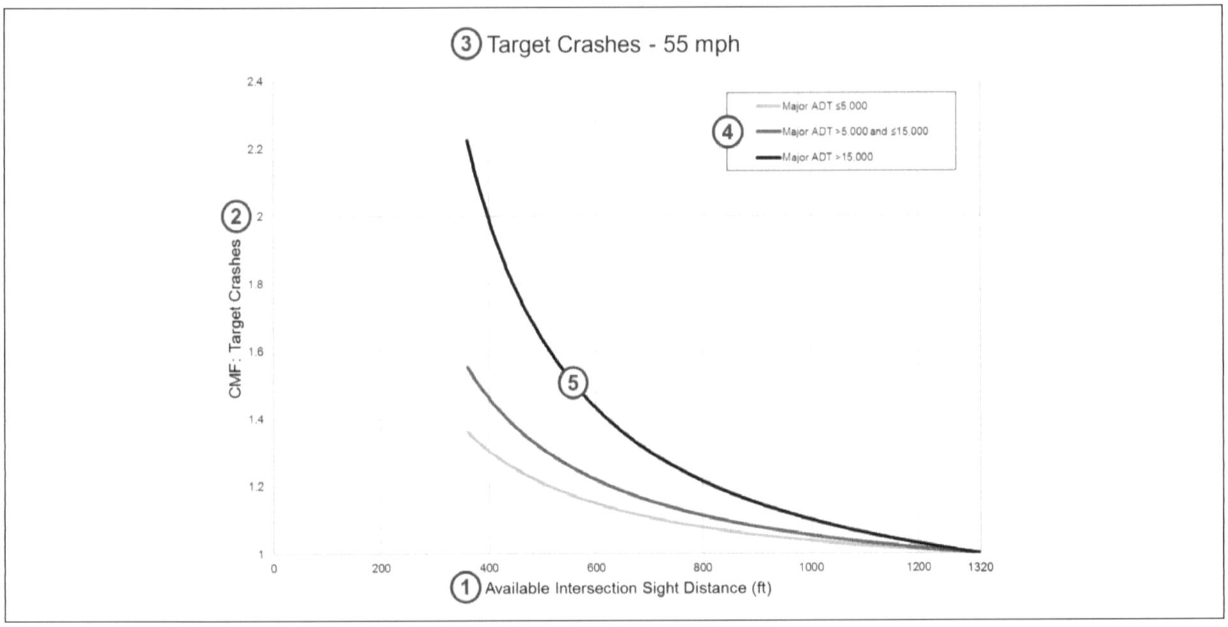

Figure 1. Sample chart.

Charts for each crash type are separated by posted speed limit, ranging from 35 to 60 mph. This yields six charts for each crash type, or 12 total charts. Caution should be used for posted speeds outside the 35 to 60 mph range.

Practitioners can apply these charts to estimate the following:

- The expected change in crash frequency at intersection approach directions (i.e., left direction or right direction), approach level (i.e., for one minor road), or intersection level (i.e., for the intersection as a whole) from changing the available ISD.
- The safety effects of changing available ISD at existing intersections.
- The safety effects of changing available ISD during the planning and design stages of the project development process.

The following are several notes of importance regarding the charts.

1. Users will select a chart based on the crash type of interest (target or target fatal and injury crashes) and the major road posted speed (e.g., 55 mph). CMFunctions may be applied directly, but caution should be used when considering the range of posted speed and available ISD. Note that the base ISD condition is 1,320 ft for the charts.
2. The output from the charts is a CMF that applies to a specific minor road approach direction, which requires the analyst to know the initial travel directions of vehicles involved in crashes at the study intersection.
3. Approach-direction CMFs can be combined using a weighted average based on the number of crashes per approach-direction to develop an intersection-level CMF. Additionally, CMFs may be combined to estimate an intersection-level CMF for total crashes.
4. The CMFunctions and charts treat major road AADT as grouped data. More precise AADTs cannot be used to further refine CMF estimates.
5. The maximum available ISD that may be used is 1,320 ft (i.e., a quarter mile). The data at collection sites were truncated at this point as the project team and panel felt no additional benefit would be realized for ISD beyond this value. The value 1,320 is used as the base condition for the charts, but direct comparisons may be made between two alternative CMFs on the chart.

Figure 1 includes a sample chart to illustrate the step-by-step instructions with corresponding numbers that align with the following steps:

- Step 1. Identify the minor road approach sight distance of interest (left-looking or right-looking) for which a new ISD is proposed. Measure the sight distance along the major road for that direction. This is the *existing* ISD for the approach and direction.
- Step 2. Identify the type of crash of interest for the analysis: target or target fatal and injury crash.
  - Target crashes are defined as a crash involving a vehicle from the major road and a vehicle entering from the minor road.

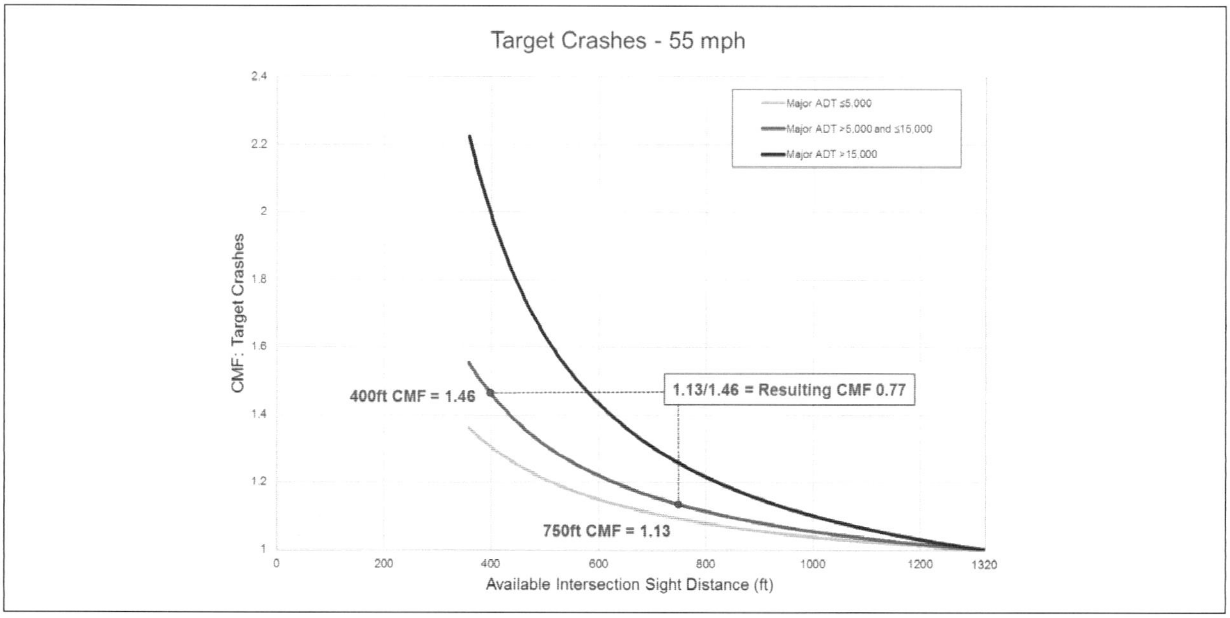

Figure 2. Example 1 – target crashes.

- Target fatal and injury crashes are a subset of target crashes that involve one or more injuries or fatalities.
- Step 3. Identify the appropriate chart type based on the type of crash and major road posted speed.
- Step 4. Identify the curve that corresponds with the appropriate major road AADT range.
  - Three ranges of major road AADT are used for target crashes.
  - Two ranges of major road AADT are used for target fatal and injury crashes.
- Step 5. Using the selected curve, plot the *existing* and the *proposed* ISD. If any ISD exceeds 1,320 ft, use 1,320 ft in the chart as this is the maximum.
- Step 6. Calculate the CMF for changing the site distance from the existing to the proposed by dividing the CMF from the chart for the *proposed* ISD by the CMF from the chart for the *existing* ISD.
- Step 7. Apply the resulting CMF to the crashes of interest associated with that direction.

Note that all 12 charts are available in the final technical report for NCHRP Project 17-59: Safety Impacts of Intersection Sight Distance.

Example: One Direction ISD Upgrade

A three-leg intersection with stop control on the minor road approach has a major road AADT of 7,000 and a major road posted speed of 55 mph. A practitioner is interested in estimating the change in average target crash frequency when increasing left-looking ISD from an existing condition of 400 ft to a proposed condition of 750 ft.

Using Figure 2, the target crashes CMF for the proposed ISD of 750 ft is 1.13. The target crashes CMF for the existing ISD of 400 ft is 1.46. Therefore, the target crashes CMF for improving ISD is 1.13/1.46, or 0.77. This factor applies to multi-vehicle crashes involving vehicles from the minor road and vehicles approaching from the left.

Conclusions

A cross-sectional study design with count regression modeling was used to quantify the relationship between safety and ISD. Target crashes were defined as those in which a vehicle on the minor road collided with a vehicle on the major road. Target crash counts were associated with specific, measured values of an approach-level ISD (i.e., one observation of target crash frequency in the database represents the target crash frequency for one minor road and major road approach combination). Estimation results for two multivariable count regression models are reported:

- Total (i.e., all severities) target crashes; and
- Target fatal and injury crashes.

Models for target fatal and incapacitating injury crashes, target angle crashes, and target daytime crashes were also explored and uncovered similar trends between the frequency of these crash types and ISD. In some cases, the statistical significance of estimated model parameters decreased due to the smaller numbers of these more refined target crash type definitions.

The analyses indicated that the expected number of target crashes are associated with available ISD. Target crash frequencies increase as available ISD decreases. The results of this research also suggest that ISD is associated with expected crash frequency in a non-linear fashion. The sensitivity of the expected number of target crashes to changes in ISD is highest when ISD is shorter, and decreases as ISD increases (i.e., the safety benefit of increasing ISD from 300 to 600 ft is substantially larger than the safety benefit of increasing ISD from 1,000 to 1,300 ft). The results also suggest that the impacts of ISD on crash frequencies vary as a function of the major road two-way AADT and the major road speed limit. The sensitivity of the expected number of crashes to changes in ISD increases as both traffic volume and speed limit increase. CMFunctions for each of the target crash types were estimated using the regression models. Applicability of the CMFunctions varies by available ISD, major road AADTs, and speed limits available in the dataset.

Future research studies should consider the safety effects of available ISD when analyzing the safety of unsignalized intersections with stop control on the minor road. The safety effects identified in this research are applicable for sites selected independent of crash history, and may differ from higher-crash intersections. The results of this research will benefit from further studies. Before–after studies of ISD are likely impractical, but additional cross-sectional studies arriving at similar results will continue to increase confidence in safety effects estimates.

Acknowledgments

This research was conducted as part of NCHRP 17-59: Safety Impacts of Intersection Sight Distance under the direction of Senior Program Officer Stephan Parker and the project panel. The authors express sincere gratitude to Mr. Parker of TRB and the entire panel for their thoughtful guidance of this research.

Author Contributions

The authors confirm contribution to the paper as follows: study conception and design: SH, KE; data collection: SH; analysis and interpretation of results: SH, RJP, KE; draft manuscript preparation: SH, RJP. All authors reviewed the results and approved the final version of the manuscript.

References

1. American Association of State Highway and Transportation Officials. *A Policy on the Geometric Design of Highways and Streets*, 6th ed. American Association of State Highway and Transportation Officials, Washington, D.C., 2011.
2. Mitchell, R. *Identifying and Improving Highway Accident Location*. Public Works, Hanley Wood, 1972.
3. Agent, K. R., N. Stamatiadis, and S. Jones. *Development of Accident Reduction Factors*. Report No KTC-96-13, Kentucky Transportation Center, Lexington, KY, 1996.
4. Harwood, D. W., F. M. Council, E. Hauer, W. E. Hughes, and A. Vogt. *Prediction of the Expected Safety Performance of Rural Two-Lane Highways*. FHWA-RD-99-207. Federal Highway Administration, Turner–Fairbank Highway Research Center, McLean, VA, 2000.
5. Golembiewski, G. A. and Chandler, B. *Intersection Safety: A Manual for Local Rural Road Owners*. Report No. FHWA-SA-11-08. Federal Highway Administration, Washington, D.C., 2011.
6. Fitzpatrick, K., K. Balke, D. W. Harwood, and I. B. Anderson. *NCHRP Report 440: Accident Mitigation Guide for Congested Rural Two-Lane Highways*. Transportation Research Board of the National Academies, Washington, D.C., 2000.
7. Pick v. Gratiot County Road Commission. *Michigan Court of Appeals*, 1993.
8. Himes, S., K. Eccles, K. Peach, C. Monsere, and T. Gates. Estimating the Safety Effects of Intersection Sight Distance at Unsignalized Intersections. *Transportation Research Record: Journal of the Transportation Research Board*, 2016. 2588: 71–79.
9. Le, T. and R. Porter. Safety Evaluation of Geometric Design Criteria for Spacing of Entrance-Exit Ramp Sequence and Use of Auxiliary Lanes. *Transportation Research Record: Journal of the Transportation Research Board*, 2014. 2309: 12–20.
10. Hauer, E., *The Art of Regression Modeling in Road Safety*. Springer International Publishing, Switzerland, 2015.

The Standing Committee on Geometric Design (AFB10) peer-reviewed this paper (18-05576), which was selected by the Design and Construction Group (AF000) to receive its 2018 Best Practical Award.

Article

TRR
JOURNAL OF THE TRANSPORTATION RESEARCH BOARD

Transportation Research Record
2018, Vol. 2672(39) 20–33

DOI: 10.1177/0361198118758394
journals.sagepub.com/home/trr
SAGE

Enhanced Design Guidelines for Interchange Loop Ramps

Darren J. Torbic[1] and Marcus A. Brewer[2]

Abstract

The objective of this research was to develop improved design guidance for interchange loop ramps. Despite their long use, there is little research on the design, safety, and operational characteristics of loop ramps. An observational field study was conducted to investigate the relationship between speed and lane position of vehicles and design elements of the ramp proper, and differences in performance between single-lane and multi-lane loop ramps. An assessment was also undertaken to determine how well crash prediction methods in the *Highway Safety Manual* (HSM) represent the safety performance of two ramp types with distinctly different geometrics: loop and diamond ramps. Based on the research results, recommendations regarding the design of the loop ramp proper, applicable at service interchanges are as follows: (a) for a given radius and design speed, recommended lane and shoulder widths for loop ramps are provided that are expected to induce speeds at or below the ramp design speed, result in similar levels of safety, and encourage drivers to stay within their intended travel lane; (b) regarding multi-lane loop ramps, outside lane widths of 12 ft for entrance ramps and 14 ft for exit ramps are sufficient to accommodate traffic comprised primarily of passenger vehicles, but if the outside lane is expected to accommodate moderate to high volumes of trucks, the outside lane width should be increased; and (c) when implementing the HSM ramp crash prediction methodology, calibration factors should be calculated separately for diamond and loop ramps.

Interchange projects are among the most complex and expensive projects constructed by highway agencies. Design decisions on such large investments should be made wisely, considering factors such as safety, operations, right-of-way and environmental constraints, and nearby traffic generators. Interchanges are comprised of individual ramps. The primary components of a ramp include the freeway mainline ramp terminal, ramp proper, and crossroad ramp terminal. Common ramp configurations can be classified as:

- diagonal (or diamond);
- loop;
- directional;
- semi-direct;
- outer connection; and
- one quadrant.

Combinations of these ramp types make up various interchange configurations. Figure 1 illustrates typical interchange types in which loop ramps are used. Despite their long use, little research has been conducted on the design, safety, and operational characteristics of loop ramps. Improved design guidelines could be developed for loop ramps based on scientifically valid research.

The objective of this research was to develop improved design guidance for interchange loop ramps. The research focused on developing improved design guidance for the ramp proper. The research was conducted in two phases. In Phase I the research team identified key unresolved issues associated with loop ramp design and prioritized these gaps in knowledge to be investigated as part of this research. The key issues investigated during Phase II included:

- The relationship between speed and lane position of vehicles and design elements of the ramp proper.
- The impact of design elements on the safety of loop ramps.
- The difference in performance on the ramp proper of single-lane and multi-lane loop ramps.

Whereas recommendations of this research are generally applicable to loop ramps on both service and system interchanges in urban and rural areas, designers should be mindful that all ramps included in the study were located at service interchanges. This paper summarizes research conducted as part of NCHRP Project 3-105; for more details on the research, refer to Part 1 of *NCHRP Web-Only Document 227* (*1*).

[1]MRIGlobal, State College, PA
[2]Texas A&M Transportation Institute, College Station, TX

Corresponding Author:
Address correspondence to Darren J. Torbic: dtorbic@mriglobal.org

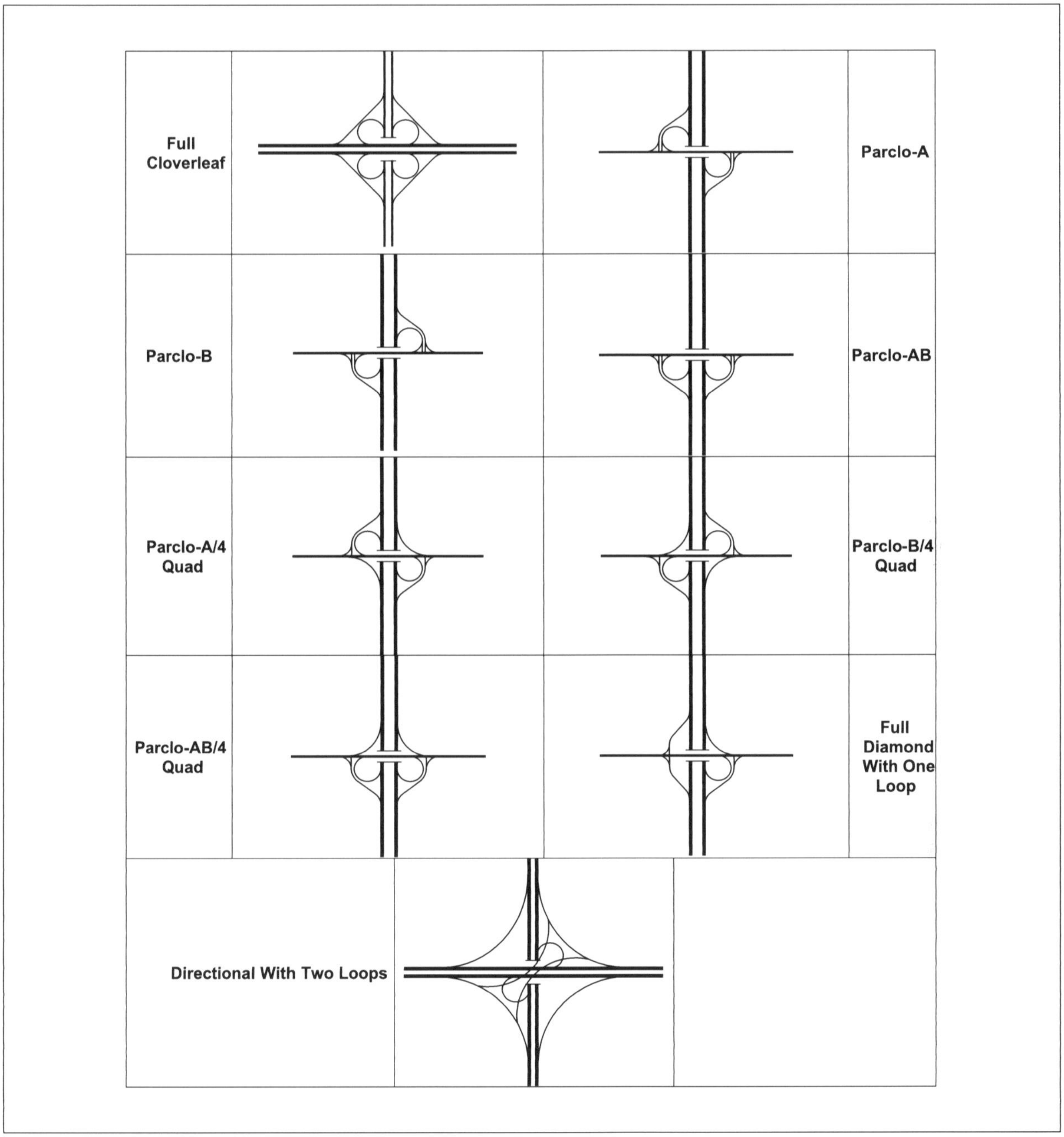

Figure 1. Interchange configurations with loop ramps.

Current Design Policy on Loop Ramps

Design guidance for loop ramps in the *Green Book* (*2*) is limited to a few sections. Chapter 10 (Grade Separations and Interchanges) provides the most guidance compared with other chapters. *Green Book* Table 10-1 (Table 1) provides guidance for selecting ramp design speeds. Ramp design speeds are specified for upper, middle, and lower range values given various conditions and ramp types. For loop ramps, minimum values usually control. For highway design speeds above 50 mph, the loop design speed preferably should not be less than 25 mph; and for freeway design speeds of 70 mph, the loop design speed preferably should not be less than 35 mph. Values in *Green Book* Table 10-1 apply to the sharpest, or controlling, curve, usually on the ramp proper. Practical radii of loop ramps are approximately 100 to 170 ft for minor movements on highways with design speeds of 50 mph or less and 170 to 250 ft for more important movements on highways with higher design speeds.

Table 1. Guide Values for Ramp Design Speed as Related to Highway Design Speed (2)

U.S. Customary										
Highway design speed (mph)	30	35	40	45	50	55	60	65	70	75
Ramp design speed (mph)										
Upper range (85%)	25	30	35	40	45	48	50	55	60	65
Middle range (70%)	20	25	30	33	35	40	45	45	50	55
Lower range (50%)	15	18	20	23	25	28	30	30	35	40
Corresponding minimum radius (ft)	See *Green Book Tables 3–7*									

Compound or spiral curves are desirable for loop ramps to obtain the desired alignment, provide a transition between through and turning roadways, and fit the natural paths of vehicles. Loop ramps located beyond a structure usually need a parallel deceleration lane due to potential sight restrictions. The speed-change lane should be developed on the near side of the structure and carried across the structure if sight distance is limited (*2*).

Although considered exceptional cases, the number of two-lane loop ramps has increased across the country. A two-lane loop ramp should not be preceded or followed by another loop ramp, and the radius of the inner edge of the traveled way of the loop ramp should not be less than 180 to 200 ft. The *Green Book* refers to the Institute of Transportation Engineers' *Freeway and Interchange Geometric Design Handbook* (*3*) for additional details on the design of two-lane loop ramps.

Review of Safety and Geometric Design Research

Several studies investigated the safety performance of loop ramps and the relationship to geometric design elements. Relevant findings from several studies are as follows:

- Speed differentials between merging vehicles and vehicles on the freeway are nearly identical for diamond and loop ramps (*4*).
- Drivers exiting loop ramps tend to reduce their speed in the freeway lane more, and decelerate along the speed-change lane at a greater rate, than drivers exiting diamond ramps (*4*).
- Comparing diamond ramps, loop ramps, and outer connection ramps, more crashes occur on exit ramps than entrance ramps by a ratio of three to two (*5*).
- Loop ramps have higher crash rates than diamond ramps (*6, 7*). The exception to this generalization appears to be urban, free-flow loop ramps, which have lower crash rates than diamond ramps (*8*).

Observational Study of Loop Ramp Proper

An observational field study of driver behavior and vehicle operations on the ramp proper of single- and multi-lane loop ramps was conducted to address two high-priority gaps in knowledge related to the design of loop ramps: (a) the relationship between speed and lane position of vehicles and design elements of the ramp proper, and (b) the difference in performance on the ramp proper between single- and multi-lane loop ramps. Data were collected at 15 entrance and 13 exit ramps in California, Kansas, Missouri, and Texas. Twenty-five ramps were located in urban and suburban areas, and three ramps were in rural areas. Table 2 presents general characteristics of the study locations and ramps. This project focused on observed operations on the ramp proper; the effects of speed-change lane design and adjacent design elements were not considered in this project. For more details on speed-change lanes and their observed effects on speed, refer to NCHRP Report 730 (*4*).

Field Data Collection

Data collection activities were similar at both entrance and exit ramps. Field data were collected on weekdays, during non-peak-hours. The primary measures collected in the field included:

Entrance Ramps:

- Speeds of vehicles at the midpoint and end (i.e., point of tangency [PT]) of the controlling curve in the direction of travel on the ramp
- Positions of vehicles within the roadway at the midpoint of the controlling curve.

Exit Ramps:

- Speeds of vehicles at the beginning (i.e., point of curvature [PC]) and midpoint of the controlling curve in the direction of travel on the ramp
- Positions of vehicles within the roadway at the midpoint of the controlling curve
- Unusual or critical behavior (e.g., braking, swerving, or use of shoulder) near the beginning of the controlling curve.

Laser guns were used to collect free-flow speeds of vehicles on the ramps; vehicles were classified as passenger vehicles (including light trucks and sport-utility vehicles) or trucks (including single-unit and tractor-semi-trailer trucks). Video recorders were used to document the lane position of vehicles at the midpoint of the controlling curve, using

Table 2. Characteristics of Study Locations and Ramp Proper

											Controlling curve							
Ramp ID	State	Freeway	Freeway design speed (mph)	Posted speed limit on freeway (mph)	Crossroad	Posted speed limit on crossroad (mph)	Type of curve	Ramp design speed (mph)	Advisory speed for ramp (mph)	Number of lanes	Radius (ft)	Length (ft)	Super (%)	Inside shoulder width (ft)	Lane width (ft)	Outside shoulder width (ft)	Overall length of curves on ramp (ft)	Overall length of ramp (ft)
Entrance ramps																		
1	MO	I-435	70	65	US 24	35	Comp	15	None	1	150	416	0.05	2.5	20.5	4.0	577	1,225
2	KS	I-435	70	70	Shawnee Mission Pkwy	55	Comp	20	25	1	250	1,030	0.05	9.5	14.6	6.8	1,318	1,418
3	KS	I-435	70	70	Shawnee Mission Pkwy	55	Comp	20	25	1	250	1,018	0.05	10.1	14.4	7.5	1,306	1,406
4	MO	I-70	70	70	MO 65	65	Comp	15	25	1	177	344	0.05	6.3	17.3	5.3	1,432	1,747
5	TX	I-30	70	60	Hampton Rd	35	Comp	20	20	1	312	1,434	0.04	6.0	14.5	4.0	1,635	2,007
6	TX	SH 360	70	60	E. Mid Cities Blvd	40	Simple	15	None	1	230	940	0.04	6.5	21.0	4.0	940	1,445
7	TX	US-67	60	60	Loop 12	45	Comp	15	20	1	150	333	0.04	1.0	16.0	1.0	910	910
8	TX	US-67	60	60	Loop 12	45	Comp	15	20	1	125	254	0.04	1.0	16.0	1.0	888	951
9	TX	I-35	75	75	US-77/SH 579	60	Comp	15	None	1	200	458	0.04	8.0	14.0	2.0	1,206	1,253
10	TX	I-35	50	60	Commerce St	30	Comp	15	None	1	110	220	0.04	1.0	20.5	3.0	564	564
11	TX	FM 2818	50	50	Wellborn Rd	45	Simple	20	25	1	200	748	0.06	6.0	14.5	2.5	748	911
12	CA	I-5	65	65	Arena Blvd	40	Comp	20	None	2	186	377	0.07	8.0	13.1	3.75	537	787
13	CA	US 50	65	65	E. Bidwell	45	Simple	20	None	2	154	698	0.07	9.5	13.3	3.5	698	1,448
14	CA	I-80	65	65	Elkhorn Rd	45	Simple	20	None	2	123	571	0.07	8.0	13.5	4.0	571	667
15	CA	I-80	65	65	Elkhorn Rd	45	Simple	20	None	2	122	566	0.07	6.5	14.0	4.0	566	766
Exit ramps																		
16	MO	I-435	70	65	US 24	35	Comp	20	25	1	181	424	0.05	3.5	18.7	7.1	706	1,384
17	KS	I-435	70	70	Shawnee Mission Pkwy	55	Comp	20	25	1	250	1,048	0.05	10.3	15.0	6.8	1,336	1,436
18	MO	I-70	65	55	US 40	35	Comp	15	25	1	160	411	0.03	4.4	20.0	5.4	890	960
19	MO	I-70	65	55	US 40	35	Comp	15	25	1	160	447	0.05	4.2	20.9	4.3	822	892
20	KS	I-435	70	70	95th St.	40	Comp	20	None	1	300	861	0.05	12.8	15.2	6.6	1,107	1,307
21	MO	I-70	70	70	MO 65	65	Comp	15	25	1	218	809	0.04	6.0	17.2	6.25	1,106	1,263
22	TX	I-35E	50	60	Commerce St	30	Comp	20	20	1	104	260	0.05	1.0	20.0	4.0	458	575
23	TX	I-635	70	60	Freeport Pkwy	35	Simple	15	20	1	200	461	0.02	7.0	17.0	2.0	461	546
24	TX	I-635	70	60	Freeport Pkwy	35	Simple	15	20	1	200	461	0.02	7.0	17.0	3.0	461	551
25	TX	I-30	70	60	Hampton Rd	35	Comp	20	20	1	246	970	0.05	6.0	14.0	2.0	1,477	1,962
26	TX	SH 360	70	60	E. Mid Cities Blvd	40	Simple	15	30	1	230	1,050	0.04	7.0	16.0	4.5	1,050	1,415
27	TX	FM 2818	50	50	Wellborn Rd	45	Simple	25	30	1	250	601	0.06	6.0	14.5	2.0	601	763
28	CA	SR 99	65	65	Sheldon Rd	45	Simple	20	30	3	171	557	0.07	8.5	14.2	3.5	557	1,719

Note: "Comp" = compound curve.

reference markers painted in the travel lane at 6-in.-intervals. Recorders were also used to record maneuvers of vehicles as they transitioned from the deceleration lane to the ramp proper of an exit loop ramp. The primary purpose was to assess unusual or critical driving behavior (e.g., braking, swerving, or use of the shoulder beyond the deceleration lane) along the first several hundred feet of the ramp proper. Video recordings were viewed in the office for post-processing.

Analysis of Observational Data

Three types of observational data from the field studies were analyzed:

- vehicle speed,
- lane position, and
- exiting maneuver.

Vehicle Speed

Vehicle speeds were analyzed to assess the effects of design elements on speed on loop ramps. The data collection procedure resulted in two speed measurements denoted as follows for each subject vehicle along a ramp:

- Entrance ramps
  - Speed 1: speed of subject vehicle at midpoint of controlling curve
  - Speed 2: speed of subject vehicle at PT of controlling curve
- Exit ramps
  - Speed 1: speed of subject vehicle at PC of controlling curve
  - Speed 2: speed of subject vehicle at midpoint of controlling curve

The design elements considered in the analyses included:

- Ramp length
- Design speed of controlling curve
- Radius of controlling curve
- Superelevation of controlling curve
- Length of controlling curve
- Inside shoulder width of controlling curve
- Lane width of controlling curve
- Outside shoulder width of controlling curve
- Freeway speed limit
- Crossroad speed limit
- Vertical profile (grade, up or down)
- Type of speed-change lane
- Type of horizontal curvature (simple or compound)
- Type of crossroad traffic control
- Vehicle type

Regression models were developed using the restricted maximum likelihood (REML) method to estimate the effect of design elements on vehicle speeds. A model for predicting speeds at the midpoint of the controlling curve was developed first, then starting with the same variables from that model, a separate model was estimated to predict the speeds at the PT of the controlling curve. Models yielding the most information and best goodness of fit for estimating the speeds of vehicles at the midpoint and end of the controlling curve on the ramp proper of an entrance loop ramp are shown in Figure 2. The models were developed using data for single- and multi-lane ramps combined.

The prediction model for estimating vehicle speeds at the midpoint of the controlling curve on an entrance loop ramp is interpreted as follows. Keeping everything else constant, vehicle speeds:

- Increase by approximately 2 mph by incrementally increasing the radius (measured to the inside of the traveled way) of the controlling curve by 50 ft.
- Increase by approximately 0.3 mph for every 1-ft incremental increase of the lane width of the controlling curve.
- Increase by approximately 0.9 mph for every 1-ft incremental increase of the outside shoulder width of the controlling curve.
- Increase by approximately 0.7 mph for every 1-ft incremental increase of the inside shoulder width of the controlling curve.

In addition, vehicle speeds at the midpoint of the controlling curve on an entrance loop ramp in Lane 2 (outside lane) of a multi-lane ramp are estimated to be approximately 2 mph faster than speeds in Lane 1 (inside lane), and truck speeds at the midpoint of the controlling curve are expected to be approximately 4.3 mph slower than speeds of passenger vehicles.

For the model predicting vehicle speeds at the end of the controlling curve, the radius (CCRadius), outside shoulder width (CC_OSW), and vehicle type (VehType[Truck]) remained statistically significant with similar coefficients to the midpoint model. This suggests that their influence is roughly unchanged between the two points where Speed 1 and Speed 2 were measured.

The same process was repeated for exit loop ramps, beginning with developing a model to estimate vehicle speed at the midpoint of the controlling curve on the ramp proper and then using the same significant predictors to model vehicle speed at the beginning of the controlling curve. The final models for exit loop ramps are provided in Figure 3.

The prediction model for estimating vehicle speeds at the midpoint of the controlling curve on an exit loop ramp is interpreted as follows. Keeping everything else constant, vehicle speeds:

- Increase by approximately 2.7 mph by incrementally increasing the radius (measured to the inside of the traveled way) of the controlling curve by 50 ft.

```
Midpoint of Controlling Curve on the Ramp Proper of an Entrance Loop Ramp
Linear mixed-effects model fit by maximum likelihood

       AIC      BIC    logLik
  9025.673 9075.068 -4503.837

 Random effects:
  Formula: ~1 | SiteID
        (Intercept) Residual
 StdDev:     1.21075 2.970354

 Fixed effects: Speed1 ~ LaneFactor + CCRadius + CC_LW + CC_OSW + CC_ISW + VehType

                     Value Std.Error   DF    t-value p-value    Significance
 (Intercept)      8.359490  3.654970 1769   2.287157  0.0223    Significant at 95% CL
 LaneFactor(Lane2) 1.978092 0.633945 1769   3.120292  0.0018    Significant at 99% CL
 CCRadius         0.039645  0.007180   11   5.521428  0.0002    Significant at 99% CL
 CC_LW            0.313094  0.186978 1769   1.674496  0.0942    Significant at 90% CL
 CC_OSW           0.912132  0.291748   11   3.126443  0.0096    Significant at 99% CL
 CC_ISW           0.681708  0.190667   11   3.575383  0.0044    Significant at 99% CL
 VehType(truck)  -4.332671  0.205156 1769 -21.118901  0.0000    Significant at 99% CL


 End of Controlling Curve on the Ramp Proper of an Entrance Loop Ramp
 Linear mixed-effects model fit by maximum likelihood

       AIC      BIC    logLik
  9717.268 9755.686 -4851.634

 Random effects:
  Formula: ~1 | SiteID
        (Intercept) Residual
 StdDev:    1.177096 3.614952

 Fixed effects: Speed2 ~ LaneFactor + CCRadius + CC_OSW + VehType

                      Value Std.Error   DF    t-value p-value    Significance
 (Intercept)      16.275717 1.1234930 1770  14.486710  0.0000    Significant at 99% CL
 LaneFactor(Lane2) 1.443538 0.4498573 1770   3.208879  0.0014    Significant at 99% CL
 CCRadius          0.054387 0.0066668   12   8.157932  0.0000    Significant at 99% CL
 CC_OSW            1.078759 0.2089150   12   5.163628  0.0002    Significant at 99% CL
 VehType(Truck)   -4.051046 0.2494547 1770 -16.239606  0.0000    Significant at 99% CL
```

Figure 2. Speed prediction models to estimate vehicle speeds at the midpoint and end of the controlling curve on the ramp proper of an entrance loop ramp.

- Increase by approximately 1 mph for every 1-ft incremental increase of the outside shoulder width of the controlling curve.

In addition, vehicle speeds at the midpoint of the controlling curve in Lane 2 or Lane 3 (outside lanes) on a multi-lane loop ramp are approximately 1.2 mph faster than speeds in Lane 1 (inside lane) of the same ramp. Truck speeds at the midpoint of the controlling curve are approximately 4.9 mph slower than speeds of passenger vehicles. Vehicle speeds at the midpoint of the controlling curve with a simple curve are approximately 3.6 mph faster than on ramps designed with a

```
Midpoint of Controlling Curve on the Ramp Proper of an Exit Loop Ramp

Linear mixed-effects model fit by maximum likelihood

        AIC      BIC    logLik
    8057.741 8116.751 -4017.871

  Random effects:
   Formula: ~1 | SiteID
          (Intercept) Residual
  StdDev:   0.8642518 3.052811

  Fixed effects: Speed2 ~ LaneFactor + CCRadius + CC_OSW + VehType + RampRadiusType + SCLType

                             Value Std.Error   DF    t-value p-value      Significance
  (Intercept)             9.511987 1.7298242 1564   5.498817  0.0000      Significant at 99% CL
  LaneFactor(Lane2&3)     1.240567 0.3808842 1564   3.257072  0.0011      Significant at 99% CL
  CCRadius                0.053491 0.0084193    6   6.353419  0.0007      Significant at 99% CL
  CC_OSW                  1.008357 0.2055809    6   4.904916  0.0027      Significant at 99% CL
  VehType(Truck)         -4.873262 0.2734472 1564 -17.821583  0.0000      Significant at 99% CL
  RampRadiusType(Simple)  3.550549 0.7998250    6   4.439158  0.0044      Significant at 99% CL
  SCLType(Drop lane)      2.911437 1.3000126    6   2.239545  0.0664      Significant at 90% CL
  SCLType(Parallel)       3.975007 0.7467030    6   5.323411  0.0018      Significant at 99% CL
  SCLType(Weave)          4.334356 1.1546617    6   3.753788  0.0095      Significant at 99% CL


  Beginning of Controlling Curve on the Ramp Proper of an Exit Loop Ramp
  Linear mixed-effects model fit by maximum likelihood

        AIC      BIC    logLik
    9032.565 9059.388 -4511.282

  Random effects:
   Formula: ~1 | SiteID
          (Intercept) Residual
  StdDev:    3.444244 4.137368

  Fixed effects: Speed1 ~  CCRadius + VehType
                       Value Std.Error   DF    t-value p-value     Significance
  (Intercept)      17.515341  4.125814 1565   4.245305   0e+00     Significant at 99% CL
  CCRadius          0.090093  0.019539   11   4.610841   8e-04     Significant at 99% CL
  VehType(Truck)   -5.967394  0.370594 1565 -16.102245   0e+00     Significant at 99% CL
```

Figure 3. Speed prediction models to estimate vehicle speeds at the midpoint and beginning of the controlling curve on the ramp proper of an exit loop ramp.

compound curve, and the type of speed-change lane significantly affects vehicle speeds at the midpoint of the controlling curve on an exit loop ramp. Vehicle speeds are approximately 2.9 mph faster following a lane drop, 4 mph faster following a parallel speed-change lane, and 4.3 mph faster following a weave area compared with vehicle speeds on an exit loop ramp with a tapered speed-change lane.

Lane Position. Field data were analyzed to assess the effects of design elements on the position of vehicles in the traveled way on loop ramps. The position of the vehicle reflects that of the right tire closest to the inside lane line (either the inside lane line separating the inside travel lane and inside shoulder, or the lane line separating the travel lanes on a multi-lane ramp). A positive value indicates the right tire is within the

Figure 4. Sample images from video to document vehicle lane position on a single-lane ramp.

intended travel lane, whereas a negative value indicates the right tire encroached onto the shoulder or inside travel lane. Figure 4 provides screen captures of video used to record lane positioning.

Researchers conducted lane position analyses similar to analyses for the speed data. Figure 5 provides the final models yielding the most information and best goodness of fit for estimating the lane position of vehicles at the midpoint of the controlling curve on the ramp proper of loop ramps.

Key findings from analyses of lane position on loop ramps are as follows:

Entrance Ramps:

- Design elements that significantly affect lane position at the midpoint of the controlling curve include lane width, outside shoulder width, superelevation, and grade.
- Vehicles in the outside lane of a multi-lane loop ramp are positioned approximately 10 in. farther away from the inside lane line than vehicles traveling in the inside lane.
- There is no significant difference in the lane positions between passenger vehicles and trucks.

Exit Ramps:

- Design elements that significantly affect lane position at the midpoint of the controlling curve are lane width and type of freeway mainline ramp terminal.
- Vehicles in the outside lane of a multi-lane loop ramp are positioned approximately 20 in. farther away from the inside lane line than vehicles traveling in the inside lane.
- Trucks are positioned approximately 6.6 in. farther away from the inside lane line than passenger vehicles.

Exiting Maneuvers. A qualitative analysis was conducted to assess patterns of unusual or critical maneuvers along the first several hundred feet of the ramp proper at nine exit loop ramps, and to determine which features of the sites could be contributing to the behavior. Approximately 7 h of video were reviewed to identify erratic or critical maneuvers of exiting vehicles, including maneuvers for approximately 1,300 passenger vehicles and 100 trucks. Figure 6 provides several screen captures illustrating critical maneuvers at one of the study ramps. In this figure, the driver makes a last-second decision to exit the freeway. The driver applied the brakes in the freeway lane and turned sharply into the painted gore area to complete the maneuver onto the exit ramp.

Table 3 provides a summary of encroachment and critical maneuvers observed on exit ramps by encroachment, maneuver type, and vehicle type. None of the maneuvers resulted in a crash. Key findings from the qualitative analysis are as follows:

- A majority of drivers entered the loop ramps in a controlled manner.
- Of the 139 encroachments and critical maneuvers observed, three involved swerving or severe braking, and 136 maneuvers involved encroachments on the shoulder.
- Given that 7.5% of the observed vehicles were trucks and trucks accounted for approximately 10% of the total number of encroachments and/or critical maneuvers observed, trucks may be slightly overrepresented in these shoulder encroachments and critical maneuvers, or both. For practical purposes, this percent

```
Midpoint of Controlling Curve on the Ramp Proper of an Entrance Loop Ramp

  Linear mixed-effects model fit by REML

        AIC      BIC    logLik
   14758.13 14812.09 -7369.066

  Random effects:
   Formula: ~1 | SiteID
          (Intercept) Residual
  StdDev:     4.07133 21.72377

  Fixed effects: LanePos ~ LaneFactor + CC_LW + CC_OSW + CCSuper100 + CCLength + Grade + VehType

                       Value Std.Error   DF   t-value p-value      Significance
  (Intercept)       42.18501 14.053540 1621  3.001735  0.0027      Significant at 99% CL
  LaneFactor(Lane2) 10.09629  3.036311 1621  3.325182  0.0009      Significant at 99% CL
  CC_LW              2.28886  0.654232 1621  3.498541  0.0005      Significant at 99% CL
  CC_OSW            -2.03742  1.009476    9 -2.018290  0.0743      Significant at 90% CL
  CCSuper100        -5.26412  1.262833    9 -4.168499  0.0024      Significant at 99% CL
  CCLength          -0.01661  0.006159    9 -2.697257  0.0245      Significant at 95% CL
  Grade(Up)        -14.59897  4.199929    9 -3.476004  0.0070      Significant at 99% CL
  VehType(Truck)    -0.55856  2.180207 1621 -0.256196  0.7978      Not significant at 90% CL


  Midpoint of Controlling Curve on the Ramp Proper of an Exit Loop Ramp
  Linear mixed-effects model fit by REML

        AIC      BIC    logLik
   14327.94 14370.84 -7155.968

  Random effects:
   Formula: ~1 | SiteID
          (Intercept) Residual
  StdDev:    6.570811 22.20448

  Fixed effects: LanePos ~ LaneFactor + CC_LW + CCSuper100 + SCLType + VehType
                           Value Std.Error   DF   t-value p-value    Significance
  (Intercept)          -42.16374 20.419250 1567 -2.064902  0.0391    Significant at 95% CL
  LaneFactor(Lane2)     20.34655  3.257214 1567  6.246611  0.0000    Significant at 99% CL
  CC_LW                  4.88004  0.954767 1567  5.111233  0.0000    Significant at 99% CL
  CCSuper100            -2.94853  1.730668   10 -1.703692  0.1193    Not significant at 90% CL
  SCLType(Drop lane)    13.74523  5.599425   10  2.454757  0.0340    Significant at 95% CL
  VehType(Truck)         6.59117  2.162167 1567  3.048412  0.0023    Significant at 99% CL
```

Figure 5. Lane position models for entrance and exit loop ramps.

difference is small enough that trucks should not, in general, be considered as overrepresented in the number of encroachments and critical maneuvers, or both that were observed.

- Combinations of narrower lane width and higher approach speed at the beginning of the controlling curve showed a higher proportion of observed encroachments and critical maneuvers.

Application of HSM Crash Prediction Method to Loop Ramps

The crash prediction methodology for ramp segments presented in HSM Chapter 19 does not separate procedures for specific ramp configurations, such as loop and diamond ramps. The same safety performance functions and crash modification factors are applied to both loop and diamond ramps. Therefore,

Figure 6. Sample images from video to document unusual or critical maneuvers on exit ramps.

researchers investigated how well these crash prediction methods represent the safety performance of these two ramp types with distinctly different geometrics. Details of this investigation, using 478 ramps in California and Washington, are provided in a recent paper by Torbic et al. (*9*). Those results indicate that the HSM crash prediction can be applied to both loop and diamond ramps; but to properly compare the safety performance, separate calibration of loop and diamond ramps is needed. If models are not calibrated separately for loop and diamond ramps, loop ramps will appear to have unrealistically high crash frequencies relative to the diamond ramps.

Design Guidance

Based on the findings from this research and existing design policies, design guidance for the ramp proper of loop ramps was developed addressing the following topics:

- recommended lane and shoulder widths for loop ramp proper,
- multi-lane ramps, and
- safety prediction of design alternatives.

Table 3. Summary Statistics of Encroachment and Critical Maneuvers Observed on Exit Ramps by Encroachment, Maneuver Type, and Vehicle Type

Ramp ID	Encroachment	Critical maneuver	Passenger vehicle	Truck	Total
16	Inside shoulder	None	3	2	5
	Outside shoulder	None	38	1	39
	None	Severe braking	0	1	1
	All	All	41	4	45
17	Outside shoulder	None	31	1	32
	All	All	31	1	32
18	Inside shoulder	None	1	0	1
	All	All	1	0	1
20	Inside shoulder	None	20	1	21
	Outside shoulder	None	1	1	2
	All	All	21	2	23
21	Inside shoulder	None	5	0	5
	Outside shoulder	None	0	2	2
	All	All	5	2	7
24	Inside shoulder	None	3	1	4
	Outside shoulder	None	0	2	2
	Outside shoulder	Swerving	1	0	1
	All	All	4	3	7
25	Inside shoulder	None	1	0	1
	Outside shoulder	None	1	1	2
	Outside shoulder	Swerving	0	1	1
	All	All	2	2	4
26	Outside shoulder	None	2	0	2
	All	All	2	0	2
27	Inside shoulder	None	18	0	18
	All	All	18	0	18
All Ramps	All	All	125	14	139

Recommended Lane and Shoulder Widths for Loop Ramps

The speed prediction and lane positioning models developed as part of this research along with the HSM crash prediction procedures were used to develop recommended lane and shoulder widths for the ramp proper of loop ramps. The recommended lane and shoulder widths were developed in three general steps:

1. Apply the speed prediction models for estimating speeds at the midpoint of the controlling curve
2. Use the HSM crash prediction procedures to compare predicted crash frequencies for different alternatives
3. Apply the lane position model to remove alternatives that result in vehicles encroaching on the shoulder or adjacent lane.

Table 4 presents recommended lane and shoulder widths for entrance loop ramps; and Tables 5 and 6 present the recommended lane and shoulder widths for exit loop ramps with simple and compound curves, respectively, for curve radii ranging from 100 to 300 ft and for design speeds from 20 to 35 mph. These recommended lane and shoulder widths are expected to induce speeds at or below the ramp design speed, result in similar levels of safety, and encourage drivers to stay within their intended travel lane. The guidelines are applicable to the controlling curve of the loop ramp. It is assumed that the same, or similar, widths will be used for all curves along the ramp proper. Alternatively, rather than using Tables 4 through 6, designers can use the speed and lane position prediction models developed as part of this research and the HSM crash prediction procedures to evaluate and assess design alternatives in more detail.

Table 4. Recommended Lane and Shoulder Width Combinations for Controlling Curve on Entrance Loop Ramps

Design speed (mph)	Radius (ft)	Lane width (ft)	Outside shoulder width (ft)	Inside shoulder width (ft)
20	100	14–16	2	2
25	150	14	2	2–6
			3	3–5
		16	2	2–5
			3	3–4
		18	2	2–4
			3	3
		20	2	2–3
25	200	14	2	2–3
		16–20	2	2
30	200	14	2	4–10
			3	3–9
			4	4–8
			5	5–6
		16	2	3–10
			3	3–8
			4	4–7
			5	5–6
		18	2	2–8
			3	3–7
			4	4–6
			5	5
		20	2	2–6
			3	3–5
			4	4
30	250	14	2	2–8
			3	3–6
			4	4–5
		16	2	2–7
			3	3–5
			4	4
		18	2	2–6
			3	3–4
		20	2	2–5
			3	3–4
30	300	14	2	2–5
			3	3
		16	2	2–4
			3	3
		18	2	2–3
		20	2	2
35	300	14	3	4–10
			4	4–9
			5	5–8
		16	2	5–10
			3	3–9
			4	4–8
			5	5–7
			6	6
		18	2	4–8
			3	3–7
			4	4–6
			5	5
		20	2	3–6
			3	3–5
			4	4

Table 5. Recommended Lane and Shoulder Widths for Exit Loop Ramps: Simple Curves

Design speed (mph)	Radius (ft)	Lane width (ft)	Outside shoulder width (ft)	Inside shoulder width (ft)
20	100	16	2	8–10
		18		6–8
		20		4–6
25	150	16	2	8–10
		18		6–8
		20		4–6
25	200	16	2	8–10
		18		6–8
		20		4–6
30	200	16	2	8–10
		18		6–8
		20		4–6
		16	3	7–9
		18		5–7
		20		3–5
30	250	16	2	8–10
		18		6–8
		20		4–6
35	300	16	2	8–10
		18		6–8
		20		4–6
		16	3	7–9
		18		5–7
		20		3–5

Multi-Lane Ramps

Based upon speeds and lane positions of vehicles on multi-lane loop ramps, no special design considerations are necessary for multi-lane loop ramps to accommodate differentials in speeds of vehicles in the outside lane compared with the inside lane, or to accommodate vehicles in the outside lane that gravitate to the inside and encroach on the inside travel lane. For multi-lane loop ramps, outside lane widths of 12 ft for entrance ramps and 14 ft for exit ramps are sufficient to accommodate traffic comprised primarily of passenger vehicles. Where the outside lane is expected to accommodate a moderate to high volume of trucks, the outside lane width should be increased.

Safety Prediction of Design Alternatives

When implementing the HSM crash prediction methodology to estimate predicted and expected crash frequencies, or both for individual ramps, separate calibration factors should be calculated for diamond and loop ramps.

Table 6. Recommended Lane and Shoulder Widths for Exit Loop Ramps: Compound Curves

Design speed (mph)	Radius (ft)	Lane width (ft)	Outside shoulder width (ft)	Inside shoulder width (ft)
20	100	16	2	8–10
		18		6–8
		20		4–6
25	150	16	2	8–10
		18		6–8
		20		4–6
		16	3	7–9
		18		5–7
		20		3–5
		16	4	6–8
		18		4–6
		20		4
25	200	16	2	8–10
		18		6–8
		20		4–6
30	200	16	3	7–9
		18		5–7
		20		3–5
		16	4	6–8
		18		4–6
		20		4
		16	5	5–7
		18		5
30	250	16	2	8–10
		18		6–8
		20		4–6
		16	3	7–9
		18		5–7
		20		3–5
		16	4	6–8
		18		4–6
		20		4
35	300	16	3	7–9
		18		5–7
		20		3–5
		16	4	6–8
		18		4–6
		20		4
		16	5	5–7
		18		5

Conclusions and Future Research Needs

General conclusions from the research are presented for entrance loop ramps, exit loop ramps, multi-lane loop ramps, and HSM crash prediction methodology for ramps. Recommended revisions to the AASHTO *Green Book* can be found in Appendix B of Part 1 in *NCHRP Web-Only Document 227* (*1*).

Entrance Loop Ramps

- Speeds of vehicles at the end of the controlling curve were found to be slightly higher than speeds at the midpoint of the controlling curve, indicating that vehicles accelerate while traversing the controlling curve on the ramp proper.
- Roadway and cross-sectional design elements that significantly influence vehicle speeds at the midpoint of the controlling curve include curve radius, lane width, inside (right) shoulder width, and outside (left) shoulder width.
- Roadway and cross-sectional design elements that significantly influence vehicle speeds at the end of the controlling curve include curve radius and outside shoulder width.
- Roadway and cross-sectional design elements that significantly influence lane position at the midpoint of the controlling curve include lane width, outside shoulder width, superelevation, and grade.
- As vehicles traverse an entrance loop ramp, the right tires of passenger vehicles and trucks are positioned approximately an equal distance from the inside lane line. Aside from offtracking issues associated with trucks, no major concerns associated with differences between the lane positions of trucks and passenger vehicles are expected.

Exit Loop Ramps

- Vehicle speeds were found to be slightly higher at the beginning of the controlling curve than at the midpoint of the controlling curve indicating that vehicles decelerate while transitioning from the freeway mainline ramp terminal along the ramp proper.
- The radius of the controlling curve is the only roadway or cross-sectional design element that significantly influences vehicle speeds at the PC of the controlling curve.
- Trucks are expected to travel slower than passenger vehicles on exit ramps.
- Roadway and cross-sectional design elements that significantly influence lane position at the midpoint of the controlling curve include lane width and type of freeway mainline ramp terminal.
- Trucks are typically positioned farther away from the inside lane line than passenger vehicles, and most passenger vehicles are positioned within the travel lane. Thus, the lane positioning of trucks should not raise

concerns about encroachment onto the inside shoulder of an exit ramp.

Multi-Lane Loop Ramps

- Vehicles in the outside lane(s) of a multi-lane loop ramp travel 1 to 2 mph faster than vehicles in the inside lane.
- Vehicles in the outside lane of a multi-lane loop ramp are positioned slightly farther from the inside lane line than vehicles in the inside lane.

HSM Ramp Crash Prediction Method

- The HSM crash prediction methodology is better at predicting crashes on diamond ramps than on loop ramps. Separate calibration factors for diamond ramps and loop ramps are needed for better comparisons between the safety performance of these two ramp types.

Future Research

Recommendations for higher priority research related to loop ramps are as follow:

- Performance of trucks on multi-lane loop ramps: Few trucks were observed in the outside lane of the multi-lane loop ramps included in the field study. Therefore, it would be desirable to expand upon this research and further evaluate differences in performance of single- and multi-lane ramps, focusing on the performance of trucks in the outside lanes.
- Capacity of loop ramps: The *Highway Capacity Manual* does not provide a methodology to estimate the capacity of a loop ramp, but a general rule of thumb is that a single-lane loop ramp can handle about 1,500 to 2,000 vph, and a two-lane loop ramp can handle up to 3,600 vph (*3*). These estimates are generally acceptable to practitioners given the overall capacity of a ramp is often limited by the capacity of the ramp terminal connections rather than the capacity of the ramp proper, but more could be learned about the capacity of a loop ramp, for example, how the capacity of a loop ramp varies with the speed of the ramp.
- Practical size of loop ramps: The *Green Book* states that, "considering all factors, experience shows that the practical size of loops resolves into approximate radii of 30 to 50 m (100 to 170 ft) for minor movements on highways with design speeds of 80 km/h (50 mph) or less and 50 to 75 m (150 to 250 ft) for more important movements on highways with higher design speeds" (*2*). Research is needed to determine the "practical size" of a loop ramp, considering issues such as tradeoffs between desired speed, distance traveled, and construction and right-of-way costs.

Acknowledgments

The research reported here was performed under NCHRP Project 3-105, "Design Guidance for Interchange Loop Ramps" as part of Phase I and Phase II. For more details on the research refer to Part 1 of the *NCHRP Web-Only Document 227* (*1*).

References

1. Torbic, D. J., L. M. Lucas, D. W. Harwood, M. A. Brewer, E. S. Park, R. Avelar, M. P. Pratt, A. Abu-Odeh, E. Depwe, and K. Rau. *NCHRP Web-Only Document 227: Design of Interchange Loop Ramps and Pavement/Shoulder Cross-Slope Breaks*. TRB, National Research Council, Washington, D.C., 2017.
2. American Association of State Highway and Transportation Officials (AASHTO). *A Policy on Geometric Design of Highways and Streets*. AASHTO, Washington, D.C., 2011.
3. Leisch, J. P. *Freeway and Interchange Geometric Design Handbook*. Institute of Transportation Engineers, Washington, D.C., 2006.
4. Torbic, D. J., J. M. Hutton, C. D. Bokenkroger, D. W. Harwood, D. K. Gilmore, M. M. Knoshaug, J. J. Ronchetto, M. A. Brewer, K. Fitzpatrick, S. T. Chrysler, and J. Stanley. NCHRP Report 730: *Design Guidance for Freeway Mainline Ramp Terminals*. TRB, National Research Council, Washington, D.C., 2012.
5. Lord, D., and J. A. Bonneson. *Estimating the Safety of Four Ramp Design Configurations*, Presented at 3rd International Symposium on Highway Geometric Design, Chicago, Illinois, Transportation Research Board, 2005.
6. Lundy, R. A. *Effect of Ramp Type and Geometry on Accidents*. California Department of Public Works, 1965.
7. Yates, J. G. Relationship Between Curvature and Accident Experience on Loop and Outer Connection Ramps. *Highway Research Record*, Vol. 312, 1970, pp. 64–75.
8. Bauer, K. M., and D. W. Harwood. *Statistical Models of Accidents on Interchange Ramps and Speed-Change Lanes*. Report FHWA-RD-97-106. Federal Highway Administration, Washington, D.C., 1998.
9. Torbic, D. J., D. W. Harwood, and K. M. Bauer. Application of the HSM Ramp Crash Prediction Method to Loop and Diamond Ramps. *Transportation Research Record: Journal of the Transportation Research Board*, 2017. 2636, 43–52.

The Standing Committee on Geometric Design (AFB10) peer-reviewed this paper (18-06447).

Article

TRR
JOURNAL OF THE TRANSPORTATION RESEARCH BOARD

Transportation Research Record
2018, Vol. 2672(39) 34–40

DOI: 10.1177/0361198118758310
journals.sagepub.com/home/trr
SAGE

A Model of the Probability of a Cross-Median Crash When a Vehicle Fully Crosses the Median

Christine E. Carrigan[1] **and Malcolm H. Ray**[1]

Abstract

The consequences of cross-median crashes are often catastrophic but crashes into median barriers can also be severe. Wide medians provide traversable space where vehicles can recover or stop but sometimes even wide medians can be crossed over. Determining if a median barrier is needed at a particular site involves balancing the risks associated with crossing completely over the median and striking a vehicle in the opposing lanes with the risks of an errant vehicle striking a median barrier. Median crossover crashes can be viewed as a conditional probability model: first the vehicle must enter the median; second, the vehicle must cross completely over the median; third, a vehicle must be in the opposing lanes where it is struck and, finally, there is a chance of a severe or fatal injury if all these conditions are met. All the elements of this conditional probability model are documented in the literature except the third; the probability of a crash with an opposing-direction vehicle given that a vehicle has completely traversed the median. Estimating this probability is the subject of this paper.

For several decades there has been interest in the roadside safety community in developing selection and placement guidance for the multiple test levels of median barriers. The large variety of median widths and terrains combined with evolving testing specifications and lack of conclusive data on median cross-over crashes have been significant obstacles to developing more realistic guidance. The current implementation of the *Manual for Assessing Safety Hardware* (MASH) (*1*) combined with new data collection efforts and the availability of new analysis tools have overcome some of these obstacles to developing median barrier guidance.

Variables like the presence or absence of a barrier, the offset from the lanes to the barrier, the type and test level of the barrier, the highway and median characteristics, and the traffic characteristics can each be assessed using updated tools such as the third version of the roadside safety analysis program (RSAPv3) by any transportation agency to develop guidance for median barriers (*2*). RSAPv3 contains a great deal of internal data used to model these relationships. While many of these supporting parameters are already available in RSAPv3, one element unique to the problem of median-related-events (MREs) that was lacking in the literature is an understanding of how an MRE becomes a cross-median-crash (CMC).

The probability of a CMC – P(CMC) – is assumed to be a conditional probability in which a vehicle first fully crosses the median and then strikes another vehicle in the opposing direction. Given an MRE has occurred and the vehicle's trajectory extends the full width of the median, the P(CMC) is a function of another vehicle being present in the opposing lanes. The crash severity distribution of head-on crashes has been found in other studies and is represented in RSAPv3 (*2*). Developing a model to represent the probability of a CMC given an MRE [i.e., P(CMC|MRE)] was the objective of this study and is documented in this paper.

Model Considerations

The words median and median barrier have minor, but different definitions among the AASHTO and state Department of Transportation literature (*3–5*). NTSB has asked FHWA and NHTSA to provide a more consistent definition for CMCs, citing a lack of consistency throughout the states (i.e., H-98-13, H-11-28). It is helpful, therefore, to first establish the definitions of terms used consistently in this paper. These definitions are provided in Table 1.

The definitions for CMEs, MREs, and CMCs are interrelated. Appreciating the importance of inter-relationship at the onset of this effort is crucial to ensuring proper data management and reliable analysis outcomes. This P(CMC) model will be used as part of the encroachment probability model,

[1]RoadSafe, LLC, Canton, ME

Corresponding Author:
Address correspondence to Christine E. Carrigan:
christine@roadsafellc.com

Table 1. Definitions of Terms

Median	The portion of a divided highway separating the traveled ways for traffic in opposite directions.
Median barrier	A longitudinal barrier system intended to reduce the risk of an errant vehicle crossing the highway median. Median barriers are designed to be impacted from either direction of travel.
Median-Related-Event (MRE)	Any event in which an errant vehicle enters the median. MREs represent all vehicles that encroach left into the median, regardless of the outcome (i.e., crash or no crash).
Cross-Median Event (CME)	An event in which an errant vehicle fully crosses the median and may or may not collide with another vehicle from the opposite direction. CMEs are a subset of MREs.
Cross-Median Crash (CMC)	A cross-median crash is one in which an errant vehicle crosses the median of a highway and strikes or is struck by a vehicle from the opposite direction. CMCs are a subset of CMEs, which are, in turn, a subset of MREs.

which is a conditional probability model. The encroachment probability model first predicts the probability of a median-related-event, P(MRE). Given an MRE, the encroachment probability model then predicts the probability that the errant vehicle's trajectory will extend to the other side of the median and become a cross-median-event (CME), P(CME|MRE). Given a CME has occurred, the encroachment probability model then determines the probability that the CME will become a CMC, P(CMC|CME).

Traffic volume in the opposing direction is thought to affect P(CMC|CME). The conditional probability of CMC given CME, P(CMC|CME) is defined by this relationship:

$$P(CMC|CME) = \frac{P(CMC \cap CME)}{P(CME)}$$

In other words, the probability of both a CMC *and* CME, $P(CMC \cap CME)$ compared with the P(CME) determines the conditional probability of a CMC, which is the object of this study. This modeling technique is founded on the assumption that P(CME) is greater than zero, a reasonable assumption considering the many observed CMCs and the definitions used to establish these variables.

Avoiding Measuring the Same Variable Twice

The effect of median width is captured through modeling P(CME), so care should be taken to not measure the effect of median width when modeling P(CMC|CME). It was desirable, therefore, to model what happens when a vehicle reaches the far edge of the median, absent the influence of median width and terrain. This was accomplished by considering cross-over crashes (COCs) on undivided roadways. An undivided roadway can be viewed as essentially a divided roadway with a median width equal to the distance between the double yellow lines (i.e., typically 1 ft) and flat median terrain. Head-on and opposite direction side-swipe crashes that occurred on undivided roadways were used to develop the data set upon which these efforts are based to remove the influence of the median width and median terrain confounders. In essence, on two-lane undivided roadways the P(CME|MRE) is essentially unity.

For roadways with a median width equal to essentially zero (i.e., undivided roadways), P(CME) = P(MRE) = 1.00 because all left encroachments cross both yellow lines at the center of the roadway at the start of the encroachment event by definition.

The conditional probability of CMC given CME, which was established above, can be rewritten for this exercise as follows:

$$P(CMC|CME) = P(COC|MRE) = \frac{P(COC \cap MRE)}{P(MRE)}$$

In other words, the probability of both a COC and MRE compared with the P(MRE) is a surrogate for the conditional probability of a CMC that eliminates the effect of median width and terrain.

P(MRE) is equal to the frequency of left-encroaching vehicles compared with the average annual daily traffic (AADT) of the segment of interest, $P(MRE) = FREQ_{enc} / AADT_{seg\,i}$, whereas the $P(COC \cap MRE)$ is equal to the frequency of a COC and left-encroaching vehicle compared with the AADT of the segment of interest, $P(COC \cap MRE) = FREQ_{coc\,\&\,enc} / AADT_{seg\,i}$. The above relationship can be further simplified as follows:

$$P(CMC|CME) = P(COC|MRE) = \frac{P(COC \cap MRE)}{P(MRE)}$$

$$= \frac{FREQ_{coc\,\&\,enc} / AADT_{seg\,ii}}{FREQ_{enc} / AADT_{seg\,i}} = \frac{FREQ_{coc\,\&\,enc}}{FREQ_{enc}}$$

The frequency of left-encroaching vehicles on divided highways is known based on other studies (*2, 6*). The frequency of COCs when a left encroachment over the centerline has occurred, however, is not known. The focus of model development is to determine the frequency of COCs using the same units the frequency of encroachment is tabulated by to develop this piece of the encroachment probability model, such that the results can be incorporated into RSAPv3. The $FREQ_{COC}$ model itself must control for the remaining confounding factors.

Method

The practice for modeling count data such as highway crashes is to fit a negative binomial model, usually with a Poisson-gamma-mixture distribution. "In statistics, count data refer to observations that have only nonnegative integer values ranging from zero to some greater undetermined value" (*7*). In highway safety, zero counts of crashes are particularly important and represent areas in which crashes were not observed. One approach to accounting for zero counts as well as the non-zero counts is to track crashes by highway segment. This approach has the added benefit of allowing the consideration of the influence of segment characteristics like horizontal curvature, grade, lane width, etc. on crash frequency.

The crash count becomes the response variable and the segment characteristics such as AADT, percent trucks (PT), highway geometrics, and area type become the explanatory variables that explain the occurrence of the crashes. Each segment is associated with each of the predictor variables and the number of crashes that occur on that segment during the study period. The characteristics of a segment are used to explain why each segment experiences more or fewer crashes than other segments.

Ideally, all possible predictor variables would be known. This ideal situation, however, remains unrealized, so it is common to consider the known predictor variables when developing a model to ensure the effect of the predictor variable of interest is not misrepresented.

Roadway characteristics that may also modify the P(CMC|CME) such as PT, highway geometrics, and land use (i.e., urban and rural) are recognized confounding factors, but are accounted for elsewhere in the encroachment probability model. Ensuring that the final representation of P(CMC|CME) when implemented in the encroachment probability model does not double-count the effect of these confounders is equally important to controlling for their effect. Controlling for these confounding variables was attempted using two different approaches: 1) explicitly modeling confounding variables, and 2) limiting the data set to segments in which the confounders contained measurements within the base conditions of the previously developed models. The latter is ultimately recommended, as discussed below.

Under both approaches, a negative binomial regression model of COCs was estimated using the COUNT package available in R (*8, 9*). The model relates the explanatory variables to the response variable using the method of maximum likelihood to quantify the magnitude of each predictor's relationship. The fit statistics are presented along with each model. The *p*-value is a measure of how probable the result observed may have occurred by chance. A low *p*-value indicates the results are statistically significant and were unlikely to have occurred by chance (e.g., $p<0.05$). A higher *p*-value only indicates that the results have not proven the null hypothesis false, not that the null hypothesis is true. The *p*-value cannot be relied on alone.

The pseudo-R^2 statistic was determined for each model. The pseudo-R^2 is not interpreted the same way as the coefficient of determination for an ordinary least squares regression. A low value pseudo-R^2 can indicate lack of fit, whereas higher values carry no such indication. There is no definition for a low value. The Akaike Information Criterion (AIC) fit statistic provides comparative information, with lower values indicating a better fitting model then the model it is compared with. The Bayesian Information Criterion (BIC) is interpreted the same way. Both are calculated from the likelihood function (*10*).

Negative binomial model parameters are estimated using maximum likelihood in which the parameters of the probability distribution that characterize the data are estimated. The log of the likelihood function (LL) is used to determine which parameters make the model most likely to be the case when the data is considered. Through an iterative process, the derivative of the LL function is taken and set to zero to estimate the parameters. When the difference between iterative values is less than a specified tolerance (e.g., 10^{-6}), the iterations stop and the values are at the maximum likelihood estimated values. The LL is also reported with the models, however, it is only useful when calculating other fit statistics (e.g., AIC and BIC).

Any measurement has uncertainty, which should be communicated. This uncertainty in statistical analyses can be conveyed through noting the standard error or the confidence interval along with the measurements. The standard error is a measure of how much the estimate could change within the model. The 95% confidence interval is essentially the same type of statistic as standard error: the 95% confidence interval limits indicate that the analysis is 95% confident the true value of the coefficient is within the stated range. It is important to note that the 95% confidence interval is equal to twice the standard error for normally distributed errors, and negative binomial models are assumed to have normally distributed errors.

Data Used for Modeling

The original intent was to use the models developed under NCHRP 794, "Median Cross-Section Design for Rural Divided Highways" (*11*). Unfortunately, there appears to be a typographical error in the model printed in NCHRP Report 794, as both the CMC+CME and CMC models shown are identical. Instead, the Highway Safety Information System database of Ohio and Washington highway crashes was used, which could be linked to highway segment information such as AADT, percentage of heavy vehicles (PT), segment length (SegL), area type, speed limit (spd_limt), number of lanes (no_lanes), lane width (lanewid), vehicle type and the crash severity distribution as represented by the KABCO scale. This database was requested for Ohio for 2002 through 2010 and for Washington from 2002 through 2007.

Table 2. Descriptive Statistics for P(CMC) Data Set

Continuous variables	Min.	1st Qu.	Median	Mean	3rd Qu.	Max.
L	0.1	0.13	0.19	0.31	0.35	2
AADT	10	1,110	2,280	3,611	4,760	68,336
PT	0.0	3.8	5.7	7.00	8.6	67
Lane width	7	10	10	10.97	12	41
Shoulder width	0	2	3	3.59	4	30
DOC	0	0	0	0.52	0	76
PG	0	0	0	1.52	0	20
PC_KABCOU	0	0	0	0.0345	0	6
HV_KABCOU	0	0	0	0.0075	0	3
MC_KABCOU	0	0	0	0.0008	0	2
KABCOU	0	0	0	0.0445	0	7

Categorical variables	Count	Categorical variables	Count	Categorical variables	Count
PSL = 20	68	PSL = 40	5,528	PSL = 60	1,764
PSL = 25	5,783	PSL = 45	20,958	PSL = 65	229
PSL = 30	392	PSL = 50	6,868	Rural	185,414
PSL = 35	30,921	PSL = 55	149,064	Urban	35,561

Note: PSL = posted speed limit (mph); LW = lane width (ft).

Preparing the Data to Develop the Model

A data set of homogenous segments was merged with cross-over and opposite direction side-swipe crashes (i.e., acctype field codes 1 and 4). The crashes were counted by crash severity and vehicle type and assigned to the appropriate homogenous segment using the recorded route and milepost of each crash. The resulting data set included a list of segments. Each segment had a field for AADT, PT, SegL, area type, spd_limt, no_lanes, and lanewid. Each segment also contained a field for the count of crashes occurring on the segment by each crash severity and vehicle type (e.g., passenger car fatal crash = PC_K; heavy vehicle serious crash = HV_A, etc.). The data set includes 1,204,084 segments, some with crashes and some without.

In some instances, the segments included fields for which the information was not available (NA) or the field contained a nonsense value (e.g., AADT = 0). Prior to working with the data set, segments with incomplete data or nonsense values were removed from consideration. The data set was filtered to remove these segments as shown here prior to any modeling. The segments remaining after each filter step are noted in parenthesis.

- Consider only segments in which the land use type (i.e., urban or rural) is known (1,202,105)
- Consider only segments in which the length in miles is $0.1 \leq L \leq 2$ (404,620)
- Consider only segments in which AADT > 0 (403,666)
- Consider only segments in which the PT is known (242,862)
- Consider only segments in which the value for the number of lanes is equal to two (221,171)
- Consider only segments in which the posted speed limit > 0 mph (220,975).

This filtering of the data set resulted in 220,975 segments being included in the modeling data set. The descriptive statistics for this data set are shown in Table 2.

Model Development

Model with Control Variables

Recall, the number of crashes per year were tabulated in the data set. A negative binomial model was fit to the data set of head-on crashes, and the log of the SegL in miles was included as an offset to allow for the frequency of opposite direction crashes to be evaluated per year per mile. The resulting parameter estimates are shown in Table 3 for the cross-over model with control variables that take this form:

$$FREQ_{COC} = \cdot e^{B1} \cdot AADT^{B2} \cdot PT^{B3} \cdot \prod_{i=4}^{N} e^{Bi \cdot Ai}$$

where

$FREQ_{COC}$ = Frequency of cross-over crashes per year per mile,

AADT = Annual Average Daily Traffic (vpd),

A_i = Control variable values for each segment under consideration,

Table 3. Negative Binomial Model for Cross-Over Crashes

Coefficients:	Parameter estimate	Standard error	*P*-value	95% confidence interval	
(Intercept)	−9.3993	0.49	< 2e–16	−10.4910	−8.5228
log(AADT)	0.9025	0.02	< 2e–16	0.8690	0.9363
log(PT)	0.0235	0.00	1.32e–15	0.0178	0.0293
Urban	0.1343	0.03	6.69e–06	0.0758	0.1928
Rural	1.0000	NA	NA	NA	NA
PSL.20	1.0000	NA	NA	NA	NA
PSL.25	0.4377	0.47	0.353	−0.3989	1.5042
PSL.30	0.5085	0.50	0.309	−0.3970	1.6172
PSL.35	0.1129	0.47	0.810	−0.7197	1.1770
PSL.40	0.0043	0.47	0.993	−0.8350	1.0725
PSL.45	−0.0708	0.47	0.880	−0.9059	0.9948
PSL.50	−0.3035	0.47	0.521	−1.1438	0.7653
PSL.55	−0.2767	0.47	0.556	−1.1104	0.7880
PSL.60	−0.7111	0.51	0.162	−1.6382	0.4095
PSL.65	−0.6755	0.73	0.357	−2.1498	0.7372
lanewid	0.0215	0.00	1.13e–06	0.0127	0.0301
DOC	0.0454	0.01	2.73e–14	0.0326	0.0569
PG	0.0454	0.01	1.39e–08	0.0187	0.0386
SHLDR_PRE	−0.0517	0.00	< 2e–16	−0.0606	−0.0429
AIC			68,272		
BIC			68,447		
Dispersion parameter (α)			1.099		
Standard error			0.0730		
LL (full)			−34,119		
Pseudo R^2			0.15		

B_i = Regression coefficients,

N = Total number of control variables considered per segment.

Limiting Data set to Base Conditions to Remove Confounders

Recall that the initial data set was filtered to remove segments with missing or nonsense values, which resulted in a data set of 220,975 segments used in the model developed that included control variables. A different approach is taken in this section when the data set only includes segments that meet the base conditions of the complementary encroachment probability model for which this P(CMC) model is being explored, such that this P(CMC) model does not account for variation in highway characteristics that are already accounted for elsewhere in the encroachment probability model.

The 220,975 segment data set was further limited for this analysis as shown here, with the remaining segments after each filter step noted in parenthesis:

- Consider only segments in which the area type is rural (185,414)
- Consider only segments in which the PT ≥ 10 (36,038)
- Consider only segments in which the posted speed limit ≥45mph (32,069).
- Consider only segments in which 10 ft ≤ lanewid ≥ 12 ft (24,690)
- Consider only segments in which the DOC = 0 (21,918)
- Consider only segments in which –2% ≤ PG ≥ +2% (17,443).

This base-condition limited data set resulted in 17,443 segments being included in the analysis. The descriptive statistics for this base-condition limited data set are shown in Table 4 using the same shorthand for the categorical variables previously discussed with respect to Table 2.

Again, a negative binomial model was fit to the data set of cross-over and opposite direction side-swipe crashes and the log of the SegL in miles was included as an offset to allow for the frequency of head-on crashes to be evaluated per year per mile. The resulting parameter estimates are shown in Table 5 for the cross-over model which explicitly limits confounders and takes this form:

$$FREQ_{COC} = \cdot e^{B1} \cdot AADT^{B2}$$

where

Table 4. Descriptive Statistics for Limited P(CMC) Data Set

Continuous variables	Min.	1st Qu.	Median	Mean	3rd Qu.	Max.
L	0.1	0.17	0.32	0.4879	0.65	2
AADT	40	1,594	2,850	3,604	5,110	20,260
PC_KABCOU	0	0	0	0.0308	0	3
HV_KABCOU	0	0	0	0.0170	0	3
MC_KABCOU	0	0	0	0.0007	0	1
KABCOU	0	0	0	0.0501	0	5
Categorical variables		Count	Categorical variables	Count	Categorical variables	Count
PSL = 45		1,143	PSL = 55	15,510	PSL = 65	107
PSL = 50		562	PSL = 60	121		

Table 5. Negative Binomial Model for Cross-Over Crashes

Coefficients:	Parameter estimate	Standard error	*P*-value	95% confidence interval	
(Intercept)	−11.3901	0.49	< 2e–16	−12.3625	−10.4393
log(AADT)	1.1050	0.06	< 2e–16	0.9928	1.2193
AIC			6,142		
BIC			6,158		
Dispersion parameter (α)			1.296		
Standard error			0.339		
LL (full)			–3,069.25		
Pseudo R^2			0.12		

$FREQ_{COC}$ = Frequency of cross-over crashes per year per mile,

AADT = Annual Average Daily Traffic (vpd),

B_i = Regression coefficients.

Recommended Model

Recall, a $FREQ_{COC}$ model that controls for highway characteristics is desired and that this model is presumed to be a function of AADT. The severity of these crashes is understood through a different model and need not be considered here, rather, this model should consider all observed crashes to allow for the conversion from a frequency model to a probability model. The simpler model in which the confounders are explicitly controlled through limiting segments considered to base-condition segments, as documented in Table 5, is preferred because (1) the AIC and BIC values are lower, (2) the model makes better engineering sense, and (3) the simpler model best satisfies the principals of parsimony (i.e., Occam's Razor).

The frequency of COCs when a left encroachment over the centerline on a two-lane undivided highway has occurred is now estimated to be:

$$FREQ_{COC \cap enc} = \cdot e^{-11.3901} \cdot AADT^{1.1050}$$

It was previously derived that $P(CMC|CME) = FREQ_{coc \cap enc} / FREQ_{enc}$. The frequency of left encroaching vehicles by direction of travel, $FREQ_{enc}$ has been documented previously *(2, 6)*.

Table 6. Recommended P(CMC|CME)

Rural Divided Highways	
For 0 < AADT < 29,000:	$0.178\, AADT^{-0.1002} e^{(0.2104+0.04128\, AADT/1000)}$
29,000 < AADT < 43,000:	$3.042\, AADT^{1.1050} \cdot 10^{-6}$
43,000 < AADT < 67,000:	$1.145 AADT^{-0.1002}$
67,000 < AADT < 90,000:	$0.117 AADT^{0.1050}$
AADT > 90,000:	$1.300\, AADT^{1.1050} \cdot 10^{-6}$
Urban Divided Highways	
For 0 < AADT < 26,000:	$0.052 AADT^{0.0162} e^{(0.2104+0.04128\, AADT/1000)}$
26,000 < AADT < 45,000:	$2.934 AADT^{1.1050} \cdot 10^{-6}$
45,000 < AADT < 90,000:	$0.335 AADT^{0.0162}$
AADT > 90,000:	$1.353\, AADT^{1.1050} \cdot 10^{-6}$

This cross-over centerline model accounts for traffic volume in the opposing lanes and is only dependent on vehicles being present in the opposing lanes. This principle can be extended to medians using the conditional probability model. As in the centerline cross-over model, the probability of a CMC, given a CME, is only a function of traffic volume in the opposing lanes and vehicles being present. Therefore, this undivided model can be used to model CMC given a CME.

The $FREQ_{enc}$ is documented by direction of travel. That is to say that vehicles originating from the primary direction of travel and encroaching into the median are counted once,

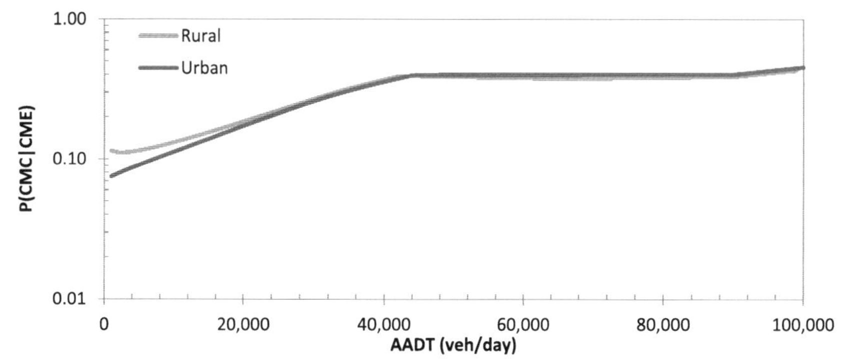

Figure 1. Recommended P(CMC|CME).

whereas those encroachments originating in the opposing direction of travel and encroaching into a median are counted separately. Until this point, the $FREQ_{COC \cap enc}$ model has considered the frequency of crashes from each direction in total for a median width of zero. This frequency model must be divided by two to allow for consideration of each direction of travel as the encroachments are considered, and as it will be used in RSAPv3. The results are shown in Table 6 in equation form, combined with the encroachment frequency model, and graphically in Figure 1.

Conclusion

The P(CMC|CME) model provides a missing piece of the conditional encroachment probability model for modeling cross-median crashes and will provide valuable insight into predicting the probability of these events and ultimately the risk of a serious or fatal (A + K) crash when these events occur. This model is proposed for use in RSAPv3 to represent the influence traffic volume in the opposing direction has on the probability of a cross-median crash, P(CMC|CME). This new P(CMC|CME) model will be added to RSAPv3 for use by any agency who wishes to consider median design on a case-by-case basis or develop agency-specific guidance.

This model is proposed to be used in conjunction with the severity model [i.e., P(A+K|CMC)] to determine the risk of an A + K for CMCs. This value is an important part of the conditional probability model for determining the frequency of severe and fatal CMCs and it is also an important aspect of comparing the severity of median barrier crashes, as median barriers should only be placed where the frequency of severe or fatal barrier crashes is less than the frequency of severe and fatal CMCs.

Acknowledgments

This work was sponsored in part by NCHRP 22-31. The authors thank the project panel and the program officer, Mr. Mark Bush, for their thoughtful comments and feedback throughout the research effort.

References

1. Technical Committee for Roadside Safety. *Manual for Assessing Safety Hardware*. American Association of State Highway and Transportation Officials, Washington, D.C., 2009.
2. Ray, M. H., C. E. Carrigan, C. A. Plaxico, and T. O. Johnson. *Engineer's Manual: Roadside Safety Analysis Program (RSAP) Update*. Roadsafe LLC, Canton, Maine. http://rsap.roadsafellc.com/RSAPv3EngineersManual.pdf. Accessed December 2012.
3. Technical Committee on Roadside Safety. *Roadside Design Guide*. American Association of State Highway and Transportation Officials, Washington, D.C., 2006.
4. American Association of State Highway and Transportation Officials (AASHTO). *A Policy on the Geometric Design of Highways and Streets*. AASHTO, Washington, D.C., 2011.
5. CALTRANS Traffic Manual. *Chapter 7, accessed online*. http://www.dot.ca.gov/hq/traffops/engineering/safety-devices/docs/Median-Barrier.pdf. Accessed February 4, 2016.
6. Ray, M. H., and C. E. Carrigan. Consideration of Roadside Features in the Highway Safety Manual. *NCHRP Project 17-54, National Cooperative Highway Research Program*. Transportation Research Board, Washington, D.C., (in progress).
7. Hilbe, J. M. *Modeling Count Data*. Cambridge University Press, Cambridge, 2014.
8. Hilbe, J. M. *COUNT: Functions, Data and Code for Count Data*. R Package Version 1.3.4. https://CRAN.R-project.org/package=COUNT. Accessed October 19, 2016.
9. R Core Team (2016). *R: A Language and Environment for Statistical Computing*. R Foundation for Statistical Computing, Vienna, Austria. https://www.R-project.org/. Accessed June 28, 2016.
10. Hilbe, J. M. *Negative Binomial Regression*, 2nd ed. Cambridge University Press, Cambridge 2012.
11. Graham, J. L., D. W. Harwood, K. R. Richard, M. K. O'Laughlin, E. T. Donnell, and S. N. Brennan. *NCHRP Report 794: Median Cross Section Design for Rural Divided Highways*. Transportation Research Board of the National Academies, Washington, D.C., 2014.

The Standing Committee on Roadside Safety Design (AFB20) peer-reviewed this paper (18-01503).

Article

TRR
JOURNAL OF THE TRANSPORTATION RESEARCH BOARD

Transportation Research Record
2018, Vol. 2672(39) 41–51

DOI: 10.1177/0361198118758676
journals.sagepub.com/home/trr
SAGE

Development of a Standardized Buttress for Approach Guardrail Transitions

Scott K. Rosenbaugh[1], Jennifer D. Schmidt[1], and Ronald K. Faller[1]

Abstract

Approach guardrail transitions (AGTs) incorporate increased post and rail sizes, reduced post spacings, and specialized buttress end geometries to smoothly transition from deformable W-beam guardrail to rigid barriers. This transition in barrier stiffness makes AGTs sensitive systems that require specific combinations of these components to function properly. Changing components, or even the removal of a curb below the rail, can negatively affect the safety performance of an otherwise crashworthy system. However, recent full-scale crash testing has indicated that a properly designed buttress at the downstream end of an AGT may be utilized with multiple AGT systems. Thus, the objective of this project was to develop a standardized buttress to reduce vehicle snag and be compatible with a wide variety of previously developed Thrie beam AGT systems, either with or without a curb. The standardized buttress was designed with a dual taper on its front upstream edge. A longer lower taper was designed to mitigate tire snag below the rail, while a shorter upper taper was designed to prevent vehicle snag and limit the unsupported span length of the rail. This buttress design was evaluated in combination with a critically weak AGT without a curb, which represented the worst-case scenario. The standardized buttress was successfully crash tested to MASH TL-3. Guidance was provided for both the attachment of the buttress to various Thrie beam AGTs as well as how to transition the shape of the buttress to adjacent bridge rails or rigid parapets downstream of the AGT.

Approach guardrail transitions (AGTs) are utilized to attach deformable W-beam guardrail to various rigid barriers, including bridge rails and reinforced concrete parapets. To smoothly transition between the different barrier stiffnesses and prevent vehicle snag, AGTs typically incorporate thicker and/or larger guardrail segments (i.e., Thrie beam), increased post sizes, and decreased post spacings. In addition, the upstream end of the rigid barrier to which the AGT is attached is often modified with various tapers and flares to minimize the risk of vehicle snag. Curbs have also been placed below the guardrail and adjacent to the rigid barrier to further reduce the likelihood of tire snag. AGTs consist of a specific combination of these components and roadside features to make them crashworthy.

Over the last couple of decades, multiple AGTs have been developed to satisfy the safety performance criteria of either the Manual for Assessing Safety Hardware (MASH) or NCHRP Report 350 (*1,2*). However, full-scale crash testing has illustrated the sensitive nature of guardrail stiffness transitions. Changing only a single component or feature of an AGT can significantly alter its safety performance. For example, the addition/removal of a curb, altering the geometry of the rigid parapet, or altering the embedment length of the transition posts, can be the difference between a failure and a successfully crash tested AGT (*3–11*). Owing to the sensitivity of stiffness transitions, AGT components and features (e.g., curb usage and rigid barrier geometry) are not interchangeable between systems.

The majority of failures observed during crash testing have been the result of excessive vehicle contact with the rigid parapet, especially for AGTs that did not utilize a curb beneath the guardrail. These tests indicated that the geometry of the rigid parapet was more critical than previously believed. Thus, the development of a concrete buttress end geometry was desired to minimize the risk of vehicle snag and to be crashworthy in combination with various Thrie beam AGTs.

Research Objective

The objective of this research project was to develop and evaluate a standardized buttress geometry for use with Thrie beam AGTs. The transition buttress needed to be compatible with all of the previously developed Thrie beam AGT systems that were successfully crash tested to the Test Level 3 (TL-3) performance criteria of either MASH or NCHRP

[1]Midwest Roadside Safety Facility, University of Nebraska-Lincoln, Lincoln, NE

Corresponding Author:
Address correspondence to Scott K. Rosenbaugh: srosenbaugh2@unl.edu

Report 350. In addition, the buttress needed to safely transition from stiffened Thrie beam to a variety of concrete parapet and bridge rail shapes. Finally, AGTs incorporating the standardized buttress needed to be crashworthy in both curbed and non-curbed installation configurations.

Standardized Buttress Design

Development of the standardized transition buttress began with a review of previous full-scale crash testing on AGTs connected to concrete parapets. Since a limited number of AGTs had been evaluated to MASH standards, the review included both MASH and NCHRP Report 350 tested systems. Thirty-nine crash tests, which were conducted on 20 different transition systems, were reviewed in order to identify tendencies between the crashworthy systems and those that failed to meet the safety criteria (*12*). Knowledge gathered from this review was utilized to guide the design of the standardized buttress.

During the literature review, it was noted that nearly all AGTs were designed with the Thrie beam end connector mounted vertically on the concrete parapet. If the parapet had a sloped face (e.g., New Jersey, F-shape, or single slope barriers), a wedge-shaped connection plate was typically utilized between the Thrie beam end connector and the parapet, which allowed the rail to remain in a vertical orientation. Only two tests were conducted on Thrie beam AGTs with the rail twisted to match the sloped face of the parapet, and both of those NCHRP Report 350 tests resulted in vehicle rollovers (*5,13*). Thus, it was desired to keep the rail element vertical throughout the AGT. To keep the AGT design simple and avoid the added components and costs associated with requiring a connection plate, the standardized buttress was designed with a vertical front face geometry. The vertical shape could then be transitioned into different parapet shapes downstream of the rail end connector.

Multiple AGTs have been designed with a rub rail placed below the rail to mitigate tire snag. However, five out of the seven tests conducted on AGTs incorporating rub rails were failures, of which four were vehicle rollovers (*14–19*). These results indicate that tire interactions with rub rails may lead to vehicle instabilities during redirection. Therefore, a rub rail was not incorporated into the design of the standardized buttress.

Without a rub rail, the front upstream corner of the buttress needed to be tapered back to reduce snag on the buttress. Previous crash testing has shown that tapering the front corner 4 to 5 in. backward was sufficient to prevent snag and often resulted in crashworthy designs (*11,20,21*). Therefore, the lateral extent of the taper on the front corner of the standardized buttress was desired to be at least 4 inches.

The slope of the taper, and the associated longitudinal extent of the taper, affects the performance of the standardized buttress in opposing ways. A shallow slope over a long

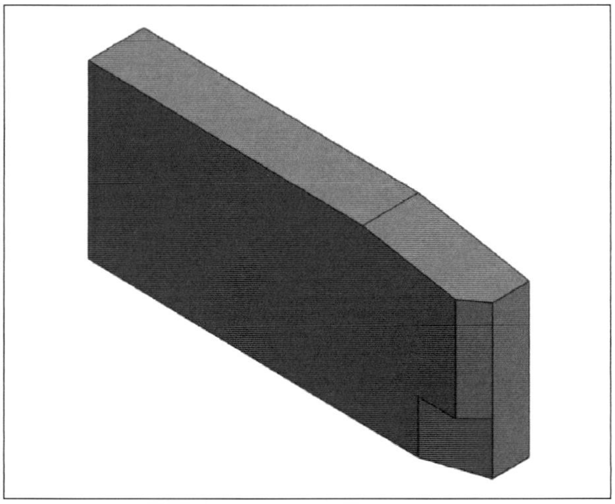

Figure 1. General shape of the standardized AGT buttress incorporating a dual tapered front edge.

distance was desired to reduce vehicle and tire snag on the buttress. However, increasing the longitudinal length of the taper also increases the unsupported length of the Thrie beam between the buttress and the adjacent transition post. Increasing the unsupported length of the rail would result in a reduction in stiffness, an increase in deflection, and increased potential for vehicle snag. Thus, a steeper taper over a shorter longitudinal distance was desired to maintain rail stiffness and prevent excessive deflections.

To balance these two effects, a dual taper design was selected, as shown in Figure 1. The lower portion of the buttress below the Thrie beam utilized a shallow taper to minimize tire snag, while the upper portion of the buttress behind the rail utilized a steep taper to limit the unsupported span length of the rail but still reduce vehicle snag. Previous MASH crash testing has demonstrated that a slope rate of 3:1 can prevent tire snag during vehicle impacts into AGTs (*22*). Thus, the lower taper on the buttress had a 3:1 slope, which resulted in a 12-in. long by 4-in. deep taper. The lower taper extended to the bottom of the Thrie beam, or a height of 11 inches. The upper taper had a 1:1 slope, resulting in a 4-in. by 4-in. taper behind the rail.

To prevent vehicle snag on the buttress above the Thrie beam, the upstream face of the standardized buttress was set at 32 in. tall, which was 1 in. above the top of the Thrie beam. However, many concrete barriers and bridge rails are installed with a 36 in. height to satisfy MASH TL-4. Thus, to match the height of adjacent TL-4 bridge rails, the height of the standardized buttress was transitioned from 32 in. up to 36 in. utilizing a 6:1 vertical slope beginning at the upstream end of the buttress.

The standardized buttress was designed with a 7-ft length to limit the total length of the AGT system while still providing enough barrier length to resist impact loads. To be compatible with adjacent TL-4 bridge rails and concrete parapets, the capacity of the buttress was designed to withstand a TL-4 impact load of 80 kips. The final configuration was 12 in. wide and was reinforced with 16 no. 4 rebar stirrups. In addition, six no. 4 longitudinal rebar were placed along each of the front and back faces of the buttress. The top edges of the buttress were chamfered 1 in.

Selection of Critical Transition Configuration

The standardized buttress needed to be compatible with a wide variety of Thrie beam AGT systems, both with and without a curb. Therefore, the buttress had to be connected to a critical AGT creating a worst-case scenario in order to properly evaluate the system. A review of existing AGTs successfully tested to TL-3 of MASH or NCHRP Report 350 was conducted to find the weakest Thrie beam AGT design (i.e., the most flexible system). This critically weak AGT would pose the greatest risk of vehicle snag on the rigid buttress. The system with the highest dynamic deflection was an AGT originally developed for the Iowa Department of Transportation (DOT). This transition utilized the smallest transition posts and the shortest embedment depths of the reviewed systems. Thus, the Iowa AGT was identified as the weakest of the reviewed transition systems.

Further, the Iowa AGT was successfully tested to both MASH and NCHRP Report 350 TL-3 criteria when used in combination with a curb (*3,4,6*). However, similar AGTs evaluated without a curb failed to satisfy either testing standard. These crash test results not only reinforce the notion that this system is the most susceptible to vehicle snagging, but also indicates that testing without a curb is more critical by allowing the vehicle tires to extend under the rail and snag on the buttress. Therefore, the AGT originally developed for the Iowa DOT, but without a curb, was selected as the critical AGT configuration for the evaluation of the standardized buttress.

To prevent altering the stiffness of the selected AGT, the rail segments and posts needed to be positioned properly relative to the buttress. The original design had an 11-in. offset between the upstream face of the buttress and the centerline of the first transition post. A 1-in. chamfer was present on the corners of the buttress creating a 12-in. span length in which the rail was unsupported in the lateral direction. Since the new standardized buttress incorporated a 4-in. × 4-in. chamfer on the front corner behind the rail, the centerline of the first transition post was placed 8 in. upstream of the buttress to maintain the 12-in. unsupported span length. These dimensions are shown in Figure 2.

Finally, the upstream end of the original AGT design, which was untested and connected to 27-in. tall guardrail, was altered to incorporate the MASH TL-3 crashworthy, Midwest Guardrail System (MGS) stiffness transition

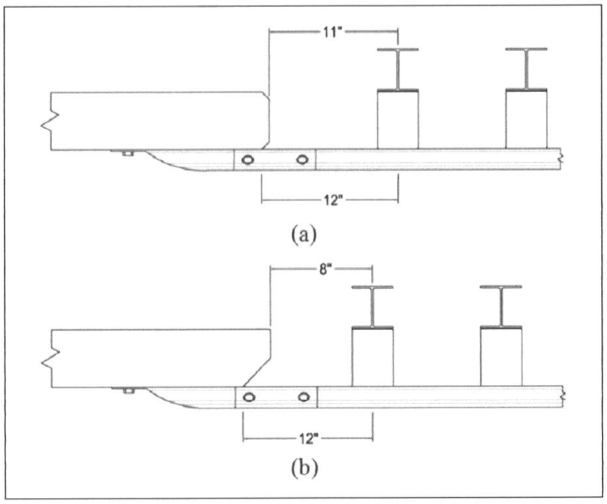

Figure 2. Buttress to transition offset: (*a*) original as-tested AGT and (*b*) AGT in combination with the standardized buttress.

(*23,24*). Both the original Iowa AGT and the critical configuration utilized to test the standardized buttress are shown in Figure 3.

Full-Scale Test AGTB-1

A full-scale crash test was conducted on the proposed standardized buttress in combination with a critical Thrie beam AGT according to MASH 3–21 criteria. The 2270P pickup truck impacted the AGT approximately 80½ in. upstream from the buttress to maximize the potential for vehicle snag on the upstream end of the buttress. The pickup was contained and redirected with moderate vehicle roll and pitch displacements of 27.3 degrees and 10.6 degrees, respectively. The maximum dynamic deflection of the Thrie beam guardrail and the support posts adjacent to the buttress was 6.0 inches. The left-front tire disengaged from the pickup as it contacted the buttress below the rail. The tire was pushed backward into the wheel well and toe pan area of the floor, causing a maximum occupant compartment deformation of 4¼ in., which was within the 9 in. MASH limit. A summary of the test results, including sequential photographs, is shown in Figure 4.

Occupant impact velocities (OIVs) and occupant ridedown accelerations (ORAs) were calculated from the onboard acceleration data. While the OIVs fell within MASH acceptable ranges, the longitudinal ORA value was −30.0 g, which exceeded the 20.49 g MASH limit. The longitudinal ORA was surprising, as longitudinal ORAs of this magnitude had not been previously observed in oblique angle crash tests, and there was no indication from the test video that vehicle decelerations were excessive. While there was some vehicle and tire snag on the tapered portions of the buttress, it did not appear to be significant

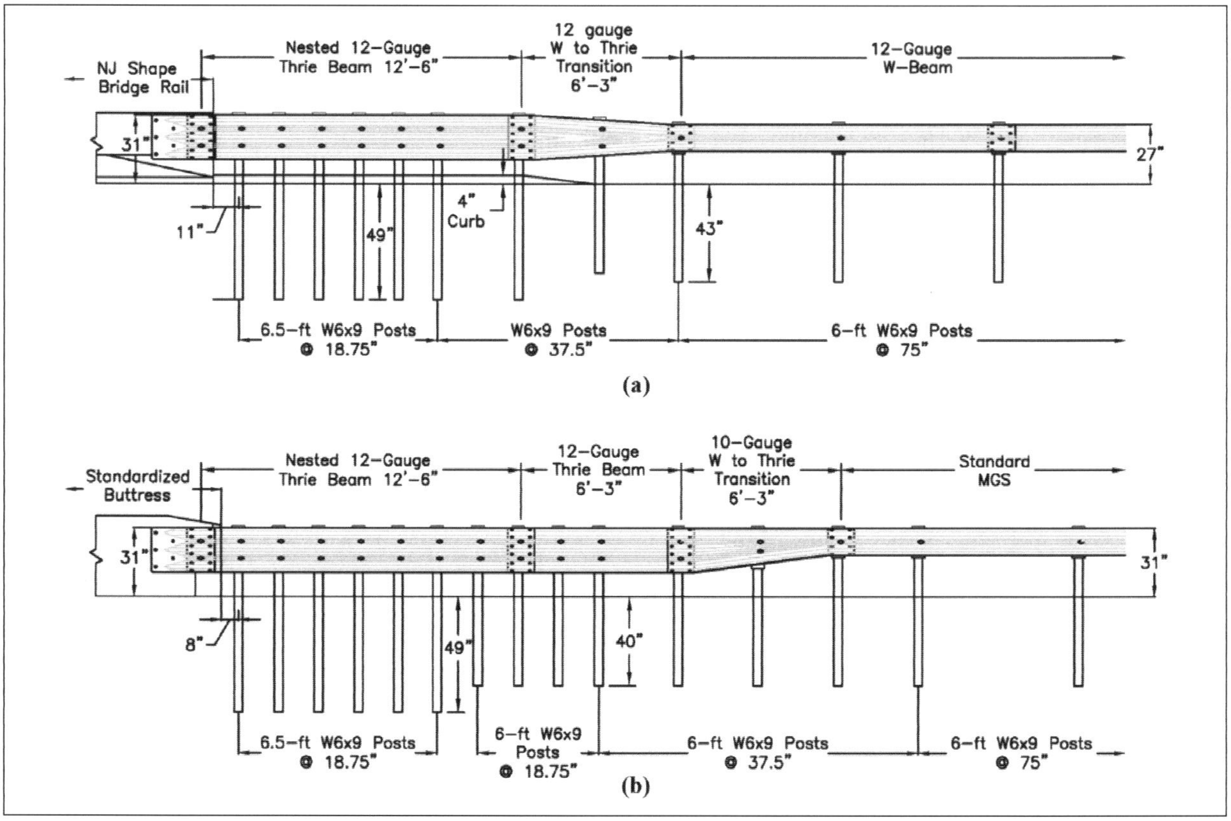

Figure 3. Selected AGT design: (*a*) original as-tested configuration and (*b*) critical configuration for evaluating the standardized buttress.

enough to cause accelerations of this magnitude. Review of the crash-tested vehicle revealed significant deformations to the floorpan and shifting of the seat frame. Unfortunately, the on-board data recorders were positioned on a mounting bracket which was attached to the seat frame. Thus, if the seat frame displaced during the test, the measured accelerations would apply only to the local acceleration of the seat frame and would not be representative of the vehicle as a whole. On-board video cameras showed significant and sudden movement of the seats beginning approximately 100 ms into the impact event, which occurred at the same time as the large deceleration spike in the data. In addition, there was an 18 g positive longitudinal acceleration spike following the −30 g spike, as shown in Figure 4, which corresponded to a 4.5 mph increase in vehicle velocity. Since the vehicle did not increase its velocity during redirection, this was further evidence that the acceleration data was compromised by the shifting seat frame. Thus, the accelerometer data was believed to be in error.

Although this large deceleration spike and resulting longitudinal ORA seemed unrealistic and was likely magnified by movement of the accelerometers relative to the vehicle, the actual ORA values for test AGTB-1 could not be obtained. Therefore, the test was determined to be a failure according to MASH evaluation criteria because of excessive longitudinal ORA. Complete details of the crash test can be found in Rosenbaugh et al. (*12*).

Buttress Redesign

After the failure experienced during test no. AGTB-1, the buttress was redesigned to reduce the amount of vehicle and tire snag. The dual taper design and reinforcement pattern of the buttress was maintained, but small changes were made to the tapers on the front edge of the buttress. To reduce the severity of tire snag below the rail, the slope of the lower taper was reduced from a 3:1 slope to a 4:1 slope. In addition, the lateral offset of the lower taper was increased by ½ in. to 4½ inches. The height of the lower taper increased to 14 in. to reduce the vehicle snag on the lower portion of the upper taper. The 14-in. height also corresponded to the height to the bottom of the transition blockout. Thus, the lower taper measured 4.5 in. deep × 18 in. long × 14 in. high.

High-speed video from test AGTB-1 showed that the pickup truck bumper and front corner were not at risk of impacting the front face of the buttress. However, a reduction to the slope of the taper may reduce snag on the taper itself. Thus, a small reduction in the lateral extent of the upper taper

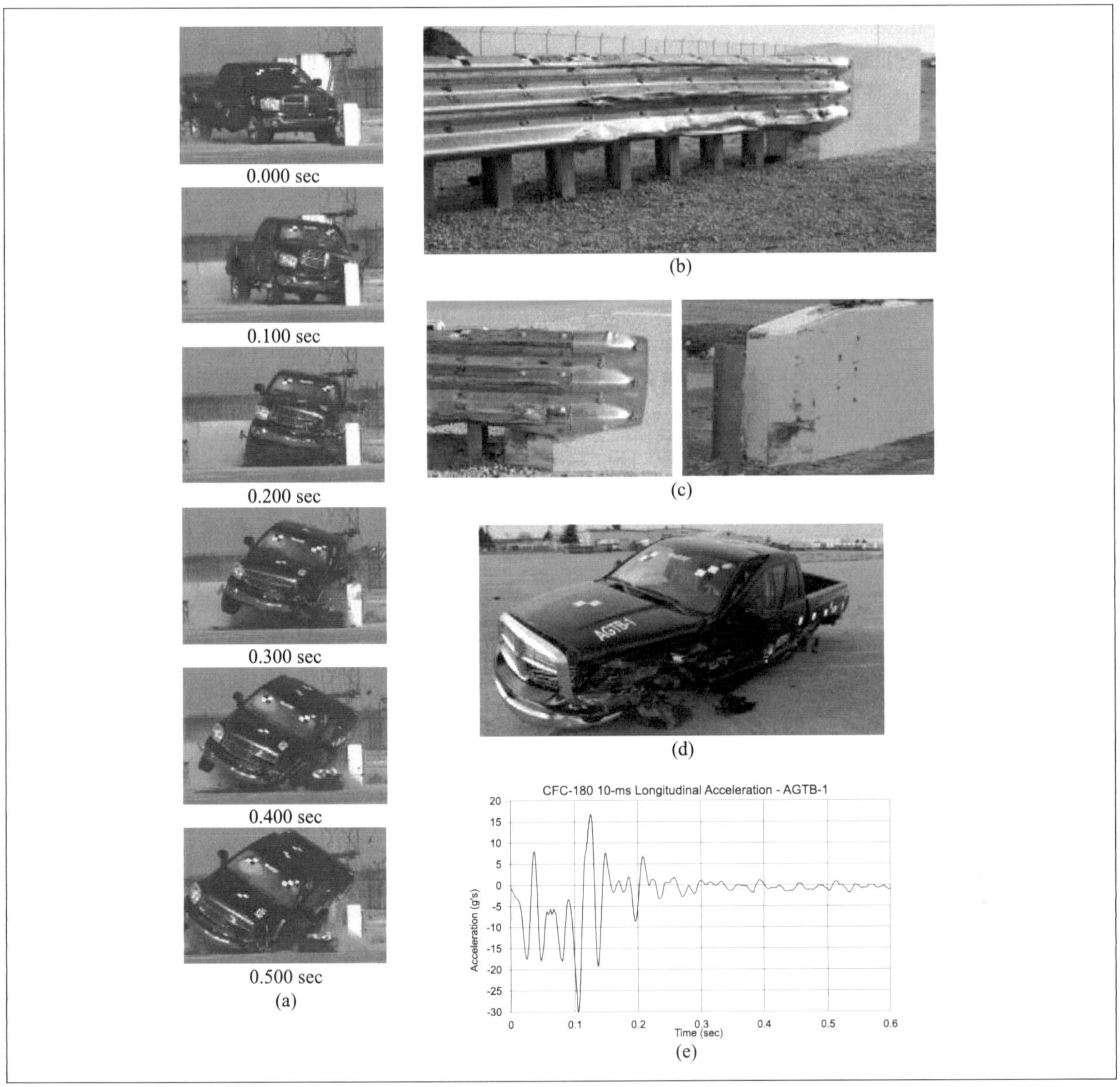

Figure 4. Test results from test AGTB-1: (*a*) sequential photos, (*b*) system damage, (*c*) AGT and buttress damage, (*d*) vehicle damage, and (e) longitudinal acceleration.

resulted in a 3 in. deep × 4 in. long chamfer. The final design for the standardized buttress is shown in Figure 5.

Full-Scale Crash Test AGTB-2

A second MASH 3–21 full-scale crash test was conducted on the redesigned standardized end buttress in combination with the critical AGT. The transition design and post-to-buttress offset distance remained the same from test AGTB-1, and only the geometry changes to the buttress were different. Photographs of the test installation for test AGTB-2 are shown in Figure 6.

The 2270P pickup truck impacted the AGT at 62.4 mph and a 25.4-degree impact angle approximately 86 in. upstream of the buttress to maximize the potential for vehicle snag on the buttress. The pickup was contained, redirected, and remained relatively stable throughout the impact event. The maximum deflection of the Thrie beam guardrail and the support posts adjacent to the buttress was 5.3 inches. The right-front tire extended under the rail and snagged on the lower taper of the buttress, which caused it to push backward into the wheel well and toe pan. However, occupant compartment deformations were all within the MASH limits. A

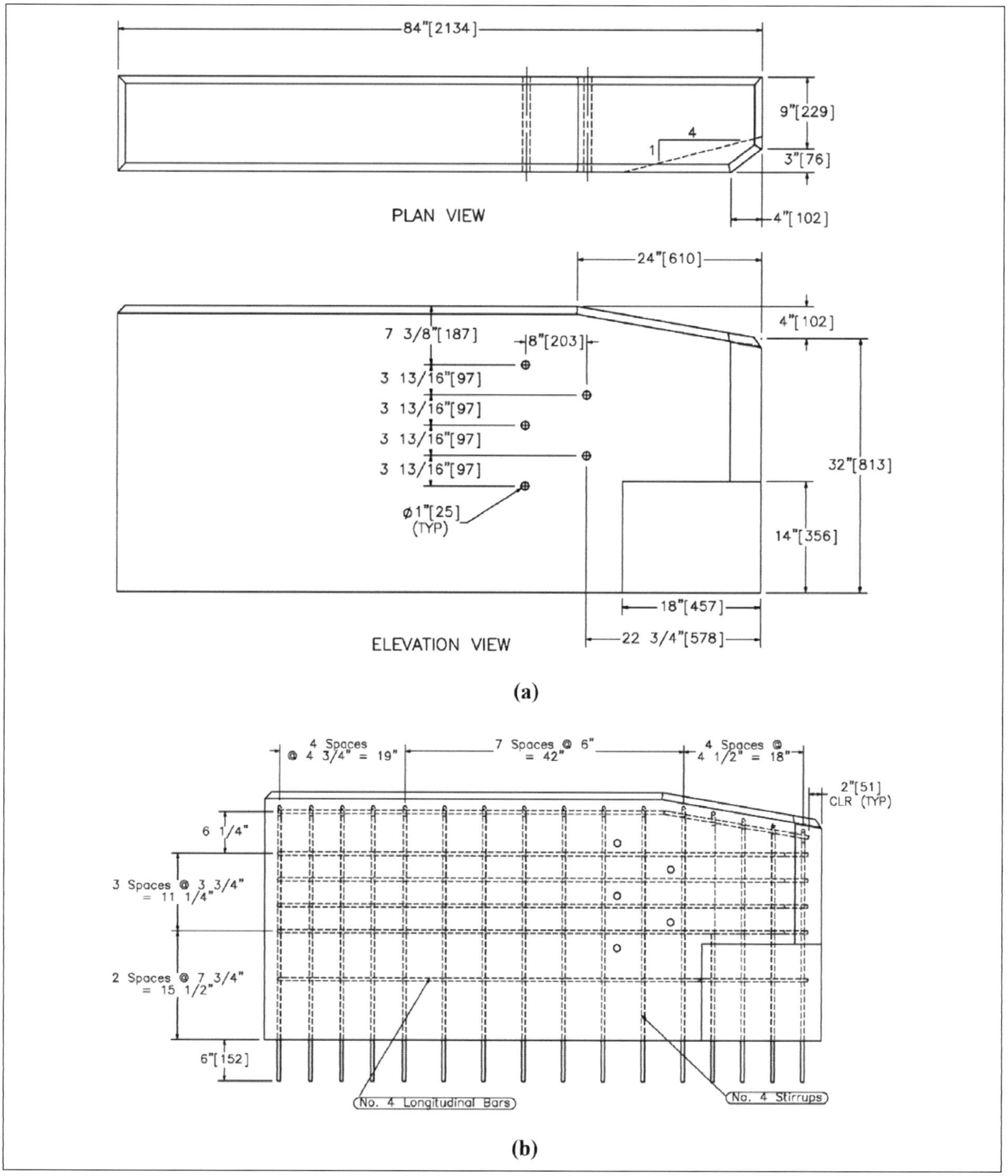

Figure 5. Final standardized buttress design with (*a*) exterior geometry and (*b*) reinforcing layout.

summary of the test results, including sequential photographs, is shown in Figure 7.

During the test, the windshield had shattered and deformed a maximum of 4⅛ inches. However, high-speed video showed the windshield was damaged as a result of airbag deployment, not from interaction with the barrier. Airbags have been shown to shatter, and even tear, windshields in previous oblique angle impacts (*25*). Similar to the previous

Figure 6. Test installation photographs: (*a*) layout, (*b*) buttress connection, (*c*) buttress overhead, and (*d*) adjacent posts.

tests, the windshield was not considered in the evaluation of test no. AGTB-2 because the windshield damage was not caused by interaction with the barrier system.

Lateral and longitudinal OIVs and ORA were calculated from the on-board accelerometers, and all values satisfied the MASH limits. The longitudinal ORA value was −7.1 g, showing a significantly reduced value than compared with test AGTB-1. Therefore, the standardized concrete buttress satisfied the evaluation criteria of MASH test 3–21. Complete details of the crash test can be found in Rosenbaugh et al. (*12*).

A comparison of the test results from both full-scale tests is shown in Table 1. In general, the OIV and ORA values recorded for AGTB-2 were lower than those observed during AGTB-1. Although the ORAs during test AGTB-1 may have occurred after the seat frame began shifting, the OIVs would have been calculated prior to the seat frame shifting. In addition, both the vehicle roll and pitch angular displacements were reduced during AGTB-2. Thus, the redesigned buttress may have resulted in reduced snag and increased stability.

Conclusions, Recommendations, and Implementation Guidance

A standardized concrete buttress was developed and successfully crash tested to MASH test 3–21 safety performance criteria. MASH test 3–20 with the small car was not considered critical since the lighter weight vehicle would result in reduced rail deflections and a reduced risk of snag on the buttress. A MASH 3–20 test was conducted on a different

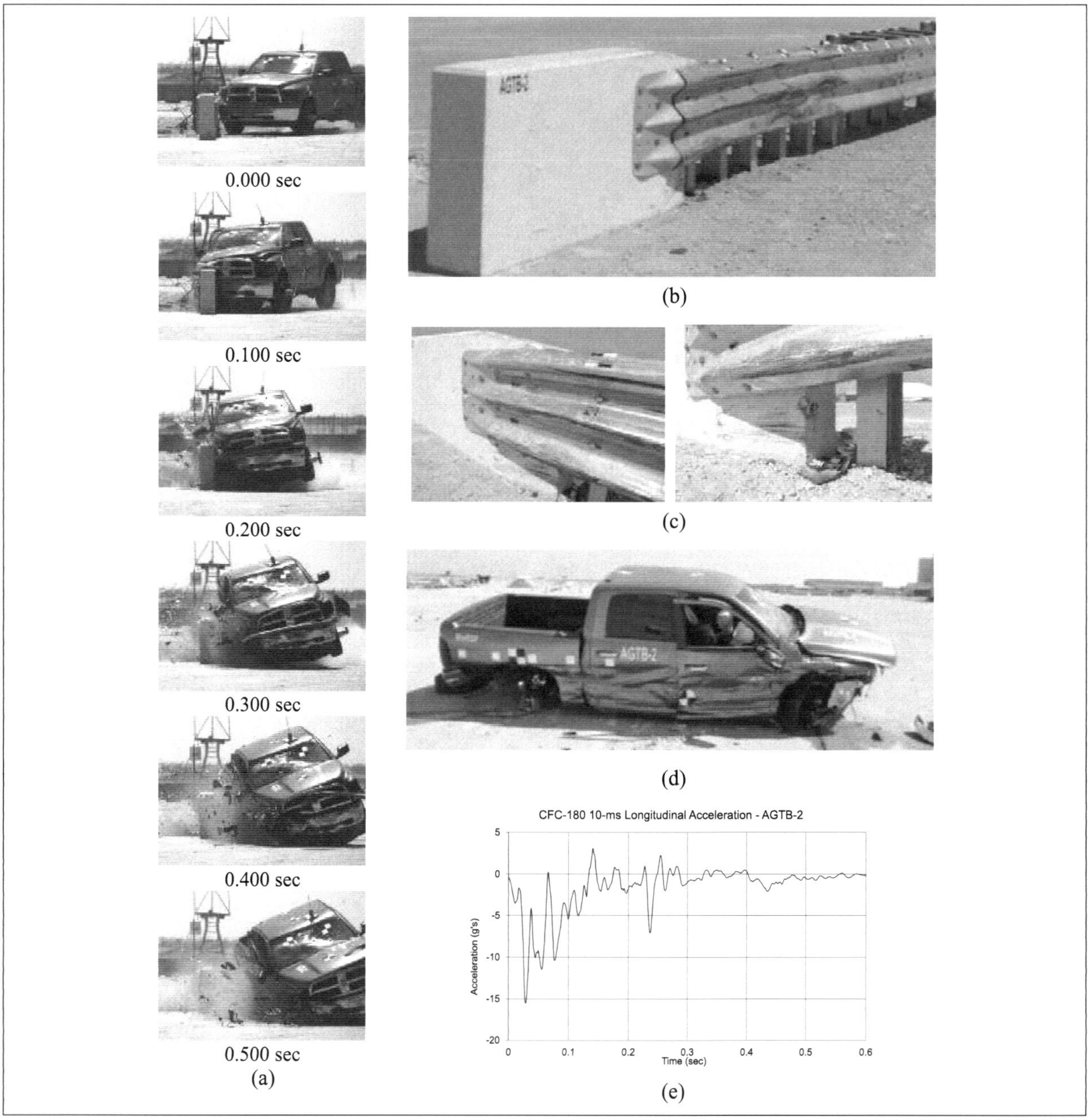

Figure 7. Test results from test AGTB-2: (*a*) sequential photos, (*b*) system damage, (*c*) AGT and buttress damage, (*d*) vehicle damage, and (*e*) longitudinal acceleration.

transition system incorporating a similar version of the standardized buttress (*26*). This AGT utilized a top rail height of 34 in., or 3 in. higher than standard transitions. Thus, there was an increased risk of the small car extending under the rail and snagging on the buttress. The 34 in. tall transition was attached to the buttress design developed here, except the buttress height and the height of the taper transition point were each increased 3 in. The full-scale crash test results on this increased height AGT satisfied MASH 3–20 evaluation criteria. Since the standard 31-in. lower rail height would reduce the risk of small car snag, the test on a 34 in. tall AGT reinforced the idea that the small car test was not critical. Therefore, the standardized transition buttress was considered to be a MASH crashworthy device.

The standardized buttress was developed to be compatible with a wide variety of Thrie beam AGT systems, both with and without a curb. As part of its evaluation, it was crash tested in combination with a critically weak approach

Table 1. Test Results from Tests AGTB-1 and AGTB-2 (*12*)

Test criteria	Test AGTB-1	Test AGTB-2	MASH limit
Impact speed	61.9 mph	62.4 mph	na
Impact angle	24.4°	25.4°	na
Maximum dynamic deflection	6.0 in.	5.3 in.	na
Maximum vehicle roll	27.3°	21.2°	75°
Maximum vehicle pitch	10.6°	6.3°	75°
OIV:			
Lateral	27.7 ft/s	24.6 ft/s	≤40 ft/s
Longitudinal	22.7 ft/s	20.3 ft/s	≤40 ft/s
ORA:			
Lateral	−10.0 g's	−10.4 g's	≤20.49 g's
Longitudinal	−30.0 g's	−7.1 g's	≤20.49 g's
Occupant compartment crush:			
Wheel well & toe pan	4¼ in.	6⅜ in.	≤9 in.
Floor pan	3⅜ in.	4 in.	≤12 in.
Side front panel	3¾ in.	6¾ in.	≤12 in.
Side door (below seat)	3 in.	2 in.	≤12 in.
Side door (above seat)	3⅞ in.	4¼ in.	≤9 in.

Note: na = not applicable.

guardrail transition without a curb. This worst-case scenario posed the greatest risk for snag on the upstream end of the buttress. Since the buttress proved crashworthy in this critical configuration, the standardized buttress should remain crashworthy when utilized with other AGTs as the stiffer systems would only reduce vehicle snag. Therefore, the standardized buttress can be used in combination with any Thrie beam AGT system that has previously been successfully tested to either NCHRP Report 350 or MASH criteria. These AGTs may be either ¼-post or ½-post spacings (i.e., 18¾-in. and 37½-in. post spacings). Further, since the standardized buttress was tested without a curb, and curbs tend to reduce tire snag, the standardized buttress can be utilized with these AGTs in either a curbed or non-curbed installation.

For the successful attachment of various AGTs to the standardized transition buttress, the same post, blockout, and rail components from the original as-tested AGT design should be utilized within the transition region. Thus, the post size, post embedment depth, post spacing, blockouts, rail thickness, rail height, and rail segment lengths should not be altered when the standardized buttress is utilized within other AGT designs. However, the offset between the buttress and the first transition post may vary. The unsupported span length of the rail, which is measured from the location where the rail is no longer laterally supported by the buttress to the centerline of the adjacent post, should remain the same as the original as-tested AGT so that the stiffness of the transition is not affected. Examples of this distance are shown in Figure 8. Because the unsupported span length varies with the flares, tapers, and post spacings utilized among various AGT designs, the offset distance from the standardized buttress to the first transition post will vary. Subsequently, the location of the Thrie beam terminal connector attachment bolts will also vary.

Until recently, most AGTs were only evaluated and crash tested near the connection between the rail and the rigid parapet. However, more recent testing has highlighted the critical nature of the upstream stiffness transition between W-beam guardrail and the stiffened Thrie beam AGT. New AGT installations should utilize a crashworthy upstream stiffness transition even if they were not originally developed with one. For installations transitioning from MGS to the standardized buttress, it is recommended to utilize the MGS stiffness transition on the upstream end of the AGT, as was done here with the Iowa AGT. Details on how to incorporate the MGS stiffness transition into a Thrie beam AGT can be found in previous reports and papers (8,9,*23,24,27*).

The standardized transition buttress was developed with a vertical face to optimize vehicle stability during impacts. However, the adjacent bridge rail or concrete parapet may not have the same geometry. Thus, the downstream end of the buttress must contain a shape transition aligned with the adjacent bridge rail or concrete parapet. Shape transitions should be gradual to prevent vehicle instabilities. Based on previous simulation efforts, transitions to the face geometry of a rigid barrier incorporating lateral slopes steeper than 10:1 may cause stability issues (*28*). Thus, it is recommended to utilize a maximum 10:1 lateral slope to transition the shape of the standardized buttress, and shape transitions may begin 6 in. downstream of the Thrie beam terminal connector, or 8 in. downstream of the attachment bolts. For drastic shape changes, the length of the buttress may need to be extended beyond its 7 ft minimum length.

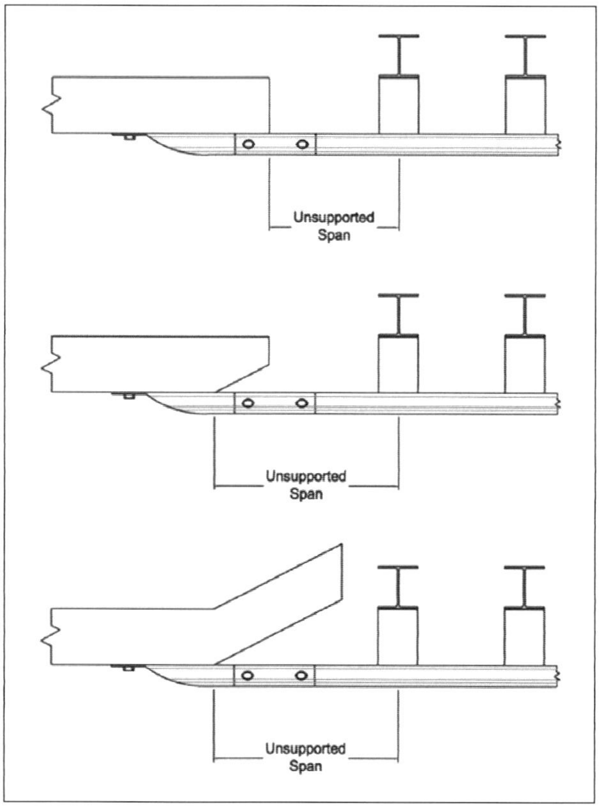

Figure 8. Examples of unsupported span lengths for various AGT configurations.

Height transitions may be necessary for attachment to taller bridge rails and concrete parapets. The upstream end of the buttress was successfully tested with a vertical taper of 4 in. over a 24-in. length. This vertical slope on the upstream end may be continued upward with the same 6:1 slope until the desired height is reached. Note, this is a steeper slope than the previous 8:1 vertical slope guidance stemming from previous testing (*29*). Finally, if the adjacent bridge rail or parapet is only 32 in. tall, the entire buttress can be installed with a constant 32 in. top height. Further implementation guidance can be found in Rosenbaugh et al. (*12*).

Acknowledgments

The authors wish to acknowledge several sources that made a contribution to this project: (1) the State DOTs comprising the Midwest States Pooled Fund for sponsoring and guiding the research project and (2) MwRSF personnel for constructing the test installations and conducting the crash tests.

References

1. American Association of State Highway and Transportation Officials (AASHTO). *Manual for Assessing Safety Hardware*, 2nd ed. AASHTO, Washington, D.C., 2016.
2. Ross, H. E., Jr., D. L. Sicking, R. A. Zimmer, and J. D. Michie. *NCHRP Report 350: Recommended Procedures for the Safety Performance Evaluation of Highway Features*. HRB, National Research Council, Washington, D.C., 1993.
3. Faller, R. K., J. D. Reid, J. R. Rhode, D. L. Sicking, and E. A. Keller. *Two Approach Guardrail Transitions for Concrete Safety Shape Barriers*. Report no. TRP-03-69-98, Midwest Roadside Safety Facility, University of Nebraska-Lincoln, Lincoln, 1998.
4. Faller, R. K., J. D. Reid, and J. R. Rhode. Approach Guardrail Transition for Concrete Safety Shape Barriers. *Transportation Research Record: Journal of the Transportation Research Board*, 1998. 1647: 111–121.
5. Bligh, R. P., W. L. Menges, and R. R. Haug. *Evaluation of Guardrail to Concrete Bridge Rail Transitions*. Report no. FHWA/TX-04/4564-1, Texas Transportation Institute, Texas A&M University, College Station, 2003.
6. Polivka, K. A., R. K. Faller, D. L. Sicking, J. R. Rhode, R. W. Bielenberg, J. D. Reid, and B. A. Coon. *Performance Evaluation of the Guardrail to Concrete Barrier Transition – Update to NCHRP 350 Test No. 3-21 with 28 in. C.G. Height (2214T-1)*. Report no. TRP-03-175-06, Midwest Roadside Safety Facility, University of Nebraska-Lincoln, Lincoln, 2006.
7. Arrington, D. R., R. P. Bligh, and W. L. Menges. *MASH Test 3-21 on TL-3 Thrie Beam Transition without Curb*. Report no. FHWA/TX-13/9/1002-12-3, Texas Transportation Institute, Texas A&M University, College Station, 2013.
8. Winkelbauer, B. J., J. G. Putjenter, S. K. Rosenbaugh, K. A. Lechtenberg, R. W. Bielenberg, R. K. Faller, and J. D. Reid. *Dynamic Evaluation of MGS Stiffness Transition with Curb*. Report no. TRP-03-291-14, Midwest Roadside Safety Facility, University of Nebraska-Lincoln, Lincoln, 2014.
9. Schmidt, J. D., S. K. Rosenbaugh, and R. K. Faller. Evaluation of the Midwest Guardrail System Stiffness Transition with Curb. *Journal of Transportation Safety and Security*, Vol. 9, No. 1, 2017, pp. 105–121.
10. Alberson, D. C., W. L. Menges, and S. K. Schoeneman. *NCHRP Report 350 Test 3-21 on the Ohio Transition from Thrie Beam to Concrete Parapet*. Report no. 401021-1, Texas Transportation Institute, Texas A&M University, College Station, 2000.
11. Alberson, D. C., W. L. Menges, and S. K. Sandars. *NCHRP Report 350 Test 3-21 on the Ohio Type 1 Transition from Thrie Beam to Concrete Parapet with Asphalt Curb*. Report no. 401021-5, Texas Transportation Institute, Texas A&M University, College Station, 2001.
12. Rosenbaugh, S. K., et al. *Development and Evaluation of Standardized Concrete End Buttress*. Draft Report no. TRP-03-369-17 (in progress), Midwest Roadside Safety Facility, University of Nebraska-Lincoln, Lincoln, 2017.
13. Strybos, J. W., J. N. Mayer, and M. E. Bronstad. *Crash Evaluation of a Thrie Beam on Wood Post Transition to a New Jersey Shaped Parapet*. Report no. FHWA-FPL-96-012, Southwest Research Institute, San Antonio, 1996.
14. Buth, C. E., W. L. Menges, and B. G. Butler. *NCHRP Report 350 Test 3-21 of the Vertical Wall Transition*. Report no. 404211-2, Texas Transportation Institute, Texas A&M University, College Station, 1998.
15. Buth, C. E., W. L. Menges, and B. G. Butler. *NCHRP Report 350 Test 3-21 of the Vertical Flared Back Transition*. Report

no. 404211-4, Texas Transportation Institute, Texas A&M University, College Station, 1998.

16. Buth, C. E., W. F. Williams, and W. L. Menges. *NCHRP Report 350 Evaluation of the Vertical Wall Transition*. Report no. 404211-12, Texas Transportation Institute, Texas A&M University, College Station, 1998.

17. Mayer, J. B. *Crash Tests of Guardrail to Bridge Rail Transitions: NCHRP Test 3-21, SwRI Test No. TBRR-1*. Report no. 06-8321, Southwest Research Institute, San Antonio, 1998.

18. Bligh, R. P., K. K. Mak, W. L. Menges, and W. F. Williams. *NHCRP Report 350 Evaluation of the Minnesota DOT Transitions*. Report no. RF473390-03, Texas Transportation Institute, Texas A&M University, College Station, 2000.

19. Buth, C. E., W. L. Menges, and S. K. Schoeneman. *NCHRP Report 350 Assessment of Existing Roadside Safety Hardware*. Report no. FHWA-RD-01-042, Texas Transportation Institute, Texas A&M University, College Station, 2000.

20. Soyland, K., R. K. Faller, D. L. Sicking, and J. C. Holloway. *Development and Testing of an Approach Guardrail Transition to a Single Slope Concrete Median Barrier*. Report no. TRP-03-47-95, Midwest Roadside Safety Facility, University of Nebraska-Lincoln, Lincoln, 1995.

21. Jewell, J., N. Clark, and R. Peter. *Vehicular Crash Tests of a Nested Thrie Beam Transition Barrier*. Report no. FHWA/CA/TL-2001/09, California Department of Transportation, Sacramento, 2002.

22. Williams, W. F., R. P. Bligh, and W. L. Menges. *MASH TL-3 Testing and Evaluation of the TxDOT T131RC Bridge Rail Transition*. Report no. FHWA/TX-13/9-1002-12-4, Texas Transportation Institute, Texas A&M University, College Station, 2013.

23. Rosenbaugh, S. K., K. A. Lechtenberg, R. K. Faller, D. L. Sicking, R. W. Bielenberg, and J. D. Reid. *Development of the MGS Approach Guardrail Transition Using Standardized Steel Posts*. Report no. TRP-03-210-10, Midwest Roadside Safety Facility, University of Nebraska-Lincoln, Lincoln, 2010.

24. Lechtenberg, K. A., M. Mongiardini, S. K. Rosenbaugh, R. K. Faller, R. W. Bielenberg, and F. D. B. Albuquerque. Development and Implementation of the Simplified MGS Stiffness Transition. *Transportation Research Record: Journal of the Transportation Research Board*, 2012. 2309: 1–11.

25. Bielenberg, R. W., J. L. Lingenfelter, J. E. Kohtz, R. K. Faller, and J. D. Reid. *Testing and Evaluation of MASH TL-3 Transition Between Guardrail and Portable Concrete Barriers*. Report no. TRP-03-335-17, Midwest Roadside Safety Facility, University of Nebraska-Lincoln, Lincoln, 2017.

26. Rosenbaugh, S. K., et al. *Evaluation of a 34-in. Tall Thrie Beam Approach Guardrail Transition*. Draft Report no. TRP-03-367-17 (in progress), Midwest Roadside Safety Facility, University of Nebraska-Lincoln, Lincoln, 2017.

27. Rosenbaugh, S. K., K. D. Schrum, R. K. Faller, K. A. Lechtenberg, D. L. Sicking, and J. D. Reid. *Development of Alternative Wood-Post MGS Approach Guardrail Transition*. Report no. TRP-03-243-11, Midwest Roadside Safety Facility, University of Nebraska-Lincoln, Lincoln, 2011.

28. Schmidt, T. L., R. K. Faller, J. D. Schmidt, J. D. Reid, R. W. Bielenberg, and S. K. Rosenbaugh. *Development of a Transition between an Energy-Absorbing Concrete Barrier and a Rigid Concrete Buttress*. Report no. TRP-03-336-16, Midwest Roadside Safety Facility, University of Nebraska-Lincoln, Lincoln, 2016.

29. Soyland, K., R. K. Faller, D. L. Sicking, and J. C. Holloway. *Development and Testing of an Approach Guardrail Transition to a Single Slope Concrete Median Barrier*. Report no. TRP-03-47-95, Midwest Roadside Safety Facility, University of Nebraska-Lincoln, Lincoln, 1995.

The Standing Committee on Roadside Safety Design (AFB20) peer-reviewed this paper (18-05386).

Article

TRR
JOURNAL OF THE TRANSPORTATION RESEARCH BOARD

Transportation Research Record
2018, Vol. 2672(39) 52–60

DOI: 10.1177/0361198118768533
journals.sagepub.com/home/trr
SAGE

Human Personality Based Investigation on Distance Threshold for Freeway Landscape Pattern

Zhonghua Wei[1], Sheng Liu[1], Shi Qiu[1], Tongyang Zhang[2], and Shaofan Wang[3]

Abstract

Freeway landscape design is closely related to driver performance. How to stimulate drivers' positive physiological condition and improve overall driving performance by adopting ideal landscape design has become a research interest in recent years. This study explores the impact of changing the landscape pattern at appropriate intervals on driver fatigue and negative mental workload. Drivers' mental performance was evaluated under three types of spatial patterns: open, semi-open, and vertical. Thirty drivers, 15 Type A Behavior Pattern (TABP) and 15 Type B Behavior Pattern (TBBP) drivers, respectively, were recruited to conduct the driving simulator experiment. Data on drivers' physiological variation trends show that the mental workload of TABP and TBBP drivers varied under different spatial patterns. The principal factor high frequency (HF) was used to investigate the distance threshold for different freeway landscape patterns based on the analysis of physiological data collected by electrocardiogram (ECG). This study reveals that changing landscape pattern at a certain interval can benefit drivers' physical and mental status. It is suggested that, when the design speed is 100 km/h, changing landscape pattern every 11 km can reduce fatigue and improve driving performance in both TABP and TBBP drivers. The conclusions of this study provide a rationale and guidance for agencies to adopt different spatial patterns in future freeway landscape design.

Freeway landscape is the aesthetic dimension of a freeway and refers to the visible features of a section of freeway, including the preservation of native plants, roadway alignment, cantilevering stone wall, and other physical elements. Freeway landscape plays an essential role in providing a safe and comfortable driving environment for drivers, providing sight guidance and increasing visual stimulus (*1,2*). As an important element of freeway landscape design, spatial pattern has received extensive attention (*3,4*). When driving on the road, drivers' focal distances become longer gradually with the diminishing of visual field, which makes drivers feel a sense of closure. The degree of closure felt by drivers, also known as the sense of space (*5*), is measured by spatial pattern and calculated according to

$$S = D/H \tag{1}$$

where:

S = classification value for spatial pattern;

D = lateral distance from the driver's line of sight to right shoulder landscape; and

H = height of right shoulder landscape.

The relationships between the spatial patterns and drivers' subjective visual perceptions are listed in Table 1 (*6*).

Based on the S value, multiple definitions of landscape classifications have been proposed (*7,8*). Freeway driving has characteristics such as high speed and changing surroundings. Using indicators such as subjective degrees of visual fields, previous research classified freeway spatial patterns into three typical enclosure types, namely, open ($S>3.0$), semi-open ($S\in[2.0, 3.0]$), and vertical ($S<2.0$) (Figure 1) (*6*).

It has been identified that the homogenous landscape and inadequate changes of landscape tend to increase drivers' distress and depressed emotions (*2,9*). However, in many cases, a single spatial pattern is commonly applied due to lack of consideration. For example, monotonous landscapes in a certain straight and long section of road can easily make drivers feel a sense of monotony and mental fatigue (*10*). In

[1]Beijing Key Laboratory of Traffic Engineering, Beijing University of Technology, Beijing, China
[2]China Highway Engineering Consulting Corporation, Beijing, China
[3]Beijing Key Laboratory of Multimedia & Intelligent Software Technology, Faculty of Information Technology, Beijing University of Technology, Chaoyang, Beijing, China

Corresponding Author:
Address correspondence to Shaofan Wang: wangshaofan@bjut.edu.cn

Table 1. Relationship between *D*/*H* and Drivers' Visual Perception

Spatial pattern	Value of *D*/*H*	Value of vision	Driver's visual perception
Open	$D/H>3.0$	$V\in[4.0, 5.0)$	Broad vision
Semi-open	$D/H\in[2.0, 3.0]$	$V\in[2.8, 4.0)$	Limited vision
Vertical	$D/H<2.0$	$V\in[1.0, 2.8)$	Enclosed vision

addition, the constant landscape pattern may even trigger harmful driving behaviors, deteriorate driver alertness, and prolong cognitive information processes, which possibly leads to unexpected crashes (*3,10*). In China, many agencies and designers neglect the rationale for transition of landscape types during the freeway design phase. Taking humanizing factors into account in landscape design is of great importance and is an urgent need (*1*). Therefore, investigation of varying distance thresholds of landscape patterns is needed to inspire drivers' best performance.

As a typical and steady psychological characteristic, human personality can be categorized in terms of risky behaviors and traffic accident involvement (*11–13*). For instance, Feng et al. (*14*) indicated that an individual's personality has an influence on their speed selection behavior. More active and extroverted drivers are prone to have speeding behaviors. Deffenbacher and Lynch (*15*) found that the high anger drivers engage in more aggressive and risky behavior on the road, and the group of high anger drivers have more accident-related outcomes than low anger drivers. Some previous studies have found that personality might interfere with attention, perception, and information processing, which affects driving behaviors and thereby can increase the likelihood of crashes (*16,17*).

Drivers can be classified into different types according to some commonly used personality tests such as the Sixteen Personality Factor (16PF) and the Type A Behavior Pattern Questionnaire (TABPQ). Drivers are divided into two typical types among these various types of human personality, namely the Type A Behavior Pattern (TABP) and the Type B Behavior Pattern (TBBP). The TABP drivers are identified as having greater tension, being hyperactive, and typically time-urgent; they are more likely to have dangerous and adverse driving behaviors (*18,19*). These adverse driving behaviors, such as anger, fatigue, and risk-taking driving behaviors, could have a negative effect on traffic efficiency and safety on freeways (*15*). On the contrary, the TBBP drivers are peaceful, patient, easy-going and not in a hurry when driving.

This paper aims to investigate drivers' mental performance and landscape distance threshold under different landscape patterns from the perspective of human personality. It is anticipated that the findings from this research will provide an appropriate reference for freeway agencies in future landscape design.

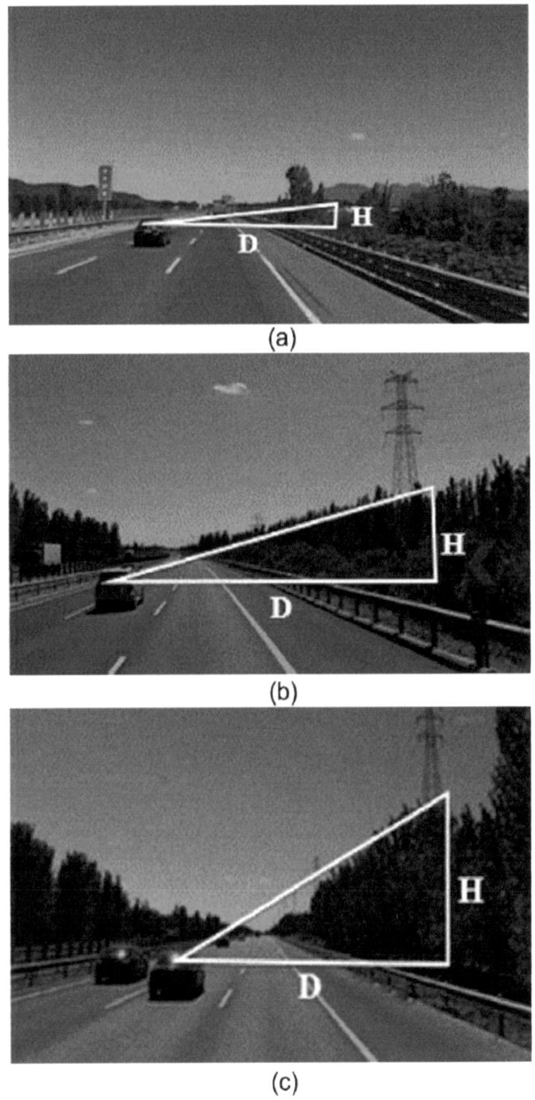

Figure 1. Freeway landscapes under different spatial patterns: (a) open spatial pattern; (b) semi-open spatial pattern; (c) vertical spatial pattern.

Design of Simulated Driving Experiment

In China the age group of drivers with the highest percentage of traffic accident frequency (29.6%) is drivers aged between 20 and 35 years (*20*). Based on the TABPQ (*21*), 30 drivers (17 males and 13 females) of this age range were recruited to engage in the experiment. Participants were asked to complete a questionnaire consisting of 60 choice questions with one point per question of choice. After completion, 15 of the questionnaires which scored over 27 points were recognized as the TABP drivers to be tested in

Table 2. Characteristics of Participants

Personality	Number	Age (years)	Driving experience (years)
A	15	27.00±6.10	5.56±1.59
B	15	25.11±3.25	4.77±2.16
All	30	26.56±4.84	5.17±1.89

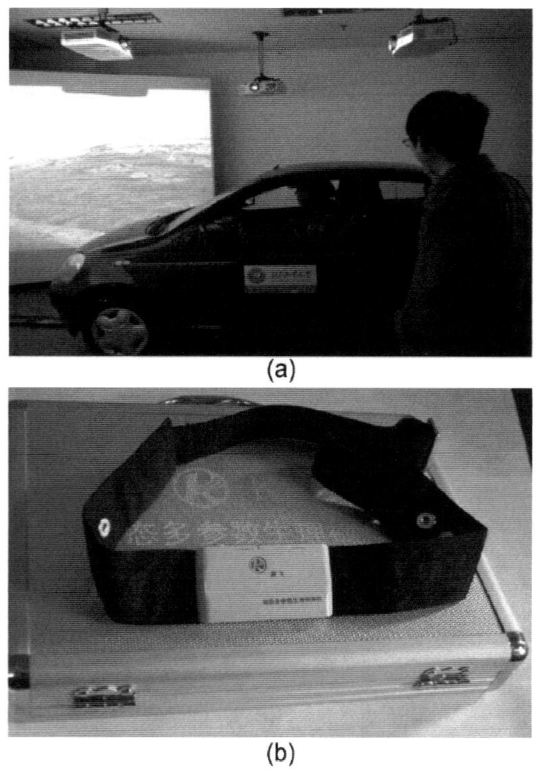

Figure 2. Apparatus for experiments: (a) the AutoSim AS1600 driving simulator system; (b) the KF-2 electrocardiograph.

this experiment while the rest were recognized as TBBP drivers, as shown in Table 2 (*22*). Significance test was used to analyze the relationship between drivers' age and driving experience, respectively. The result is $F\ (1,\ 16) = 0.67$, $p = 0.42 > 0.05$ and $F\ (1,\ 16) = 0.75$, $p = 0.39 > 0.05$, indicating no significant difference between the TABP and TBBP drivers.

The simulated driving experiment was conducted by the AutoSim AS1600 driving simulator system (Figure 2*a*). As shown in Figure 3, 3DsMax was used to generate experimental scenarios, which were loaded into the driving simulator. To avoid driver fatigue, the distance of the simulated freeway section was set as 15 km (*4,9*). Two lanes were set in each direction based on the Chinese national standard CJJ 37-2012 (*23*). Each lane was 3.75 m wide with 2 m right shoulder; the post-speed is 100 km/h.

Figure 3. Landscape spatial patterns simulated by 3DsMax: (a) open spatial pattern; (b) semi-open spatial pattern; (c) vertical spatial pattern.

All the participants were asked to maintain sufficient rest without consumption of any alcohol or drugs two hours before the experiment. To assure their calmness during the experiment process, the participants were given 10 minutes to rest after putting the KF-2 electrocardiograph (Figure 2*b*) around their chests and after each experiment. All participants were asked to drive at the post-speed.

Data Collection

Physiological data were collected through the simulator study, as recorded by the KF-2 electrocardiograph. ECG parameters from physiological data were used to reflect drivers' mental status and physical workload (*23–25*). The HR and HRV are two indicators of physiological status which are most commonly used in research (*10*). HR is the heartbeat speed measured by the number of heart contractions per minute, which reflects the body's physiological performance and is an indirect measure of physical workload. As a physiological phenomenon of variation in the time interval between heartbeats, HRV presents an interaction effect balanced by the individual's sympathetic nervous system (SNS) and parasympathetic nervous system (PSNS) (*26*). Imbalance between SNS and PSNS may trigger adverse reactions when the balance between the SNS activity and PSNS activity is broken, like arrhythmia and metabolic hyperthyroidism (*27*).

Time domain method and frequency domain method are the two most widely used methods for HRV analysis. Time domain method analyzes the R-R interval—the time interval between two successive heartbeats by statistical discrete trend analysis. Root mean square of successive difference (RMSSD), an important indicator of R-R interval in the time domain method, inflects the change of HRV. The indicators of the frequency domain method are: high frequency (HF), low frequency (LF), LF/HF ratio, and very low frequency (VLF). HF reflects the situation of the parasympathetic nervous system (PSNS), a high level of PSNS may inspire driver fatigue. LF reflects the situation of the sympathetic nervous system (SNS), which may trigger driver anxiety when SNS is dominant. The value of the LF/HF ratio indicates a balanced state between SNS and PSNS. VLF represents the variation tendency of body's temperature (*10,26*).

In total, three sets of simulator freeway scenarios were applied to experimental drivers in this study and physiological data were collected. Electrocardiography was used to investigate participants' physiological performances under different spatial patterns. ECG parameters consist of seven indicators as described in the data analysis section.

Data Analysis

Principal Component Analysis

Due to the possible correlation existing among the multidimensional data, an effort should be devoted to reduce redundancy in analysis (*28,29*). To remove correlated indicators in ECG parameters, principal component analysis (PCA) was adopted in this research, using a few linearly irrelevant factors to represent the characteristics of the seven parameters.

Kaiser–Meyer–Olkin (KMO) and Bartlett tests were used to examine the correlations in the physiological data and the results are shown in Table 3. An obvious correlation between ECG parameters ($p = 0.00 < 0.05$) suggests the necessity of PCA. Then, two components, HR and HF, whose eigenvalues exceed 1 were extracted based on PCA, which can reflect the most information on the primary seven ECG parameters (*30*). For the sake of achieving dimension reduction and ensuring the affection of HR and HF, Varimax rotation was applied to seven indicators. The loading distribution, obtained from the Varimax rotation, was used to determine the category of the seven indicators. As shown in Table 4, after rotation, two components correspond to the time domain and frequency domain indicators respectively (*10*). It is recognized that an indicator is representative when the factor load is higher than 0.4 or the factor dual-loads are lower than 0.4. Finally, the two indicators HR and HF, respectively related to time and frequency domain indicators, were extracted as the principal factors in this study based on PCA. In addition, the accumulated variance explained by these two components reaches 84.169%, as shown in Table 5. It means that these two principal indicators can effectively reflect the physiological status of TABP and TBBP drivers.

Table 3. Results of KMO and Bartlett's Test

Kaiser–Meyer–Olkin Measure of Sampling		0.70
Bartlett's Test of Sphericity	Approx. Chi-Square	16782.02
	Df	21
	Sig.	0.00

Table 4. Rotated Component Matrix

		Components		
No.	Indicators	1	2	Category
1	HR	0.935	0.035	Time domain indicators
2	RMSSD	0.284	0.160	
3	LF	0.469	0.807	Frequency domain indicators
4	HF	0.196	0.914	
5	LF/HF	0.400	0.786	
6	VLF	0.426	0.571	
7	LFNU	0.522	0.769	

The performances of the HR and HF were further analyzed. For HR, the descriptive statistics were first examined, as shown in Table 6. As repeated measurement was conducted in this experiment, it was necessary to detect whether the data are inter-correlated. Mauchly's test of sphericity is conducted to identify the characteristics of participants' HR in different scenarios. The results from Mauchly's test of sphericity indicate that the statistics do not meet the spherical distribution hypothesis [$p = 0.00 < 0.05$, χ^2 (2) = 119.807]. Thus, the method of Pillai's Trace was used to conduct a multivariate analysis of variance (MANOVA), and to examine the relationship between HR under different spatial patterns. The results of MANOVA suggest that there is no significant difference between drivers' HR under different spatial patterns [F (2, 177) = 2.512, $p = 0.069 > 0.05$]. The results indicate that TABP and TBBP drivers have no significant difference in physical workload under different spatial patterns. Hence, HR is not suitable to be selected as the physiological characteristic indicator of drivers.

Table 5. Total Variance Explained

	Initial eigenvalues			Extraction sums of square loading			Rotating sum of square loading		
Components	Total	% of variance	Cumulative %	Total	% of variance	Cumulative %	Total	% of variance	Cumulative %
1	3.52	73.241	73.241	3.52	73.241	73.241	2.98	60.272	60.272
2	1.43	10.928	84.169	1.43	10.928	84.169	1.97	23.897	84.169
3	0.45	5.068	89.237	—	—	—	—	—	—
4	0.32	4.937	94.174	—	—	—	—	—	—
5	0.17	3.847	98.021	—	—	—	—	—	—
6	0.15	1.030	99.051	—	—	—	—	—	—
7	0.05	0.949	100	—	—	—	—	—	—

Table 6. Descriptive Statistics of Principal Factors

					95% confidence interval	
Principal factor	Spatial pattern	Personality	Mean	Standard deviation	Lower bound	Upper bound
HR (beat/min)	Open	A	75.89	7.21	61.76	90.03
		B	71.51	7.21	60.43	82.60
	Semi-open	A	70.99	8.66	54.02	87.96
		B	72.56	6.06	60.69	84.44
	Vertical	A	74.68	5.93	63.07	86.30
		B	74.36	5.94	62.71	86.00
HF (MHz)	Open	A	283.53	30.21	224.33	342.74
		B	194.27	27.76	139.85	248.68
	Semi-open	A	295.02	24.60	246.80	343.23
		B	151.27	23.60	105.01	197.53
	Vertical	A	226.14	20.66	185.65	266.63
		B	123.07	18.58	86.65	159.49

The multiple comparison of HR was carried out to gain further understanding of the significant effect of personality and spatial patterns on the experiment participants' HR values. As shown in Tables 7 and 8, significant difference exists in drivers' HR under the interaction of personality and landscape patterns [$p = 0.00 < 0.05$]. It is found that TABP drivers have significant difference in HR with different landscape patterns while TBBP drivers have no difference. This means that TABP drivers' physical workload can more easily be influenced by a monotonous landscape. Among the three landscape patterns, the open pattern causes the strongest tension for TABP drivers.

The descriptive statistics of HF are shown in Tables 7 and 8. Mauchly's test of sphericity was used to detect characteristics of participants' HF in different scenarios. The results of Mauchly's test of sphericity suggest that the statistics do not meet the spherical distribution hypothesis [$p = 0.00 < 0.05$, $\chi^2(2) = 65.410$]. Thus, the analysis method Pillai's Trace was used for MANOVA. The results of the MANOVA indicate that significant difference exists between experimental participants under different spatial patterns [$F(2, 174) = 27.140$, $p = 0.000 < 0.05$]. Thus, HF is a proper indicator of a driver's mental workload.

The results of multiple comparisons are shown in Tables 7 and 8. It can be seen that both TABP and TBBP drivers have significant differences under different spatial patterns. In addition, TABP drivers experienced higher mental workload under an open landscape pattern than TBBP drivers, which means that negative performance from fatigue is more easily triggered in TABP drivers.

Investigation of Distance Threshold

HF reflecting the variation of TABP and TBBP drivers' mental workload were used to investigate freeway landscape distance threshold under a single spatial pattern. Travel distance and mental workload is denoted as L and V, respectively. The distribution of physiological indicators is shown in Figure 4, where each participant's mental workload variation tendency is explicitly exhibited in 15-km simulator scenarios. To achieve best fit, various types of curve fitting were compared, such as the exponential fitting, logarithmic fitting, and polynomial curves fitting. The coefficient of determination, R^2, was used to evaluate the goodness of fit. Among these types of curve fitting, polynomial curves fitting with third orders had the best goodness of fit for this series of data

Table 7. Results of Multiple Comparison for Personalities

Principal factor	Spatial pattern	(I) Personalities	(J) Personalities	Mean difference (I–J)	Standard error	Sig.	95% confidence interval	
							Lower bound	Upper bound
HR	Open	A	B	4.38*	0.98	0.00	2.45	6.31
	Semi-open	A	B	−1.58	1.13	2.16	−3.80	0.65
	Vertical	A	B	0.33	0.89	0.72	−1.44	2.09
HF	Open	A	B	89.27*	35.91	0.01	18.40	160.13
	Semi-open	A	B	143.75*	36.14	0.00	72.42	215.07
	Vertical	A	B	103.07*	25.30	0.00	53.15	153.00

Table 8. Results of Multiple Comparison for Landscape Patterns

Principal factor	Personalities	(I) Spatial pattern	(J) Spatial pattern	Mean difference (I–J)	Standard error	Sig.	95% confidence interval	
							Lower bound	Upper bound
HR	A	Open	Semi-open	4.91*	1.01	0.00	2.91	6.90
			Vertical	1.21*	0.46	0.01	0.30	2.13
		Semi-open	Vertical	−3.70*	0.93	0.00	−5.53	−1.86
	B	Open	Semi-open	−1.05	1.03	0.31	−3.08	0.98
			Vertical	−2.84	0.47	0.07	−3.77	−1.91
		Semi-open	Vertical	−1.79	.95	.059	−3.657	0.072
HF	A	Open	Semi-open	−11.48	40.86	0.78	−92.12	69.16
			Vertical	57.39*	28.52	0.05	1.10	113.69
		Semi-open	Vertical	68.87*	33.18	0.04	3.38	134.37
	B	Open	Semi-open	43.00	41.56	0.30	−39.02	125.01
			Vertical	71.20*	29.01	0.02	13.94	128.46
		Semi-open	Vertical	28.20	33.75	0.41	−38.41	94.81

points. In other words, the cubic polynomial function can be used to assess the relationship between HF and travel distance:

$$V(P) = a_3P^3 + a_2P^2 + a_1P + a_0 \quad (2)$$

where:

P = travel distance in the simulator scenarios;

V = HF value; and

a_i = ith polynomial coefficient of the parameter L in jth power (i = 1, 2, 3, 4, and j = 4−i).

Accordingly, the polynomial coefficients a_i and the coefficients of determination R^2 for TABP and TBBP drivers under different spatial patterns are shown in Table 9.

Figure 4 shows the relationship between HF value and travel distance. It presents the physiological responses of TABP and TBBP drivers to different freeway landscape scenarios explicitly. To understand the fitting curve in Figure 4, several key parameters are defined in advance. Particularly, P_1 is the travel distance to reach the point V_1, which is the local maximum value of HF; accordingly, P_2 is the travel distance to reach the point V_2, which is the local minimum HF value. In the distance interval $[0, P_1)$, the driver is in the procedure of vehicle starting, which is less affected by the spatial pattern. Hence, the travel distance in this range was not investigated in this study. The variation trend of HF in the open spatial pattern was analyzed as follows:

- In the distance interval $[P_1, P_2)$, the HF value decreases when the travel distance increases. This phenomenon indicates that driver's activity of PSNS weakens while SNS is dominant. In this circumstance, drivers are in positive and active situations. The V value reaches minimum when arriving at P_2, indicating that the driver is in the most active status. In this phase, the freeway landscape with open spatial pattern has a positive influence on the driver.
- In the distance interval $[P_2, 15]$, the HF value increases with the increase of travel distance, which indicates that the driver's PSNS is dominant. This is a signal that the driver is in a negative and depressive situation. The freeway landscape with open spatial pattern has a negative impact on the driver in this phase.

The variation trends of HF for the other two spatial patterns, which are similar to the open spatial pattern, are omitted. As the variation trend analyzed above, the critical distance interval of driving in a constant landscape is ideal as it turns to $[P_1, P_2)$; exceeding that distance an individual's mental status is changed from positive to negative. In

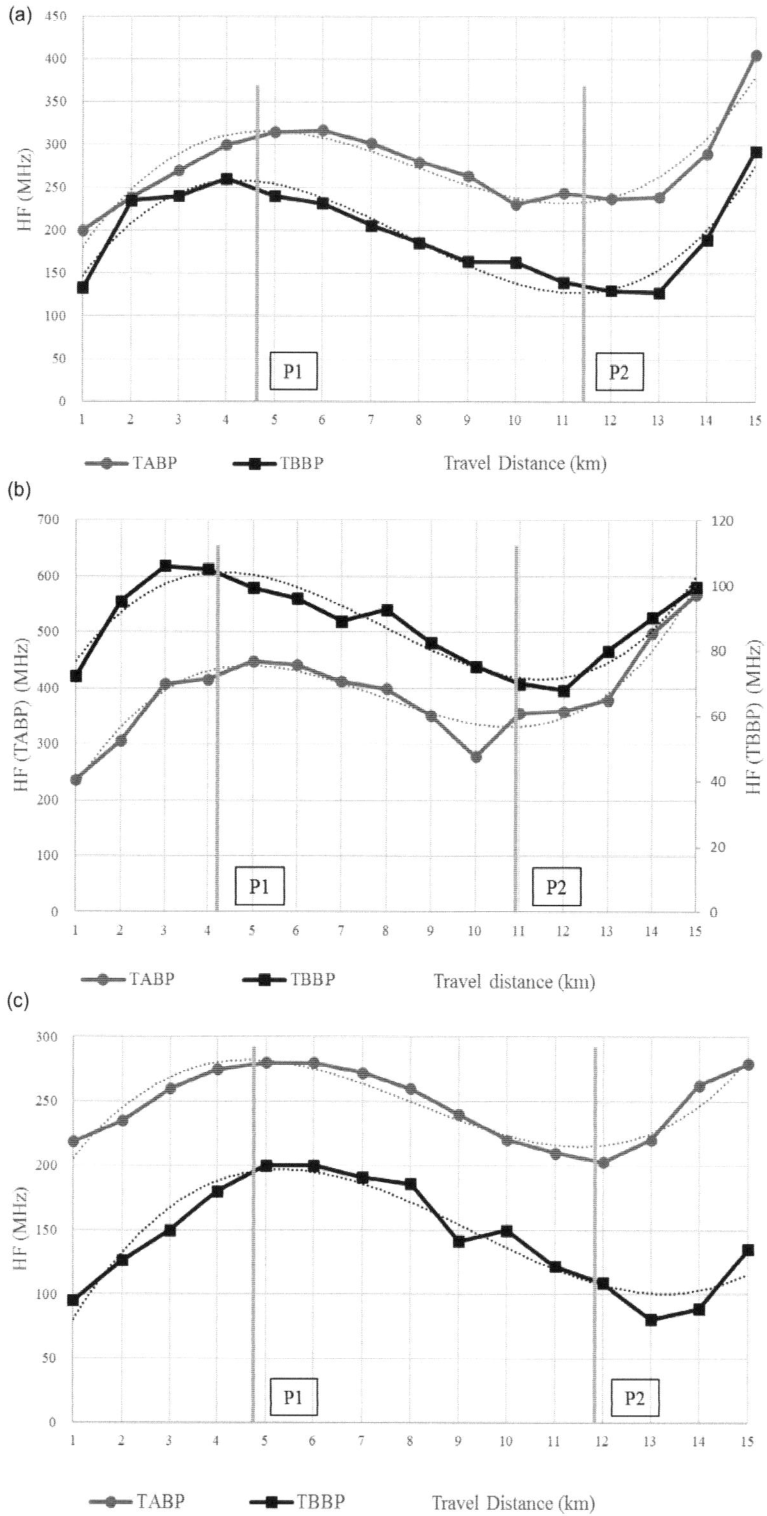

Figure 4. Variation tendency of HF under different spatial patterns: (a) the open pattern's fitting curves; (b) the semi-open pattern's fitting curves; (c) the vertical pattern's fitting curves.

Table 9. Results of the Cubic Polynomial Curve Fittings

Personality	Spatial pattern	Polynomial coefficients a_3	a_2	a_1	a_0	R^2	Distance threshold (km)
A	Open	0.7062	−16.786	112.67	83.81	0.91	11.02
	Semi-open	1.0962	−25.828	174.87	74.034	0.93	10.77
	Vertical	0.4094	−9.9419	65.861	149.62	0.90	12.35
B	Open	0.7546	−17.602	108.99	54.651	0.92	11.28
	Semi-open	0.1771	−4.1928	26.268	54.472	0.90	11.47
	Vertical	0.3828	−10.686	81.278	9.3983	0.90	13.28

addition, both TABP and TBBP drivers' mental workload increases significantly, which indicates that the constant spatial pattern has made drivers tense and depressive (*31*). Based on these results, it is suggested that appropriate stimulus, such as transition of the landscape spatial pattern, should be delivered, which would help increase driver alertness and reduce harmful effects from the negative physiological status. In order to specify the distance threshold of different spatial patterns, the following derivation formula is applied in this study:

$$\begin{cases} \dfrac{d\left(V(P)\right)}{dP} = 0 \\ P > 0 \end{cases} \tag{3}$$

The value of freeway distance P_2, which is equal to the distance threshold, can be calculated while the derived function is 0. Based on the analysis of the fitting curves under different spatial patterns, the variation trend is similar. The calculation results of distance threshold for TABP drivers under open, semi-open and vertical spatial pattern are 11.02 km, 10.77 km, and 12.35 km, while for TBBP drivers they are 11.28 km, 11.47 km, and 13.28 km, as shown in Table 9.

The change of HF value reflects the physiological mental workload of drivers. A higher HF value indicates a more stressed mental status, which should be avoided in freeway landscape design as far as possible. Based on the results in Table 9, it can be seen that the TABP drivers could maintain a shorter distance in good condition under constant landscape pattern than TBBP drivers (i.e., the distance threshold under vertical spatial pattern of TABP and TBBP is 12.35 km and 13.28 km, respectively), which means TABP drivers are more prone to driver fatigue and negative performance than TBBP drivers under the same constant landscape pattern. This can be explained by the TABP personality, which is time-urgent and more likely to have underlying risk-taking driving behaviors. In addition, the distance threshold under an open landscape pattern is significantly shorter for them, which means TABP drivers have higher mental workload under an open landscape pattern than the other two patterns.

Conclusions

To develop ideal design of freeway landscape, the effect of different freeway landscape spatial patterns on TABP and TBBP drivers' mental variation trends are examined in this study. The suitable freeway landscape distance threshold in different spatial patterns for TABP and TBBP drivers was obtained through driving simulator experiment and ECG statistics.

Based on the physiological status investigated in this study, it is concluded that a constant spatial pattern leads to an increase of HF when driving on the freeway for a certain distance with post-speed, and that it is easier to trigger driver fatigue and negative physiological status in TABP drivers than TBBP drivers. In addition, an open spatial pattern has the most adverse effect on drivers' mental workload which means a worse mental status, such as driver fatigue, for both the TABP and TBBP drivers compared with the other two spatial patterns.

The ECG statistics show the freeway distance threshold is affected by different spatial patterns. The results of this study indicate that when the post-speed is 100 km/h, the distance threshold for TABP drivers under open, semi-open and vertical spatial pattern is 11.02 km, 10.77 km, and 12.35 km, whereas for TBBP drivers it is 11.28 km, 11.47 km, and 13.28 km, respectively. It is suggested that a change in landscape spatial pattern is needed when the constant freeway landscape distance reaches 11 km, in order to avoid driver fatigue and negative physiological status.

However, both design and assessment of landscape also involve legal and political issues. The value of this study is to provide a technical rationale for other parties and guidance to take the freeway landscape distance threshold into account. In this study, the research experiment was conducted with a straight freeway and fixed post-speed scenario, which could be varied in future research. Exploring the influence of geometric design of the freeway and different post-speeds would be interesting and meaningful.

Acknowledgments

The authors acknowledge the financial support for this study provided by the National Natural Science Foundation of China (Grant No. 51608019), Beijing Excellent Talent Program (Grant No.

2015000020124G035), Beijing Municipal Science and Technology Project (No. Z171100000517003 and No. Z171100000517004), and the Ri-Xin Talents Project of Beijing University of Technology (2017-RX(1)-16).

References

1. Mok, J. H., H. C. Landphair, and J. R. Naderi. Landscape Improvement Impacts on Roadside Safety in Texas. *Landscape and Urban Planning*, Vol. 78, No. 3, 2006, pp.263–274.
2. Horst, R. V. D., and S. D. Ridder. Influence of Roadside Infrastructure on Driving Behavior: Driving Simulator Study. *Transportation Research Record Journal of the Transportation Research Board*, 2007. 2018: 36–44.
3. Antonson, H., S. Mardh, M. Wiklund, and G. Blomqvist. Effect of Surrounding Landscape on Driving Behavior: A Driving Simulator Study. *Journal of Environmental Psychology*, Vol. 29, No. 4, 2009, pp. 493–502.
4. Yeung, J. S., Y. D. Wong, and H. Xu. Driver Perspectives of Open and Tunnel Expressways. *Journal of Environmental Psychology*, Vol. 36, 2013, pp. 248–256.
5. McCluskey, J. *Road Form and Townscape*. China Architecture & Building Press, Beijing, China, 1992.
6. Zhao, X., Z. H. Wei, Z. Li, Y. Zhang, and X. Feng. Threshold Research on Highway Distance under Typical Landscape Patterns Based on Drivers' Physiological Performance. *Discrete Dynamics in Nature & Society*, Vol. 2015, Article 753732, 2015, pp.1–15.
7. Ashihara, Y. *External Space Design*. China Architecture & Building Press, China, Beijing, 1985.
8. Zhu, X. G. *Research on Highway Visual Space and in the Application of Landscape Planning*. Report presented to South China University of Technology, at Guangzhou, China, in partial fulfillment of the requirement for the degree of Doctor of Engineering, 2010 (in Chinese).
9. Simonds, J. O. *Landscape Architecture*. China Architecture & Building Press, Beijing, China. 2000.
10. Li, X. H., S. L. Zhu, and C. H. Qi. Experiment Research on Influence about the Different Road Environment on Drivers' Physiology Reaction. *Journal of Inner Mongolia Agricultural University* (*Natural Science Edition*), Vol. 31, No. 4, 2010, pp. 172–176 (in Chinese).
11. Brown, T. D. Personality Traits and Their Relationship to Traffic Violations. *Percept Mot Skill*, Vol. 42, No. 2, 1976, pp. 467–470.
12. Pal, U., and R. Torbjorn. Personality, Attitudes and Risk Perception as Predictors of Risky Driving Behavior Among Young Drivers. *Safety Science*, Vol. 41, 2003, pp. 427–443.
13. Cai, H., and Y. Z. Lin. Modeling of Operators' Emotion and Task Performance in a Virtual Driving Environment. *International Journal of Human-Computer Studies*, Vol. 69, 2011, pp. 571–586.
14. Feng, Z. X., H. Z. Yuan, J. Liu, W. H. Zhang, and H. C. Liu. Influence of Driver's Personal Characteristics on Vehicle Velocity. *Journal of Traffic and Transportation Engineering*, Vol. 12, No. 6, 2012, pp. 89–96.
15. Deffenbacher, J. L., and R. S. Lynch. Anger, Aggression, Risky Behavior, and Crash-Related Outcomes in Three Groups of Drivers. *Behavior Research and Therapy*, Vol. 41, 2003, pp. 333–349.
16. Deffenbacher, J. L., E. R. Oetting, and R. S. Lynch. Development of a Driving Anger Scale. *Psychological Reports*, Vol. 74, No. 1, 1994, pp. 83–91.
17. Gunilla, M. B. Driver Irritation and Aggressive Behavior. *Accident Analysis and Prevention*, Vol. 40, No. 3, 2008, pp. 1069–1077.
18. Boyce, T. E., and E. S. Geller. An Instrumented Vehicle Assessment of Problem Behavior and Driving Style: Do Younger Males Really Take More Risks? *Accident Analysis and Prevention*, Vol. 34, No. 1, 2002, pp. 51–64.
19. Venables, N. C., C. J. Patrick, J. R. Hall, and E. M. Bernat. Clarifying Relations Between Dispositional Aggression and Brain Potential Response: Overlapping and Distinct Contributions of Impulsivity and Stress Reactivity. *Biological Psychology*, Vol. 86, No. 3, 2011, pp. 279–288.
20. Traffic Management Bureau of the Public Security Ministry. *2013 Statistical Annual Report on Road Traffic Accidents of the People's Republic of China*. China Communication Press, Beijing, 2014 (in Chinese).
21. Zhang, B. Y. Psychophysiological Reaction in Cardiovascular Disease: II. A Study on the Behavior Pattern of Coronary Heart Disease Patients. *Acta Psychologica Sinica*, Vol. 1985, No. 3, 1984, pp. 314–321 (in Chinese).
22. Reason, J., A. Manstead, S. Stradling, J. Baxter, and K. Campbell. Errors and Violations on the Roads: A Real Distinction? *Ergonomics*, Vol. 33, No. 10–11, 1990, pp. 1315–1332.
23. MHURD. *CJJ37-2012: Code for Design of Urban Road Engineering*. Ministry of Housing and Urban-Rural Development of the People's Republic of China, Chinese Department of Transportation, 2012.
24. Mulder, G. Mental Load and the Measurement of Heart Rate Variability. *Ergonomics*, Vol. 16, 1973, pp. 69–84.
25. Lee, H. B., J. S. Kim, Y. S. Kim, H. J. Baek, M. S. Ryu, and K. S. Park. The Relationship Between HRV Parameters and Stressful Driving Situation in the Real Road. *Proc., 6th International Special Topic Conference*, Tokyo, Japan, ITAB, 2007, pp. 198–210.
26. Guo, S. D. *Cardiology Dictionary*. China Medical Science Press, 1998 (in Chinese).
27. Yang, Y. S., Z. Q. Yao, Z. Y. Li, and K. Jiao. Investigation on Correlation Between ECG Indexes and Driving Fatigue. *Machine Design and Manufacture*, Vol. 5, 2002, pp. 94–95 (in Chinese).
28. Brookhuis, K. A. Measuring Physiology in Simulators. *Handbook of Driving Simulation for Engineering Medicine & Psychology*, Vol. 17, 2011, pp. 1–10.
29. Horenko, I. On Simultaneous Data-Based Dimension Reduction and Hidden Phase Identification. *Journal of the Atmospheric Sciences*, Vol. 65, No. 6, 2008, pp. 1941–1954.
30. Zhao, G. R. A Method of Dimensionality Reduction for Large Scale Data Using PCA. *Computer Knowledge and Technology*, Vol. 10, No. 8, 2014, pp. 1835–1837.
31. Taylor, J. G., E. H. Zube, and J. L. Sell. Landscape Assessment and Perception Research Methods. *Methods in Environmental and Behavioral Research*, Vol. 18, No. 4, 1987, pp. 361–393.

The Standing Committee on Landscape and Environmental Design (AFB40) peer-reviewed this paper (18-02896).

Article

TRR
JOURNAL OF THE TRANSPORTATION RESEARCH BOARD

Transportation Research Record
2018, Vol. 2672(39) 61–67

DOI: 10.1177/0361198118758049
journals.sagepub.com/home/trr
SAGE

Mitigation of Impacts to Groundwater Quality from Highway Runoff in a Karst Terrain

Michael E. Barrett[1]

Abstract

There is a concern that urban development and the construction and operation of highways in particular represent a substantial threat to groundwater quality in areas underlain by karst geology. The objective of this paper is to determine whether common stormwater treatment practices are effective at preventing adverse impacts. The Barton Springs portion of the Edwards Aquifer, a karst system in Central Texas, presents a unique opportunity to address this issue, in that stormwater treatment requirements have been in place over a period of 20 years, when a substantial amount of development, including new highways, has been undertaken. This paper analyzes water quality data from Barton Springs over this period to identify constituents that exhibit changes in concentration and identifies potential causes of these changes. This analysis indicates that the constituents for which water quality has exhibited degradation appear to be due mainly to increases in pumping for water supply and onsite disposal of wastewater. No changes in constituent concentrations were observed for those most closely associated with highway operation, including total suspended solids, heavy metals, pesticides, and bacteria. It was found that sand filters produce an effluent quality equal to or better than runoff from undeveloped land. Consequently, it can be concluded that sand filtration, or an equivalent treatment, is an effective strategy for preventing groundwater degradation from highway operation in karst terrains.

The Edwards Aquifer is a karst system that extends from north of Austin, Texas through San Antonio and beyond. A karst aquifer consists of soluble rock such as limestone or dolomite in which dissolution of these materials results in caves and conduits that allow rapid water flow. There is a concern that this aquifer is particularly vulnerable to non-point source pollution from highways and other land development activities, because of the lack of a significant soil layer to filter stormwater runoff and the rapid rates of transport observed in groundwater. In addition, various endangered species are present within the aquifer and at a number of the springs discharging from this system. In the Austin area, the primary discharge point is at Barton Springs. A number of highway expansions and construction of a new highway over the Barton Springs portion of the aquifer have raised concerns about the impact of highway runoff on the aquifer and the listed species.

The listing documents for the Barton Springs Salamander (*Eurycea sosorum*) and the Austin Blind Salamander (*Eurycea waterlooensis*) express substantial concerns about the potential impacts of urban development on these two species (*1, 2*). In addition, the listing document for the Barton Springs Salamander discusses at length the potential adverse impacts of highways in particular, citing Barrett et al. (*3*) among others. Barrett et al. (*3*) identified a number of previous studies where impacts on soil water and groundwater were observed near highways (*4–7*). On the other hand, Barrett et al. (*1*) also documented a number of studies where no impact was observed (*8–10*), which suggests that site-specific factors may have a substantial effect on the degree to which highway derived constituents are transported to groundwater. Consequently, local data likely provides the best opportunity to assess potential water quality impacts on the Edwards Aquifer.

It has been almost 20 years since the Barton Springs Salamander was placed on the endangered species list, so it would seem appropriate to revisit the issue of the potential impacts of urban development and highway operation on the two listed species at Barton Springs. During the years since

[1]Center for Water and the Environment, University of Texas, Austin, TX

Corresponding Author:
Address correspondence to Michael E. Barrett: mbarrett@mail.utexas.edu

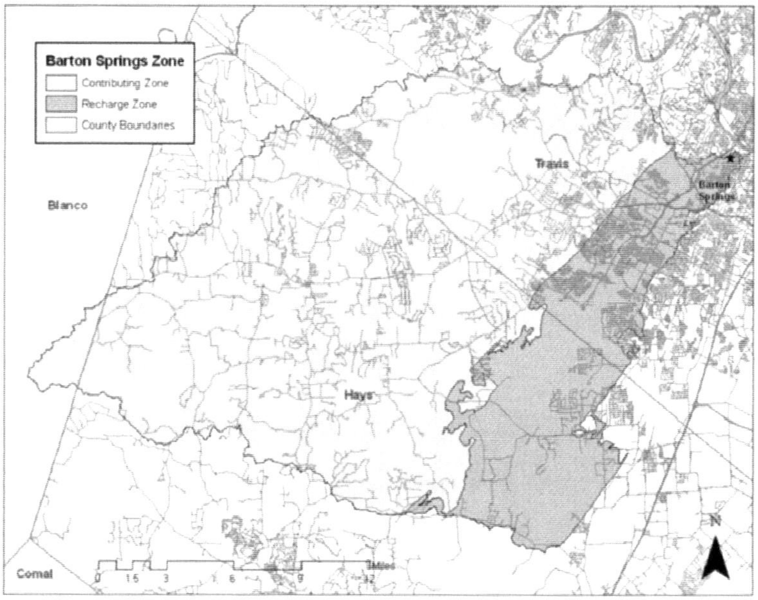

Figure 1. Map of Barton Springs Zone (*11*).

the listing there has been substantial development in the Barton Springs Zone, and there has been continuous water quality monitoring at various springs that has allowed an assessment of water quality impacts and the effectiveness of the various stormwater best management practices (BMPs) required by the City of Austin and the Texas Commission on Environmental Quality (TCEQ).

The Barton Springs Zone (shown in Figure 1), which is the source of water for the springs, has undergone a substantial amount of development over this same period. According to Herrington et al., the population in this area has increased by about 2.5 times during the period 1990–2010, from less than 60,000 individuals to more than 143,000 (*11*). The precise amount of impervious cover created by this growth is not well known. Herrington et al. estimated that about 2,100 acres of structures were added in the Barton Springs Zone based on property roles (*11*); however, this value does not include driveways, sidewalks, or publicly owned transportation infrastructure. An estimate of the increase in total impervious cover from 1991 to 2008 in just one of the five watersheds subject to the City of Austin Save Our Springs Ordinance by Sung et al. (*12*) indicated almost 1,400 acres of new impervious cover in this area alone.

In addition, a number of major highway projects were completed in the early 1990s including:

- Loop 1: from US 290 to Slaughter, completed 1990
- Loop 1: from Slaughter to Hannon (now South Bay), completed 1992
- Loop 1: from Hannon Ln to SH 45, completed in 1994
- SH 45: from Loop 1 to RM 1826, completed in 1994.

This growth has occurred over the same interval that water quality monitoring has been underway. As most of the water that enters the aquifer in this zone is ultimately discharged at Barton Springs, any changes in water quality observed there would be the best possible measure of the cumulative impacts of land development activities.

To limit the impact of land development on water quality, the City of Austin and TCEQ have adopted requirements for the management of both construction and postconstruction stormwater runoff. These requirements have been in place since 1992 for the City and 1999 for TCEQ, and they describe a variety of practices effective for reducing project impacts on water quality. Consequently, monitoring at the springs over this same period of time provides the opportunity to assess the overall effectiveness of these measures.

Data Analysis

The impact on groundwater quality resulting from urbanization as well as highway operation was evaluated using historical water quality data collected at Barton Springs by the City of Austin and the U.S. Geological Survey (https://austintexas.gov/department/barton-springs). This location was selected for several reasons. First of all, it is a known habitat of two endangered species, the Barton Springs Salamander and the Austin Blind Salamander, so any changes in water quality at this location have the potential to impact on these listed species. This location has also been a focal point for water quality monitoring in the Austin area. There is an approximate 20-year record of water quality data collected at the springs for a wide variety of constituents, many of which are commonly observed in highway runoff.

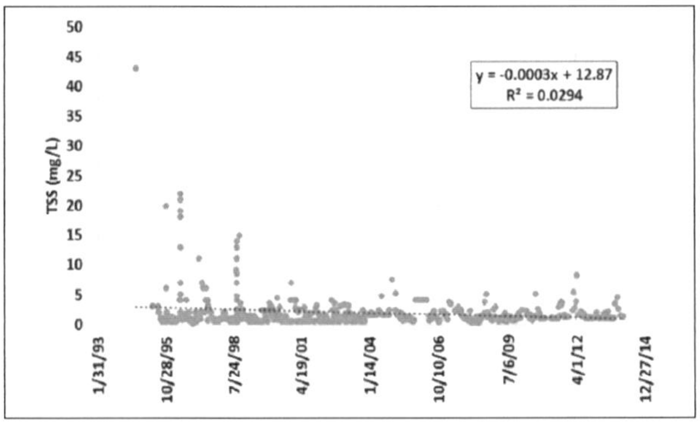

Figure 2. TSS concentrations observed at Barton Springs.

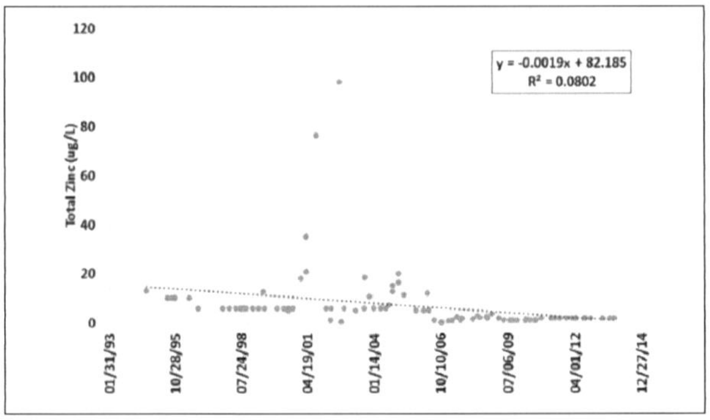

Figure 3. Zinc concentrations observed at Barton Springs.

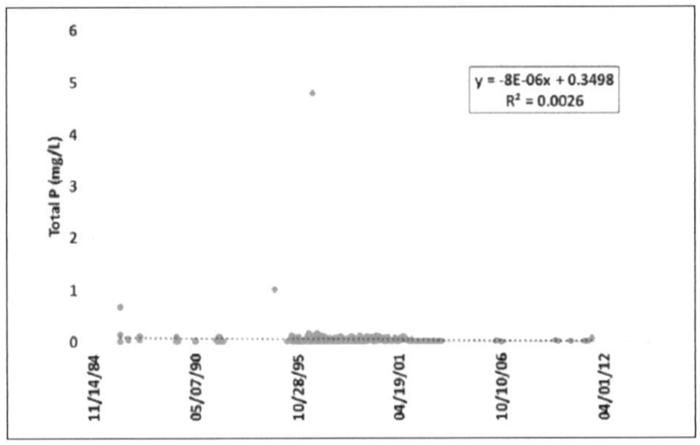

Figure 4. Total phosphorus concentrations observed at Barton Springs.

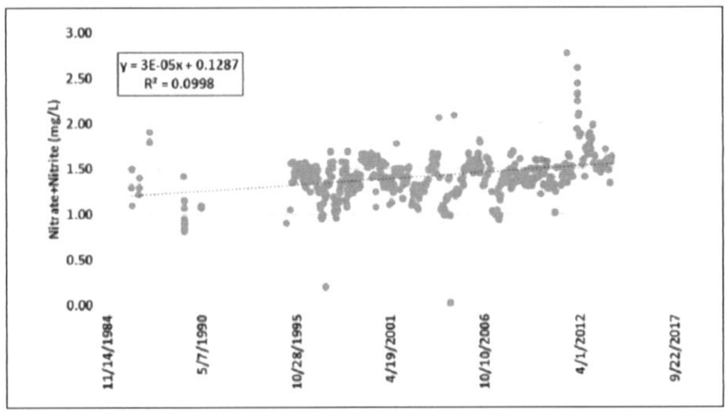

Figure 5. Nitrate concentrations observed at Barton Springs.

The available water quality data at Barton Springs were analyzed for temporal trends for a selected list of constituents commonly observed in highway runoff including solids, heavy metals, and nutrients. Total suspended solids (TSS) was selected as the constituent to represent solids in spring discharge. A graph of the observed TSS concentration for the period of record is presented in Figure 2. It is readily apparent that there is no trend of increasing TSS through time at Barton Springs. Surprisingly, the slope of the regression line shown in this figure is negative, which would indicate that concentrations are actually declining, and this trend is statistically significant. The discussion of BMP performance later in this document will show why this is the expected outcome.

To evaluate the potential for increases in heavy metal concentrations at the springs, zinc was selected as a representative constituent. It is always found in highway runoff, with a primary source being the zinc used in tire manufacturing during the vulcanization process. Tires are typically 1–2% zinc. Another source of zinc in urban areas is galvanized metal, which is found in a number of applications including fences, guardrails, metal roofs, and sign posts. Zinc concentrations for the period of record are presented in Figure 3. It is also obvious for zinc that no trend of increasing concentration at the springs is occurring. Like TSS, the slope of the regression line is negative, indicating a gradual improvement in water quality. This apparent improvement is likely an artifact of lowered detection limits achieved for the newer data. It is important to note that no zinc concentration above the laboratory quantification limit has been observed in over 8 years, going back to 2006 and covering more than 40 individual samples. Similar results were found for copper, although these are not presented in this report.

Total phosphorus and nitrate nitrogen were selected to determine whether there are temporal trends in nutrient concentrations. The graphs for these two constituents are presented in Figure 4 and Figure 5. Total phosphorus concentrations are stable; however, there is a statistically significant trend of increasing concentrations for nitrate.

Nitrogen concentrations in stormwater runoff are similar to those observed at Barton Springs, so that source is unlikely to be responsible for the observed trend. Conversely, nitrogen concentrations in wastewater are substantially higher and could be responsible for the increase if the volume increase is sufficient. Barrett and Charbeneau predicted the effect of additional onsite wastewater disposal on the concentration of nitrogen at Barton Springs (*13*), as shown in Figure 6. The point labeled "current conditions" refers to the number of systems and the water quality at Barton Springs as of 1996. To account for the observed increase in nitrate at

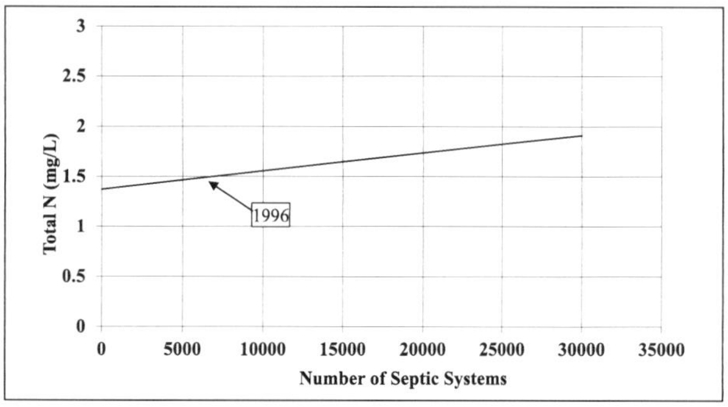

Figure 6. Predicted impact of septic systems on nitrogen concentrations at Barton Springs (*11*).

Barton Springs, an increase in effluent discharge equivalent to that produced by about 15,000 additional septic systems would be required. At 2.3 persons/household, and with a documented increase in population of about 80,000, an additional 35,000 septic systems would be required, so there has been more than enough increase in wastewater discharge to account for the observed trend.

Figure 7 presents a map showing changes through time in the number of septic systems and permitted facilities for land application of treated wastewater in the Barton Springs Zone. It is obvious from this series of maps that the discharge of wastewater has increased substantially in the last 20 years. In addition, Mahler et al. examined the isotopic composition of nitrate in surface water systems that recharge the aquifer in the Barton Springs Zone and found that much of the nitrate was of human or animal origin (*14*). They went on to conclude that:

> During 2001–10, the estimated densities of septic systems and domesticated dogs and cats (number per acre) increased in the watersheds of all five creeks, and the rate of land application of treated waste-water (gallons per day per acre) increased in the watersheds of Barton, Bear, and Onion Creeks. Considering the timing and location of the increases in the three sources, septic systems were considered a likely source of increased nitrate to Bear Creek; land application of treated wastewater a likely source to Barton, Bear, and Onion Creeks; and domestic dogs and cats a potential source principally to Williamson Creek. (*14*)

The City of Austin also recognized the threat to nitrogen concentrations in the aquifer and now requires all new septic systems on the Recharge Zone to include a nitrogen reduction system. Consequently, one can be very confident that the increasing trend for nitrate is related to increases in wastewater disposal rather than stormwater runoff.

The City of Austin has also conducted independent, extensive analysis of water quality trends at karst springs in the Austin area (*15*). Table 1 presents a list of all the constituents with worsening concentrations (generally increasing, except for dissolved oxygen, which is decreasing). Note that there is a lot of double counting on this list. For instance, the sum of calcium and magnesium is what is known as hardness, and alkalinity in natural systems is primarily provided by carbonate, as in calcium carbonate or limestone.

The striking thing about this collection of constituents is that stormwater is not a substantial source of any of them, rather they are common aquifer components. As stormwater recharge occurs and flow increases what is typically observed is a decrease in concentration for the constituents in Table 1. This is exemplified by the relationship between Barton Springs discharge and conductivity presented in Figure 8, in which higher discharge (and recharge) is associated with lower conductivity. Consequently, urban and highway runoff is not responsible for the observed trends for these constituents.

If not runoff, then what is the source of the constituents? The water in the Edwards Aquifer gradually becomes saltier to the east. In addition, there are also documented increases in concentration for most of the other constituents in Table 1, such as sulfate, fluoride, and strontium among others. Figure 9 provides a map of the Barton Springs segment of the Edwards Aquifer showing the Saline Water Zone to the east and the Saline Water Zone Boundary (also known as the "bad water line") along I-35, approaching Barton Springs at the northern boundary of the aquifer. Drought as well as pumping for water supply reduces the water surface elevation in the aquifer, causing water from the saline zone to move west and increase its relative discharge at Barton Springs.

Mahler et al. report that dye-tracing studies have indicated that one prominent aquifer flow path is near the saline zone boundary, which results in elevated sodium, chloride, and sulfate concentrations in water along that flow path relative to concentrations along other flow paths (*16*). They report that numerous studies of water chemistry indicate that Barton Springs receives some of its water from the saline-zone influenced flow path.

Consequently, it is unlikely that it is urbanization per se that is causing the observed trends, but additional pumping from the aquifer for water supply that is causing water from the saline zone to become a larger fraction of discharge at the springs. Groundwater is not used for highway operation, so the projects proposed by the Texas Department of Transportation and the Central Texas Regional Mobility Authority will not contribute to the worsening trends for the constituents listed in Table 1.

The list of constituents identified by the City of Austin as having no temporal trend is quite extensive. Table 2 contains a subset of this list, which includes constituents commonly associated with highway and urban runoff. These include sediment, nutrients, heavy metals, an insecticide, an herbicide, and bacteria. This list of constituents would be typically analyzed for when characterizing the pollutant discharge in stormwater runoff. No constituent associated with stormwater runoff has shown any increase in concentration despite adding 80,000 residents and the associated transportation and other land development infrastructure. This suggests that the stormwater treatment requirements

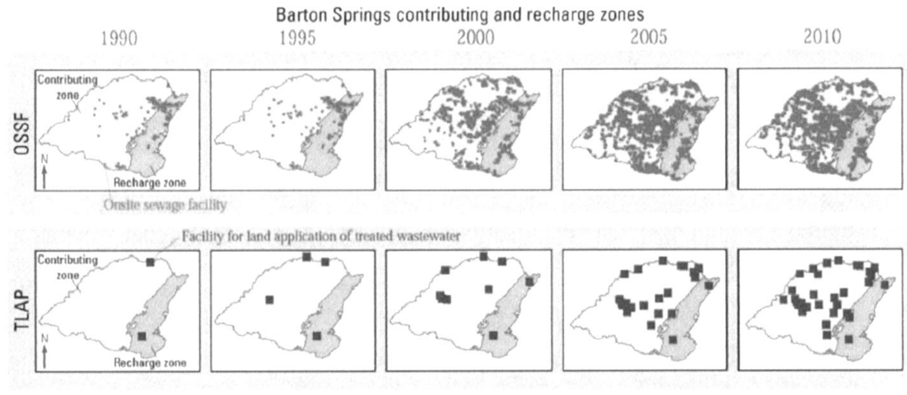

Figure 7. Changes in number of onsite sewage systems (OSSF) and permitted facilities for land application of treated wastewater [Texas Land Application Permit (TLAP)system] in the contributing and recharge zones of the Barton Springs Zone, south-central Texas, 1990–2010 (*14*).

Table 1. Constituents with Worsening Concentrations (*15*)

Conductivity	Silica
Dissolved oxygen	Calcium
Nitrate	Magnesium
Sodium	Potassium
Chloride	Barium
Alkalinity	Sulfate
Hardness	Fluoride
Strontium	

adopted by Austin and TCEQ are effective at preventing adverse effects to the subject salamanders by increased urban and highway runoff.

Discussion

The City of Austin recently completed an analysis of the water quality in runoff from upland undeveloped sites and for selected BMPs currently used in the Austin area (*17*). The purpose of the analysis was to determine the need for impermeable liners under stormwater BMPs. A comparison of the runoff concentrations is provided in Table 3 for undeveloped sites and sand filters, which are commonly used to meet the TSS reduction requirements of the TCEQ Edwards Rules. Green colored cells indicate concentrations significantly lower than undeveloped runoff, and clear cells indicate no significant differences in concentration. As shown in Table 3, sand filters produce runoff with lower or equivalent concentrations than runoff from undeveloped sites for these constituents regulated under the City of Austin Save Our Springs Ordinance.

Despite the remarkable water quality performance of sand filters they are not considered to meet non-degradation standards. The reason for this is that both the City of Austin and TCEQ rules are written around pollutant load (load = volume × concentration), rather than pollutant concentration alone. In other words, if the discharge at Barton Springs were to double as a result of an extended period of high rainfall, but discharge concentrations were to remain exactly the same, under both Austin and TCEQ rules the pollutant discharge (e.g., load) would be said to double. Similarly, a sand filter is not recognized as a non-degradation alternative for stormwater treatment solely because the volume of runoff increases as a result of development, not because of degradation in water quality. Salamanders and other aquatic organisms are affected by the concentrations of various constituents not the pollutant load; consequently, pollutant load is not an appropriate standard to apply when considering potential impacts on aquatic species.

Conclusions

In summary, a very substantial amount of development, including the construction of a number of major highways, has occurred in the area providing recharge to the Barton Springs portion of the Edwards Aquifer over the last 20 years. None of the water quality data analyzed by the City of Austin at the springs indicates any degradation at this location due to changes in stormwater runoff. This observation supports the contention that the water quality regulations adopted by the City of Austin, the TCEQ, and other entities are protective of groundwater quality in the Edwards Aquifer and have resulted in non-degradation in water quality at Barton Springs despite the high level of urbanization that has occurred in the watershed. Consequently, it can be concluded that highway runoff when managed by sand filters or equivalent practices has no detrimental impact to the salamanders found at Barton Springs and other locations in this karst aquifer.

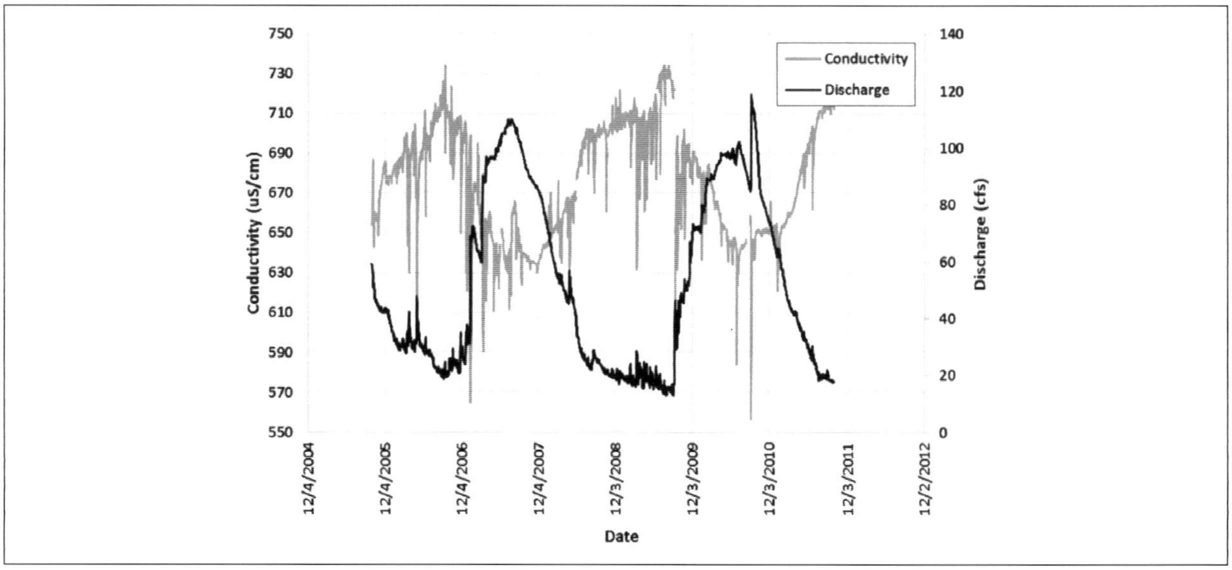

Figure 8. Relationship between conductivity and Barton Springs discharge.

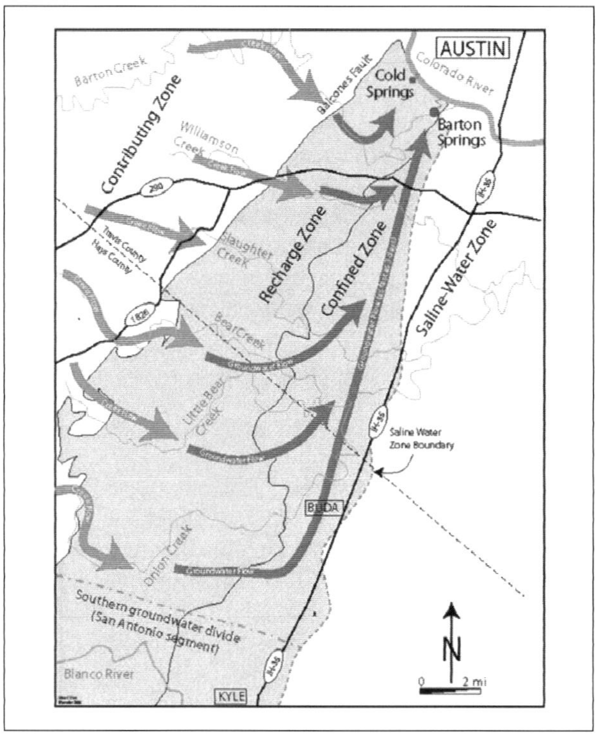

Figure 9. Map of Barton Springs segment of the Edwards Aquifer (Barton Springs/Edwards Aquifer Conservation District).

Table 2. Selected Constituents Exhibiting No Temporal Trend (9)

Total suspended solids	Dissolved phosphorus
Total volatile solids	Total phosphorus
Zinc	Diazinon
Copper	Atrazine
Lead	Turbidity
Cadmium	*E. coli*

Table 3. Comparison of Undeveloped Runoff Concentrations with Sand Filter Discharge (17)

Constituent	Upland undeveloped	Sand filter
Total suspended solids (mg/l)	30.9	11.4
T. phosphorus (mg/l)	0.056	0.078
T. nitrogen (mg/l)	0.89	0.96
Zinc (μg/l)	11.9	18.6
E. coli (Most probable number/100 ml)	6892	5312
Chemical oxygen demand (mg/l)	25.4	18.7
Total organic carbon (mg/l)	8.4	5.9

References

1. U.S. Fish and Wildlife Service. Final rule to list the Barton Springs salamander as endangered. *Federal Register* 62:23377-23392. 1997.
2. U.S Fish and Wildlife Service. Endangered and Threatened Wildlife and Plants; Designation of Critical Habitat for the Austin Blind and Jollyville Plateau Salamanders; Final Rule. *Federal Register* 78: 51328-51379. 2013
3. Barrett, M., R. Zuber, and E. Collins. A Review and Evaluation of Literature Pertaining to the Quality and Control of Pollution from Highway Runoff and Construction. Center for Research in Water Resources, Austin, TX, 1995. http://txcoastalbmp.org/wp-content/uploads/2017/05/rpt95-5.pdf. Accessed Sept. 12, 2016.

4. Yousef, Y. A., M. P. Wanielista, H.H. Harper, and T. Hvitved-Jacobsen. *Best Management Practices – Effectiveness of Retention/Detention Ponds for Control of Contaminants in Highway Runoff, FL/DOT/BMR-86-307*. Florida Department of Transportation, Tallahassee, FL, 1986.
5. Schiffer, D. *Effects of Highway Runoff on the Quality of Water and Bed Sediments of Two Wetlands in Central Florida, USGS Water Resources Investigations*. Report no. 88-4200, A.K.A. Wetlands for Stormwater Treatment, FL/DOT/SMO/90-377. Florida Department of Transportation, Office of Materials and Research, Gainesville, FL, 1988.
6. Schiffer, D. *Effects of Three Highway Runoff Detention Methods on Water Quality of the Surficial Aquifer System in Central Florida, USGS Water Resources Investigations*. Report no. 88-4170. A.K.A. "Impact of Stormwater Management Practices on Groundwater," FL/DOT/SMO/90-378. Florida Department of Transportation, Office of Materials and Research, Gainesville, FL, 1989.
7. Milligan, J. D., and R. P. Betson. The Effects of Carbonate Geology on Urban Runoff: Part 2 – Water Quality Aspects. *Perspectives on Nonpoint Source Pollution. Proc., National Conference, Kansas City, MO. United States Environmental Protection Agency*, p. 281–284. 1985.
8. McKenzie, D. J., and G. A. Irwin. *Effects of Two Stormwater Management Methods on the Quality of Water in the Upper Biscayne Aquifer at Two Commercial Areas in Dade County, Florida*. USGS Water Resources Investigations. Report no. 88-4069. United States Geological Survey, Reston, VA, 1988.
9. Ku, H. F. H., and D. L. Simmons. *Effect of Urban Stormwater Runoff on Ground Water Beneath Recharge Basins on Long Island, New York*. USGS Water-Resources Investigations. Report no. 85-4088. United States Geological Survey, Reston, VA, 1986.
10. Bell, J. H., and M. P. Wanielista. *Use of Overland Flow in Storm-Water Management on Interstate Highways*. Transportation Research Record 736, TRB, National Research Council, Washington, D.C., 1979, pp. 13–21.
11. Herrington, C., M. Menchaca, and M. Westbrook. *Wastewater Disposal Practices and Change in Development in the Barton Springs Zone*. Report no. SR-11-01. City of Austin Watershed Protection Department, Austin, TX, 2010.
12. Sung, C.Y., Y. Young-jae, and L. Ming-Han. Impervious Surface Regulation and Urban Sprawl as its Unintended Consequence. *Land Use Policy*, Vol. 32, 2013, pp. 317–323.
13. Barrett, M., and R. Charbeneau. A Parsimonious Model for Simulation of Flow and Transport in a Karst Aquifer, 1996. http://txcoastalbmp.org/wp-content/uploads/2017/05/rpt96-3.pdf. Accessed Sept. 12, 2016.
14. Mahler, B., M. Musgrove, C. Herrington, and T. Sample. *Recent (2008–10) Concentrations and Isotopic Compositions of Nitrate and Concentrations of Wastewater Compounds in the Barton Springs Zone, South-Central Texas, and Their Potential Relation to Urban Development in the Contributing Zone*. USGS Scientific Investigations Report no. 2011–5018. United States Geological Survey, Reston, VA, 2011.
15. Herrington, C., and S. Hiers. *Temporal Trend Analysis of Long-term Monitoring Data at Karst Springs, 2009*. Report no. SR-10-062010. City of Austin Watershed Protection Department, Austin, TX, 2010.
16. Mahler, B. J., B. D. Garner, M. Musgrove, A. L. Guilfoyle, and M. V. Rao. *Recent (2003–05) Water Quality of Barton Springs, Austin, Texas, with Emphasis on Factors Affecting Variability*. U.S. Geological Survey Scientific Investigations Report no. 2006–5299. U.S. Geological Survey, Reston, VA, 2006.
17. Richter, A., and E. Peacock. *Evaluation of Potential for Water Quality Impacts from Unlined Stormwater Basins in the Barton Springs Recharge Zone*. Report no. SR-15-13. City of Austin Watershed Protection Department, Austin, TX, 2015.

The Standing Committee on Stormwater (AFB65) peer-reviewed this paper (18-00038).

Article

TRR
JOURNAL OF THE TRANSPORTATION RESEARCH BOARD

Transportation Research Record
2018, Vol. 2672(39) 68–78

DOI: 10.1177/0361198118758029
journals.sagepub.com/home/trr
SAGE

Full-Scale Performance Evaluations of Various Wire-Backed Nonwoven Silt Fence Installation Configurations

J. Blake Whitman[1], Wesley C. Zech[1], Wesley N. Donald[1], and Jeffrey J. LaMondia[1]

Abstract

Silt fences have long been a key component for controlling construction stormwater runoff; nonetheless, many silt fence installations fail to perform in the field as intended. Silt fences are temporary sediment control measures used to retain sediment by impounding runoff and allowing for sedimentation on-site, while simultaneously discharging stormwater runoff at a controlled rate. This study evaluated the performance of eight alternative configurations of the Alabama Department of Transportation (ALDOT) standard wire-backed, nonwoven silt fence. Standard installation parameters associated with the ALDOT silt fence include (1) 32-in. (81.3-cm) high fence, (2) 0.95 lb/ft (1.4 kg/m) support T-post, and (3) 10 ft (3.0 m) T-post spacing. Throughout the series of configurations tested, these standard parameters were varied individually and jointly in efforts to improve overall performance. Variations to the standard parameters include (1) 24-in. (61.0-cm) high fence, (2) 1.25 lb/ft (1.9 kg/m) support T-post, (3) 5 ft (1.5 m) T-post spacing, and (4) trench offsetting. Performance analyses were conducted on each configuration and results were evaluated to determine the best overall configuration to enhance the in-field performance of the ALDOT silt fence. Ultimately, the offset 24 in. (61.0 cm) fence with 1.25 lb/ft (1.9 kg/m) T-post spaced 5 ft (1.5 m) on-center resulted in the best overall improvement, retaining an average of 93% of sediment and deflection of only 0.18 ft (0.004 m) over the course of three simulated storm events.

Construction site stormwater runoff has long been recognized by many federal, state, and local agencies as a significant and detrimental nonpoint source (NPS) pollutant of receiving waters (*1*). Earth disturbing activities (i.e., clearing, grubbing, and top soil removal) create erosive conditions by exposing soils, which typically consist of a mixture of sands, silts, and clays. During a rain event, these soils are subjected to erosive forces caused by water droplet impacts, as well as shear forces from water flowing across the soil surface (*2*). Due to these forces, soil particles become dislodged and suspended within stormwater runoff. When stormwater runoff velocities decrease, the process known as sedimentation occurs (*3*). If not properly treated on-site, sediments can be transported and discharged off-site causing a multitude of environmental concerns (i.e., fish kills, water degradation, flooding, etc.).

Silt fences are temporary measures commonly implemented by industry and government agencies for intercepting and treating sediment-laden stormwater prior to site discharge (*4, 5*). Unfortunately, the effectiveness of silt fence installations can be jeopardized due to improper installation and structural instability (*6*). Common structural components associated with silt fence installations include geotextile filter fabric, reinforcement, and vertical support structures. When installed correctly, each of these components work jointly to impound stormwater and promote the removal of solids by means of sedimentation and to a lesser degree, filtration (*7*). Ideally, when used as a perimeter control to intercept sheet flow, silt fence installations need to follow the same contour to promote even flow distribution; however, this installation method is not always achievable. When the toe of the silt fence is not at the same elevation across the entire installation, low points are created along the flow path. These points become areas of concentrated impoundments that subject silt fence installations to increased hydrostatic loading. When this occurs, vertical support structures fail resulting in overtopping and discharge of sediment-laden stormwater.

Based on the findings of Bugg et al. (*8*), the Alabama Department of Transportation (ALDOT) *standard reinforced silt fence* (*9*) yielded no improvement in water quality and

[1]Department of Civil Engineering, Auburn University, AL

Corresponding Author:
Address correspondence to J. Blake Whitman: jbw0012@auburn.edu

failed structurally when subjected to repeated 2-year, 24-h storm events. Thus, the Auburn University Erosion and Sediment Control Testing Facility (AU-ESCTF) research team were tasked with investigating alternative installation configurations that would enhance the overall performance of the standard ALDOT silt fence. In total, eight alternate installation configurations were tested in the AU-ESCTF sediment barrier testing apparatus.

Literature Review

Typically, stormwater runoff is conveyed through one or more on-site sediment control devices (e.g., inlet protection, sediment barriers, sediment basins, vegetative barriers, etc.) prior to off-site discharge into receiving waters and adjacent property (*10*). Nonetheless, these devices can be overloaded due to inadequate design, improper installation, and insufficient maintenance, which can lead to NPS pollution. Due to a wide acceptance of silt fence within industry, several studies have focused on the sediment removal performance of various silt fence installations (*7, 11–14*). Barrett et al. indicated that sediment-trapping efficiency is not a function of filtration, rather the ability to create an upstream impoundment area, which in turn promotes particle settlement (*7, 11*). Robichaud et al. illustrated an effective method for installing silt fence and further reported a trapping efficiency of 73–100% for hillside installations (*14*). Risse et al. performed sediment removal tests on the Silt-Saver® Belted Strand Retention Fence™ (BSRF), as well as the Georgia Soil and Water Conservation Commission Type-C silt fence (*13*). Results indicated that the BSRF reduced turbidity and retained sediment more effectively than the Type-C silt fence. Troxel evaluated six sediment control devices: Type-A silt fence, Type-C silt fence, mulch berm, straw bales, and two compost socks with diameters of 18 in. (45 cm) and 12 in. (30 cm) (*15*). Results indicated total suspended solids (TSS) removal efficiencies of 98.4%, 97.6%, 95%, 91.2%, 92.9%, and 88.2%, respectively. However, due to the upstream sampling point being located at the tank discharge, no determination could be made whether the reduction was a result of upstream impoundment generated by the device, or filtration.

Breaches in silt fences are common on construction sites, however, possible modifications to traditional installation practices may result in increased performance. Typical installation failures observed include undercutting, overtopping, flow bypass, post deflection, fabric sag, fabric detachment, and fabric decomposition (*16*). Donald et al. evaluated the performance of nonwoven wire-backed silt fence installations used as a ditch check and determined that (1) cutting a weir into the filter fabric helps control discharge so flows are contained within the channel, (2) placing a splash pad downstream reduces scour, and (3) pinning filter fabric to the channel eliminates the need for trenching (*17*). Perez et al. conducted experiments on nonwoven wire-backed inlet protection practices (*18*). Ultimately, a design enhancement that incorporated 2 in. × 4 in. (5.1 cm × 10.2 cm) lumber bracing, as well as a dewatering board, were implemented to achieve a structurally sound installation. Unfortunately, a lack of published research on the structural performance of silt fence, used as a perimeter control, exists. Installation details and guidelines provided by government agencies and manufacturers are typically based on rules of thumb, field evaluations, and trial-and-error (*19*). Therefore, a need exists for evaluating silt fence installation improvements to gain an understanding of individual aspects affecting performance.

The goal of this study was to develop and evaluate alternative silt fence installation methods that focus on improving structural stability. The research presented in this paper exhibits the effect small design and installation changes have on structural performance of silt fence when exposed to a replicable 2-year, 24-h design storms. A statistical analysis was conducted on deflection data to determine individual aspect functionality as it relates to structural performance. Sediment retention rates and water quality analyses are also presented.

Methodology

A series of full-scale silt fence experiments were performed to evaluate eight alternative wire-back, nonwoven silt fence installation configurations. For each installation configuration, three replicable 2-year, 24-h design storm tests were performed. Each design storm was replicated by pumping water into a 300-gal water tank that uses a series of valves and orifices to control flow over a calibrated weir. Water discharged through the weir enters a mixing trough in which sediment is introduced at a controlled rate and thoroughly mixed. The sediment-laden water exiting the bottom of the mixing trough is converted to sheet flow using slotted diversion vanes. Each test introduced sediment-laden water into the system at a constant rate of 0.22 cfs (0.03 m^3/s) for a duration of 30 minutes. The detailed methodology and procedures pertaining to sediment barrier performance testing is further outlined in Bugg et al. (*19*). The standard test method referenced for the development of the testing methodology was ASTM D7351-13: *Standard Test Method for Determination of Sediment Retention Device Effectiveness in Sheet Flow Application* (*20*).

Sediment Barrier Testing Apparatus

Testing was accomplished by using the full-scale sediment barrier testing apparatus located at the AU-ESCTF. The apparatus is comprised of a permanently stabilized 20-ft (6.1-m) wide 3H:1V slope, and a 12 ft (3.7 m) × 20 ft (6.1 m) earthen section. The permanently stabilized slope is plated with galvanized sheet metal in which diversion vanes are installed to generate sheet flow conditions. The earthen test area has 3-ft (0.9-m) retaining walls running parallel to the flow direction on either side, allowing for full height silt fence installations and adequate water impoundment. The installation area is designed so that structural failures and installation deficiencies, on a wide array of sediment barrier practices, can be easily evaluated. Before each test, the earthen test area is prepared until moisture content is within ± 5% of the optimum. The area is then leveled using a 12 ft (3.7 m) aluminum screed, and mechanically compacted using a jumping jack compaction plate. The soil used during testing, as well as for fill within the earthen test area, is native to the state of Alabama and is classified by the Unified Soil Classification System as a well-graded sand. A maximum density of 113.1 lb/ft^3 (17.77 kN/m^3) and optimum moisture content of 15.0% were determined by the method described in ASTM D698-12: *Standard Test Methods for Laboratory Compaction Characteristics of Soil Using Standard Effort* (*21*). To ensure repeatability of tests, in-place density samples were taken upstream and downstream of silt fence installations using thin wall Shelby tubes and a density drive hammer to verify 95% standard Proctor compaction was achieved prior to testing.

Silt Fence Installation Materials

The following outlines the materials used during performance testing.

- *Filter fabric:* 3.5 oz/yd^2 (118 g/m^2), nonwoven, 48-in. (121.9-cm) wide fabric that conforms to the American Association of State Highway and Transportation Officials (AASHTO) M288 standard (*22*). Fabric was attached along the top of wire reinforcing using c-ring clips approximately 2 ft (0.61 m) on-center. Fabric was placed into a 6 in. × 6 in. (15.2 cm × 15.2 cm) trench and backfilled.
- *Wire reinforcing:* 17-gauge steel woven wire reinforcement with maximum vertical spacing of 6 in. (15.2 cm) and horizontal spacing of 12 in. (30.5 cm). Wire reinforcing was used to support filter fabric.
- *Studded T-post:* 5 ft (1.5 m) and 4.3 ft (1.3 m) studded T-post, 0.95 lb/ft (1.4 kg/m) and 1.25 lb/ft (1.9 kg/m), driven into ground 24 in. (61 cm), spaced 10 ft (3.0 m) and 5 ft (1.5 m) on-center. T-posts were used as vertical supports for reinforcing wire and filter fabric.
- *Wire ties:* three 6.5 in. (15.6 cm), 11-gauge aluminum wire ties were used to attach reinforcing wire to each studded T-post.
- *C-ring clips:* 11/16 in. (1.75 cm), 16-gauge galvanized steel c-ring clips were used to secure filter fabric to reinforcing wire.

To accurately evaluate the performance of each silt fence installation configuration, the filter fabric type and weight were kept consistent throughout testing.

Control Test: Standard ALDOT Silt Fence Installation

The ALDOT standard wire-reinforced, nonwoven, trenched silt fence configuration, as illustrated in the ALDOT Standard Drawing ESC-200-4 (*9*) shown in Figure 1, *a* and *b*, was evaluated by Bugg et al. (*8*). The findings establish the performance baseline for which results of the installation modifications are compared. The standard ALDOT silt fence installation specifies constructing a silt fence that is: (1) a minimum of 32 in. (81.3 cm) above the ground surface, (2) supported by studded metal T-posts spaced 10 ft (3 m) on-center, and (3) entrenched 6 in. × 6 in. (15.2 cm × 15.2 cm) into the ground. Results from the standard ALDOT silt fence installation tests indicated that: (1) structural failure was due to T-post deflection, (2) no improvement in water quality downstream of the installation was observed, and (3) 82.7% of sediment introduced was retained upstream of the installation.

Alternative Silt Fence Installations

The sediment barrier test apparatus was prepared in accordance with the experimental specifications outlined above for each alternative installation configuration to minimize inconsistencies between tests. Eight installation configurations were evaluated to determine overall performance. Each configuration was installed in the same manner as the standard ALDOT installation, but minor modifications were implemented, as noted below.

- *Modification 1 (M1):* 0.95 lb/ft (1.4 kg/m) T-posts were replaced with 1.25 lb/ft (1.9 kg/m) T-posts.
- *Modification 2 (M2):* 0.95 lb/ft (1.4 kg/m) T-posts spacing was reduced from 10 ft (3.0 m) on-center to 5 ft (1.5 m) on-center.
- *Modification 3 (M3):* 0.95 lb/ft (1.4 kg/m) T-posts were replaced with 1.25 lb/ft (1.9 kg/m) T-posts and T-posts spacing was reduced from 10 ft (3.0 m) on-center to 5 ft (1.5 m) on-center.
- *Modification 4 (M4):* fence height was reduced from 32 in. (81.3 cm) to 24 in. (61.0 cm).
- *Modification 5 (M5):* fence height was reduced from 32 in. (81.3 cm) to 24 in. (61.0 cm) and T-post spacing was reduced from 10 ft (3.0 m) on-center to 5 ft (1.5 m) on-center.
- *Modification 6 (M6):* fence height was reduced from 32 in. (81.3 cm) to 24 in. (61.0 cm) and 0.95 lb/ft (1.4 kg/m) T-posts were replaced with 1.25 lb/ft (1.9 kg/m) T-posts.
- *Modification 7 (M7):* fence height was reduced from 32 in. (81.3 cm) to 24 in. (61.0 cm), 0.95 lb/ft (1.4 kg/m) T-posts were replaced with 1.25 lb/ft (1.9 kg/m) T-posts, and T-post spacing was reduced from 10 ft (3.0 m) on-center to 5 ft (1.5 m) on-center.

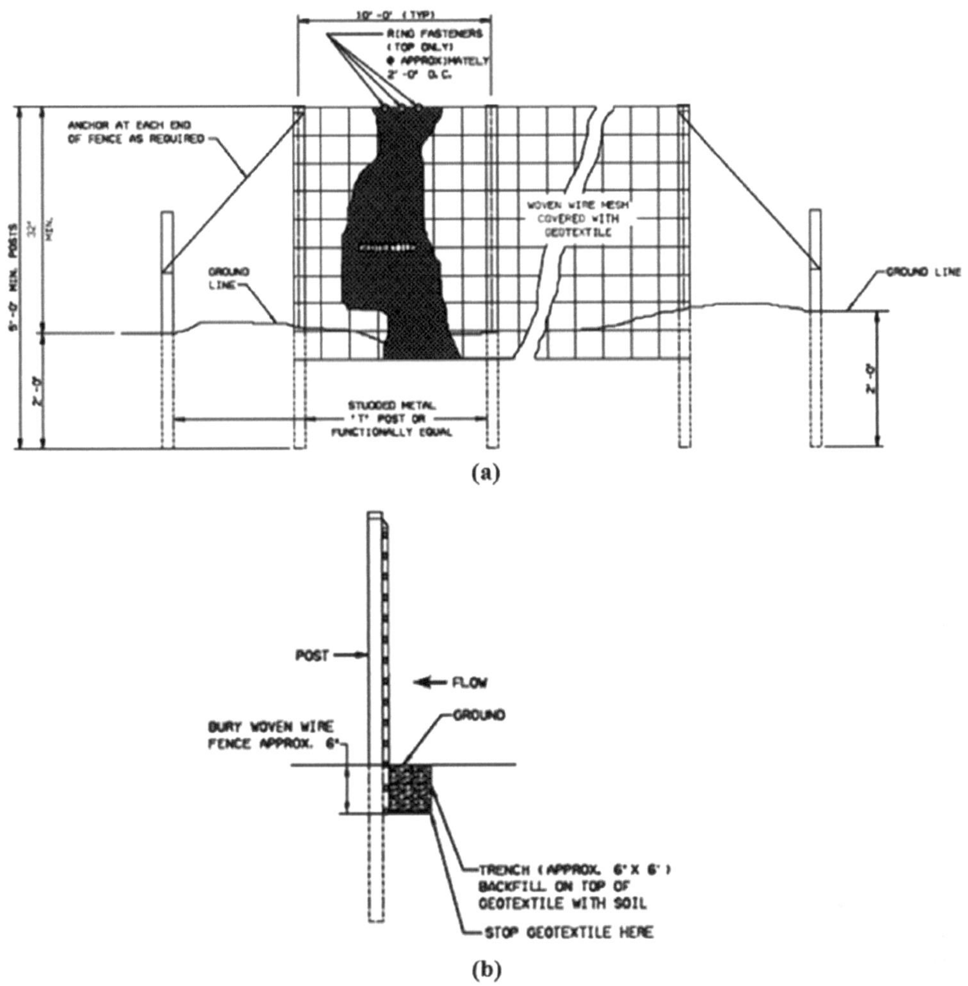

Figure 1. ALDOT standard silt fence installation (9): (*a*) standard ALDOT silt fence – front elevation view; (*b*) standard ALDOT silt fence – side elevation view.

- *Modification 8 (M8):* mimics Modification 7; however, T-post were offset 6 in. (15.2 cm) downstream of the trench.

A summary of the variances between each installation configuration is provided in Table 1.

Evaluation Criteria

The eight installation modifications were evaluated against the results reported by Bugg et al. for the standard ALDOT installation (*8*). The criteria used for evaluation were (a) structural performance, (b) sediment retention, (c) water quality, and (d) statistical relevance. Structural performance analyses were based on observations for potential failure modes. Sediment retention rates were determined by volumetrically comparing pre- and post-test topographical surveys of the test

Table 1. Summary of Silt Fence Installation Modifications

Installation	Fence height (in.)	T-Post weight (lb/ft)	T-Post spacing (ft)	Trench offset (in.)
STD	32	0.95	10	0
M1	32	1.25	10	0
M2	32	0.95	5	0
M3	32	1.25	5	0
M4	24	0.95	10	0
M5	24	0.95	5	0
M6	24	1.25	10	0
M7	24	1.25	5	0
M8	24	1.25	5	6

Note: STD = standard ALDOT installation; M = modification to standard ALDOT installation.
1 in. = 2.54 cm; 1 lb/ft = 1.5 kg/m; 1 ft = 0.3 m.

area. Water samples were taken upstream and downstream of the silt fence installation to evaluate the effect each installation had on water quality.

The fourth criteria, statistical relevance, was achieved by developing a traditional multiple linear regression model. Installation components (e.g., fence height, post weight, post spacing, and trench offset) were first recoded into unique binary independent variables that took values of 1 or 0, depending on whether the installation modified the component or not. The dependent variables were coded as deflection lengths, which ranged between 0.03 ft (0.01 m) and 0.72 ft (0.22 m). The objective for conducting the regression analysis was to determine the relative impact of each component on final fence deflection, independent of other components. It is important to recognize that because some installations were only evaluated once (e.g., M1, M2, M3, M4, M5, and M6), model results are not sufficiently statistically significant to predict deflections. However, the model does provide valid quantifiable measures to support the remaining evaluation criteria described in the following section, as previously seen in work completed by Donald et al. (*23*). The model equation can be written as

$$f(x) = \beta_0 + \beta_1 x_1 + \beta_2 x_2 + \beta_3 x_3 + \beta_4 x_4 \quad (1)$$

where

$f(x)$ = dependent variable (e.g., silt fence deflection)

β_0 = coefficient intercept

β_i = ordinary least squares coefficients

x_i = independent variables (e.g., fence height, post weight, post spacing, offset trench).

Results and Discussion

The following is a summary of the results and observation made from the experiments that use a constant sheet flow of 0.22 cfs (0.006 m^3/s) for 30 min during full-scale test evaluations of the eight installation modifications described above. For each modification, individual installations (i.e., I1, I2, and I3) were subjected to three replicated storm events (i.e., P1, P2, and P3). Based on performance tests conducted by Bugg et al. on the standard ALDOT silt fence installation, alternative construction materials and various installation methods were developed for evaluation (*8*). Throughout the investigation, precedence for improvement was placed in the following order: (1) structural integrity, (2) sediment retention, (3) water quality, and (4) statistical relevance.

Structural Performance

As outlined previously, the efficiency of a silt fence installation in removing sediment is largely dependent on its ability to impound stormwater. To achieve the desired efficiencies, two common failure modes must be addressed. First, the placement of a silt fence should not allow stormwater runoff to bypass the installation (e.g., flow should not go around the end of the silt fence). Second, the silt fence should be able to structurally withstand the hydrostatic pressure imposed by stormwater that impounds upstream of the installation.

Table 2 summarizes the structural performance of each of the installation configurations. Failure mechanisms observed throughout testing were: post deflection, fence sagging, overtopping, and undermining. Post deflections were measured for each installation configuration. The maximum and minimum post deflections for test P3 were 2.67 ft (0.81 m) (standard ALDOT installation) and 0.15 ft (0.05 m) (M8), respectfully. Each installation using 0.95 lb/ft (1.9 kg/m) T-post and/or 10 ft (3 m) T-post spacing, experienced significant post deflection, which resulted in overtopping (Figure 2, *a–f*). Excessive fence sag was observed in each installation using a 10 ft (3 m) T-post spacing (Figure 2, *a*, *b*, *d*, and *f*). Undermining was observed at multiple T-post locations during several tests (Figure 2, *g* and *h*). Although a definitive reason for this occurrence could not be determined, it was speculated that lack of compaction around T-posts due to their placement within the trench resulted in these failures.

In addition to increasing T-post size and decreasing T-post spacing, improvements to the standard installation were analyzed. While conducting tests on M1, it was noted that securing the nonwoven fabric to the T-post by cutting a slit in the fabric and looping it over the T-post (Figure 3*a*) decreased fence sag caused by hydrostatic pressure between T-post, as shown in Figure 3, *b* and *c*. This installation method also reduced pressure applied to the c-ring fasteners (Figure 3*d*) along the top of the fence, which failed during the "no loop" test.

Although undermining was not a significant factor affecting sediment retention performance for each configuration, installation improvements for reducing the reoccurrence of undermining were tested. Figure 3, *e* and *f* show the offset trench installation implemented. Even though a justifiable metric that indicates the benefits of the offset trench in regards to undermining was not obtained, a slight increase in impoundment depth [approximately 0.12 ft (0.04 m)] was noted when compared with direct trenching method. This observation indicates that the offset trench, which was mechanically compacted, may minimize flow under the installation compared with direct trenching, which requires hand compaction to avoid damaging the installation.

Sediment Retention

Topographical surveys of the test area were performed using a total station to gather elevation points pre- and postsimulated events. The data points were used to develop three-dimensional surface models of sediment deposition caused by the impoundment of the silt fence installations. Pre- and post-test surfaces for each simulated event were compared and the volumetric difference between the two was calculated. These volumes, along with the volumes of soil introduced as

Table 2. Silt Fence Failure Modes

Description	Installation	Test	Overtopping time (min:s)	Structural failure
STD (Bugg et al., 2017)	I1	P1, P2	–	No failure
		P3	15:15	Post deflection, fence sag, overtopping
	I2	P1,P2	–	No failure
		P3	14:30	Post deflection, fence sag, overtopping
	I3[a]	P1	–	No failure
		P2	15:30	Post deflection, fence sag, overtopping
M1	I1[a]	P1	–	No failure
		P2	18:45	Post deflection, overtopping, fence sag, undermining
M2	I1	P1	–	No failure
		P2	–	Undermining
		P3	26:40	Post deflection, overtopping
M3	I1	P1, P2, P3	–	No failure
M4	I1[a]	P1	–	Undermining
		P2	16:28	Post deflection, fence sag, overtopping
M5	I1	P1	–	Undermining
		P2	–	No failure
		P3	26:00	Post deflection, overtopping
M6	I1	P1, P2	–	No failure
		P3	13:10	Post deflection, fence sag, overtopping
M7	I1	P1, P2, P3	–	No failure
	I2	P1, P2, P3	–	No failure
	I3	P1, P2, P3	–	No failure
M8	I1	P1, P2, P3	–	No failure
	I2	P1, P2, P3	–	No failure
	I3	P1, P2, P3	–	No failure

Note: STD = standard ALDOT installation; [a] = test P3 was not conducted due to test P2 failure; – = overtopping did not occur.

sediment, were analyzed to determine a retained volume. Average sediment retention rates for installations that did and did not fail due to overtopping (indicated by [a] in Table 3) were 82% and 95%, respectively. The sediment retention rates for each installation are shown in Table 3.

These findings are consistent with the result reported by Donald et al. for sediment retention rates of silt fence used as ditch checks (*17*). The majority of particle sedimentation occurred along the impervious slope, creating a cliff-like deposition as shown in Figure 2*f*. This sediment consisted of large granular particles that settled rapidly when velocity was reduced by the impoundment. Sediment particle size gradually decreased, as did sedimentation depth along the flow direction. Sediment not retained by the silt fence was most likely smaller than the silt fence pores, thus were discharged in the effluent.

Water Quality

Throughout each simulated storm event, water samples were taken to evaluate the effect each installation had on water quality. Samples were taken every 5 min at five sample locations: along the impervious slope (SL1), upstream of silt fence on the surface of the impoundment (SL2), upstream of silt fence along the bottom of the impoundment via sampling pump (SL3), downstream of the silt fence (SL4), and as water discharged into the catch basin (SL5). Figure 4*a* illustrates each of the sample locations. As each installation used the same geotextile fabric, the results obtained were very similar between tests. As shown in Figure 4*b*, the difference in upstream-top of water (SL2) and downstream (SL4) water quality is negligible. As the test progresses, the water quality at each of these locations consistently improves (i.e., turbidity decreases). This improvement is most likely due to impoundment depth increasing as blinding of the fabric occurs along the face of the silt fence, creating a longer impoundment, resulting in a longer time period for sediment particles to fall out of suspension prior to reaching the silt fence fabric. This phenomenon was also observed in Bugg et al. (*8*). When comparing upstream-top of water (SL2) and downstream (SL4) to upstream-bottom of water (SL3), it is further evident that impoundment depth directly affects water quality. On average, for installations that did not fail due to overtopping, a 56% reduction in turbidity was measured between SL3 and SL4 30 min into each test. Extreme variations in water quality were only observed when failures, such as overtopping or undermining, occurred.

Figure 2. Silt fence installation configurations and failure modes: (*a*) standard ALDOT installation (*8*); (*b*) modification 1; (*c*) modification 2; (*d*) modification 4; (e) modification 5; (*f*) modification 6; (g) undermining of modification 2; (*h*) undermining of modification 4.

Statistical Relevance

To statistically determine the effects of different installation configurations; a multiple linear regression model was developed. Each installation had a corresponding combination of independent variables considered in the analysis: (1) fence height, (2) T-post weight, (3) T-post spacing, and (4) trench offset. For this regression model, the standard ALDOT installation was considered the base installation, against which each installation variation was compared. The dependent variable selected for the analysis, which is directly affected by each independent variable, was T-post deflection. Deflections obtained from P1 tests were used within the model owing to consistent initial conditions (i.e., unclogged filter fabric pores and plumb T-posts). A brief summary of T-post deflections used within the model is shown in Figure 5. The R^2 of the estimated model was 0.93, indicating a well-fitted linear model when compared with measured observations. Results of the analysis, along with statistical significances, are shown in Table 4.

Based on the statistical significance generated by the model, the following conclusions were drawn: (1) each installation component independently reduces fence deflection relative to the standard ALDOT installation, as evident by the

Table 3. Sediment Retention

Description	Installation	Sediment retained
STD	I1	87%
ALDOT	I2	87%
Installation	I3	75%
M1	I1	53%
M2	I1	76%
M3	I1[a]	87%
M4	I1	90%
M5	I1	95%
M6	I1	96%
M7	I1[a]	100%
	I2[a]	100%
	I3[a]	96%
M8	I1[a]	90%
	I2[a]	91%
	I3[a]	98%

Note: STD = standard ALDOT installation; [a] = overtopping did not occur.

negative coefficients (i.e., positive values indicate increased deflections, therefore negative values indicate decreased deflections); (2) each coefficient is statistically significant at a 90% confidence level, as indicated by p-values less than 0.1, thus signifying a positive effect on installation performance; (3) fence height and trench offsetting have the least effects on performance; and (4) T-post spacing and T-post weight have the greatest effects on performance. These statistical conclusions correlate to the structural failure mode observations outlined in Table 2, as well as to the sediment retention rates outlined in Table 3. When comparing measurable performance standards of each installation modification to the standard ALDOT installation, it is evident that each alteration facilitates a performance improvement.

Conclusions

Current wire-backed, nonwoven silt fence installation practices implemented by ALDOT lack the structural ability to

Figure 3. Silt fence improvements: (*a*) T-post loop over; (*b*) without T-post loop over; (*c*) with T-post loop over; (*d*) c-ring fasteners; (*e*) offset trench; (*f*) offset silt fence installation.

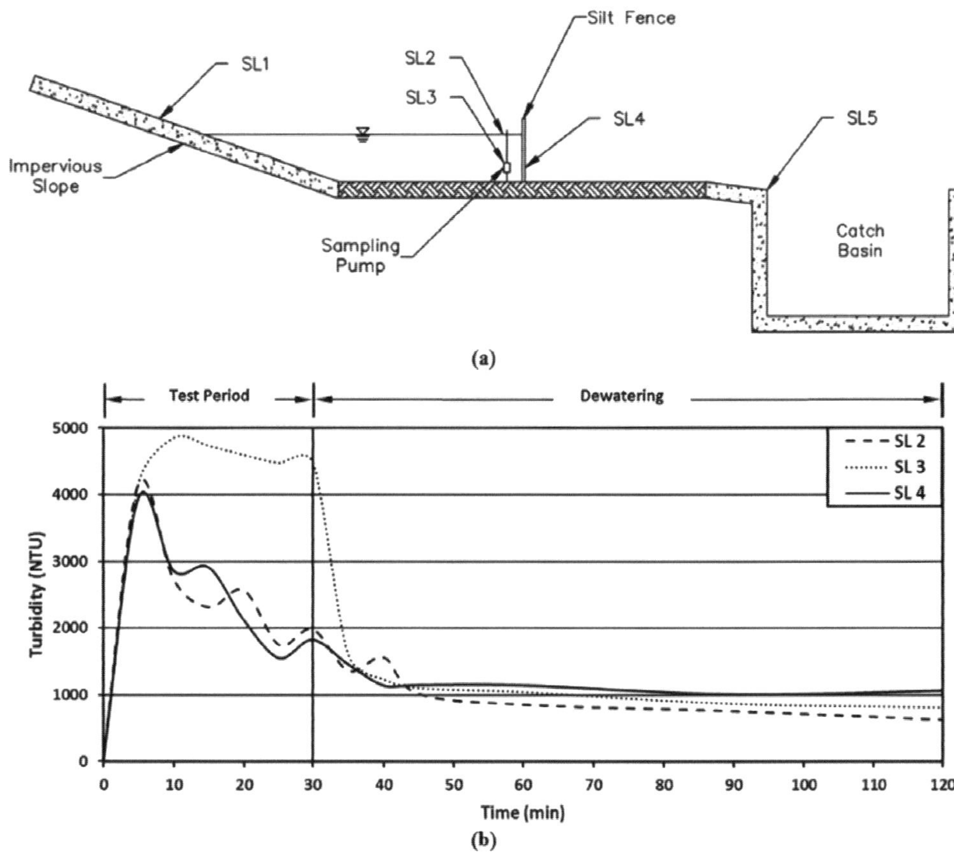

Figure 4. Water quality: (*a*) water sample locations, and (*b*) typical turbidity graph for nonwoven silt fence.

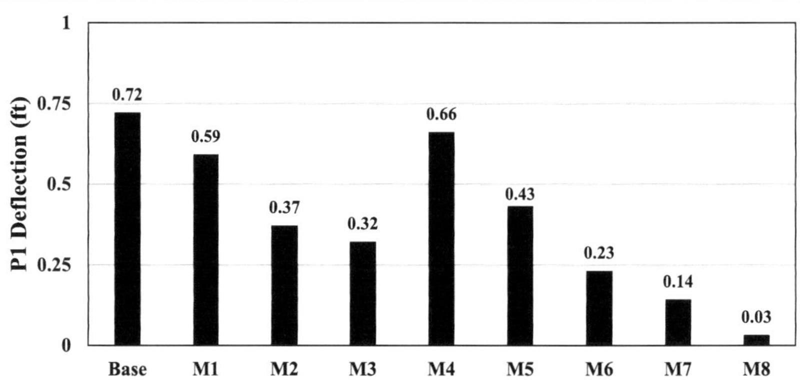

Figure 5. T-post deflection summary.

create and sustain impoundments required to promote sedimentation. The hydrostatic loading imposed on an installation by an impoundment may cause structural failures, thus resulting in untreated sediment-laden stormwater discharges into the surrounding environment. The research team at AU-ESCTF evaluated the structural performance of eight silt fence installation configurations and demonstrated that a structurally sound silt fence practice is achievable.

The information obtained through this study shows that increasing T-post weight and decreasing T-post spacing

Table 4. Statistical Relationship of Installation Components

	Statistical significance	
Installation component	Coefficients	p-Value[a]
Base	0.73	na
Fence height	−0.13	0.024
T-Post weight	−0.22	0.001
T-Post spacing	−0.23	0.000
Trench offset	−0.11	0.096

Note: na = not applicable; [a] = comparison to effects of ALDOT Standard Silt Fence at 90% confidence interval and *p*-values <0.10.

greatly improves the structural integrity of silt fence installations. Additionally, reducing fence height and implementing an offset trench only provided slight structural improvements. However, from an installation standpoint, offset trenching allows for mechanical compaction, which ultimately has the potential to reduce the occurrence of undermining. Observations during testing suggest additional filter fabric support can be achieved by looping the fabric over each T-post. The volumetric analysis conducted on retained sediment shows that structurally sound silt fence installations have a retention rate of 95% as opposed to 83% for those that overtop. Water quality data indicated that as impoundment depth increases, water along the surface of the impoundment and downstream of the silt fence decreases in turbidity. Based on these findings, modification 8 had the best overall performance characteristics of the installation variations tested. As this installation method only varies in approach and not necessarily in equipment and effort needed, this method can easily be applied in the field with only minimal training of field personnel to improve silt fence structural performance. Finally, the structural performance and retention efficiency of an installation in the field can be effectively increased by implementing routine inspections and conducting preventive maintenance by removing accumulated sediment after storm events.

Acknowledgments

The AU-ESCTF would like to thank the Alabama DOT for its sponsorship of this research.

References

1. Fennessey, L. A., and A. R. Jarrett. Influence of Principal Spillway Geometry and Permanent Pool Depth on Sediment Retention of Sedimentation Basins. *Transaction of the ASCE*, Vol. 40, No. 1, 1997, pp. 53–59.
2. Pitt, R., S. E. Clark, and D. W. Lake. *Construction Site Erosion and Sediment Controls: Planning, Design and Performance*. DEStech Publications, Lancaster, Pa., 2007.
3. Hunt, C. L., and J. K. Grow. Impacts of Sedimentation to a Stream by a Non-Compliant Construction Site. Bridging the Gap: Meeting the World's Water and Environmental Resources Challenges, Proceeding of the World Water and Environmental Resources Congress, Orlando, Florida, American Society of Civil Engineers, 2001. https://doi.org/10.1061/40569(2001)20.
4. Zech, W.C., J. L. Halverson, and T. P. Clement. Development of Silt Fence Tieback Design Methodology for Highway Construction Installations. *Transportation Research Record: Journal of the Transportation Research Board*, 2007. 2011: 21–28.
5. Zech, W.C., J. S. McDonald, and T. P. Clement. Field Evaluations of Silt Fence Tieback Systems at a Highway Construction Site. Practice Periodical on Structural Design and Construction. *American Society of Civil Engineers*. Vol. 14, No. 3, 2009, pp. 105–112.
6. Stevens, E., S. Yeri, B. Barfield, J. Arjunan, and A. Selvakumar. A Design Aid for Conventional Silt Fence Technology. *Impacts of Global Climate Change, Proceeding of the World Water and Environmental Resources Congress*, Anchorage, Alaska, American Society of Civil Engineers, 2005. https://doi.org/10.1061/40792(173)171.
7. Barrett, M. E., J. E. Kearney, T. G. McCoy, and J. F. Malina. An Evaluation of the use and Effectiveness of Temporary Sediment Controls. *Center for Research in Water Resources Technical Report 261*, University of Texas at Austin, Tex., 1995.
8. Bugg, R. A., W. N. Donald, W. C. Zech, and M. A. Perez. Performance Evaluations of Three Silt Fence Installations Using a Full-Scale Testing Apparatus. *Water*, Vol. 9, No. 502, 2017.
9. Alabama Department of Transportation (ALDOT). *Standard Drawing ESC-200 (Sheet 4)*. Montgomery, Ala., 2017.
10. Perez, M. A., W. C. Zech, X. Fang, and J. G. Vasconcelos. Methodology and Development of a Large-Scale Sediment Basin for Performance Testing. *Journal of Irrigation and Drainage Engineering*, Vol. 142, No. 10, 2016.
11. Barrett, M. E., J. F. Malina, and R. J. Charbeneau. An Evaluation of Geotextiles for Temporary Sediment Control. *Water Environment Research*, Vol. 70, No. 3, 1998, pp. 283–290.
12. Keener, H. M., L. B. Faucette, and M. H. Klingman. Flow-Through Rates and the Evaluation of Solids Separation of Compost Filter Socks versus Silt Fence in Sediment Control Applications. *Journal of Environmental Quality*, Vol. 36, No. 3, 2007, pp. 742–752.
13. Risse, L. M., S. A. Thompson, J. Governo, and K. Harris. Testing of New Silt Fence Materials: A Case Study of a Belted Strand Retention Fence. *Journal of Soil and Water Conservation*, Vol. 63, No. 5, 2008, pp. 265–273.
14. Robichaud, P. R., D. K. McCool, C. D. Pannkuk, R. E. Brown, and P. W. Mutch. Trap Efficiency of Silt Fences Used in Hillslope Erosion Studies. Soil Erosion Research for the 21 Century. *Proc., International Symposium*, Honolulu, Hawaii, USA, 2001, pp. 541–543.
15. Troxel, C. F. Life Cycle Analysis of Sediment Control Devices. MS thesis. Georgia Institute of Technology, Atlanta, Ga., 2013.
16. Stevens, E., S. Yeri, B. Barfield, K. Gasem, and M. Matlock. On and Off Site Sediment Control Using Silt Fence. Critical Transitions in Water and Environment Resources Management, *Proceeding of the World Water and Environmental Resources Congress*, Salt Lake City, Utah, American Society of Civil Engineers, 2004. https://doi.org/10.1061/40737(2004)69.
17. Donald, W. N., W. C. Zech, M. A. Perez, and X. Fang. Evaluation and Modification of Wire Backed, Nonwoven

Filter Fabric Silt Fence for use as a Ditch Check. *Journal of Irrigation and Drainage Engineering*, Vol. 142, No. 2, 2016.
18. Perez, M. A., W. C. Zech, W. N. Donald, and X. Fang. Installation Enhancements to Common Inlet Protection Practices Using Large-Scale Testing Techniques. *Transportation Research Record: Journal of the Transportation Research Board*, 2015. 2521: 151–161.
19. Bugg, R. A., W. N. Donald, W. C. Zech, and M. A. Perez. Improvements in Standardized Testing for Evaluating Sediment Barrier Performance: Design of a Full-Scale Testing Apparatus. *Journal of Irrigation and Drainage Engineering*, Vol. 143, No. 8, 2017.
20. ASTM Standard D7351. *Standard Test Method for Determination of Sediment Retention Device Effectiveness in Sheet Flow Applications*. ASTM International, West Conshohocken, Pa., 2017.
21. ASTM Standard D698. *Standard Test Methods for Laboratory Compaction Characteristics of Soil Using Standard Effort*. ASTM International, West Conshohocken, Pa., 2017.
22. AASHTO Standard M288. *Standard Specification for Geotextile Specification for Highway Applications*, AASHTO, Washington, D.C., 2017.
23. Donald, W. N., W. C. Zech, M. A. Perez, and X. Fang. Ditch Check Installation Evaluations of Wheat Straw Wattles Used for Velocity Reduction. *Transportation Research Record: Journal of the Transportation Research Board*, 2013. 2358: 69–78.

The Standing Committee on Stormwater (AFB65) peer-reviewed this paper (18-01215).

The findings, opinions, and conclusions presented are those of the authors and do not necessarily reflect the view of the sponsor.

Article

TRR
JOURNAL OF THE TRANSPORTATION RESEARCH BOARD

Transportation Research Record
2018, Vol. 2672(39) 79–87
© National Academy of Sciences:
Transportation Research Board 2018
Article reuse guidelines:
sagepub.com/journals-permissions
DOI: 10.1177/0361198118758679
journals.sagepub.com/home/trr
SAGE

Assessing Roadway Contributions to Stormwater Flows, Concentrations, and Loads with the StreamStats Application

Adam J. Stonewall[1], Gregory E. Granato[2], and Tana L. Haluska[1]

Abstract

The Oregon Department of Transportation (ODOT) and other state departments of transportation need quantitative information about the percentages of different land cover categories above any given stream crossing in the state to assess and address roadway contributions to water-quality impairments and resulting total maximum daily loads. The U.S. Geological Survey, in cooperation with ODOT and the FHWA, added roadway and land cover information to the online StreamStats application to facilitate analysis of stormwater runoff contributions from different land covers. Analysis of 25 delineated basins with drainage areas of about 100 mi^2 indicates the diversity of land covers in the Willamette Valley, Oregon. On average, agricultural, developed, and undeveloped land covers comprise 15%, 2.3%, and 82% of these basin areas. On average, these basins contained about 10 mi of state highways and 222 mi of non-state roads. The Stochastic Empirical Loading and Dilution Model was used with available water-quality data to simulate long-term yields of total phosphorus from highways, non-highway roadways, and agricultural, developed, and undeveloped areas. These yields were applied to land cover areas obtained from StreamStats for the Willamette River above Wilsonville, Oregon. This analysis indicated that highway yields were larger than yields from other land covers because highway runoff concentrations were higher than other land covers and the highway is fully impervious. However, the total highway area was a fraction of the other land covers. Accordingly, highway runoff mitigation measures can be effective for managing water quality locally, they may have limited effect on achieving basin-wide stormwater reduction goals.

The Oregon Department of Transportation (ODOT) and other state departments of transportation (DOTs) need quantitative information about the percentages of different land cover categories above any given stream crossing in the state. Such information is needed to assess the potential contribution of state highways and other sources to the quality of receiving streams. The ability to assess contributions from highways and other sources is needed to address responsibilities for managing Environmental Impact Statements, National Pollutant Discharge Elimination System permits, Municipal Separate Storm Sewer System (MS4) permits, and efforts to establish total maximum daily loads (TMDLs) (*1–4*). Information about highway contributions is needed to inform analyses of the quality and quantity of runoff, potential effects on receiving waters, and the potential effectiveness of mitigation measures. Knowledge of highways and surrounding and upstream land use/land cover percentages in a stream basin can inform decisions about the potential cost and effectiveness of on-site or watershed-scale best management practices (BMPs) to minimize adverse effects of runoff on receiving water quality.

TMDLs are used as a method to calculate allowable loads of the water-quality constituents that will reduce or eliminate water-quality impairments in receiving waters. The results of TMDL analyses are used to determine load allocations for point and non-point sources of the constituents of concern. Currently (2017), the U.S. Environmental Protection Agency (EPA) reports that there are 42,527 impaired water bodies nationwide and 71,193 TMDLs have been approved since 1995 (*5*). Many of the constituents of concern for these water-quality impairments, including pathogens, nutrients, trace metals, organic carbon, and sediment, have also been identified as highway and urban runoff quality constituents (*3,4,6–8*). Research indicates that DOTs need information, tools, and techniques to address

[1]U.S. Geological Survey, Oregon Water Science Center, Portland, OR
[2]U.S. Geological Survey, New England Science Center, Massachusetts Office, Northborough, MA

Corresponding Author:
Address correspondence to Gregory E. Granato: ggranato@usgs.gov

increasing concerns about their roles and responsibilities in the TMDL processes (*1,3,4,6,9,10*).

In 2013, the U.S. Geological Survey (USGS), in cooperation with the FHWA, published the Stochastic Empirical Loading and Dilution Model (SELDM) to provide a tool that can be used to simulate stormwater concentrations, flows, and loads from highways and other land uses (*2,3,11*). SELDM is a lumped-parameter Monte Carlo model that can be used to simulate long-term average annual runoff loads from highways and other land use areas that can be used for TMDL analyses. SELDM also has a stormwater BMP module to simulate potential effects of stormwater control measures on runoff flows, concentrations, and loads from the site of interest (*11,12*). In SELDM, each storm that is generated for an analysis is identified by sequence number and annual-load accounting year. When the time between event mid points of a series of simulated events exceeds 365 or 366 days, those events are lumped into one annual-load accounting year. SELDM generates each storm randomly; there is no serial correlation. The order of storms does not reflect seasonal patterns. The annual-load accounting years, which are random collections of events generated with sums of inter-event times less than or equal to a year, are used to generate annual highway flows and loads for the TMDL analyses. The population of annual runoff and BMP discharge loads from the site of interest (a highway or other land use area) indicates the potential contribution from the site of interest, the potential for reducing loads to meet any proposed load allocations, and the uncertainty of such estimates.

Purpose and Scope

The purpose of this paper is to describe and demonstrate methods for assessing roadway contributions to stormwater flows, concentrations, and loads by using the StreamStats application to obtain basin properties and by using SELDM to simulate the quality and quantity of runoff. The Oregon StreamStats application was used as an example because the USGS has implemented land cover and roadway information in cooperation with ODOT. Results of a spatial analysis with many midsize basins was used to demonstrate the process for estimating land cover categories and the length and area of minor roads, major roads, and state highways above any point along a stream in Oregon. Total phosphorus (TP) concentrations from the literature for different land cover types were used to demonstrate application of stormwater runoff yields with derived land cover areas to assess contributions of different land uses in the Willamette basin above USGS streamgage 14198000 Willamette River at Wilsonville, Oregon, because TP is a constituent of concern for many TMDLs nationwide (*5*). This large basin was used for the TMDL analysis because it represents an area-weighted average of the properties of its subbasins.

Methods

The methods used for the geographic and stormwater-quality analyses used in this paper are based on previous efforts to provide hydrologic and water-quality information and stormwater simulation methods for the nation (*2,3,11,13*). In this paper, however, these methods were combined to demonstrate new practice-ready capabilities developed by the USGS that can be used by other DOTs, nationwide.

Geographic Analysis Methods

The USGS, in cooperation with the ODOT, added land cover and roadway data sets to the previously existing Oregon state StreamStats application (*14,15*). The land cover data sets added to this application were derived from the 2011 National Land Cover Datasets (*16–18*). These data sets included the impervious percentage, natural, agricultural, and developed land covers. The natural/undeveloped land covers included the aggregated categories for water, wetlands, herbaceous, shrub land, forest, and barren land. The developed land use categories included open space, low-intensity, medium-intensity, and high-intensity developed areas. This data set provides the percentage imperviousness and the percentage of each land cover upstream of any point on any stream within any drainage area that is contained within Oregon. The land cover data set is implemented as a gridded raster data set with a 30-m pixel resolution. A data set of three road classes, including non-state minor roads, non-state major roads, and state highways, was developed as a vector data set from the Oregon state-road class geographic information system (GIS) coverage (*19*). This data set provides the length of all roads in each road class within each delineated basin. It provides the length of each side of divided highways. These data sets were incorporated into the application by using standard USGS StreamStats processing methods.

For this paper, the area of the roadways was calculated by using the length of each type of roadway times the width. The width of the roadways was estimated by multiplying a standard lane width (12 ft) to an estimated average number of lanes for each type of road. It was assumed that the average width of non-state minor roads is 2.25 lanes, non-state major roads is 3 lanes, and state highways was 3.5 lanes (on each side of the divided highway, including ramps and breakdown lanes).

Two methods were used for obtaining drainage areas, land cover percentages, and roadway lengths from StreamStats. The batch-processing method was used to obtain basin properties to analyze the distributions of land cover percentages in stream basins within the Willamette Valley in Oregon. A GIS was used to analyze stream segments in the National Hydrography Dataset to locate coordinates of points along the stream network that represent drainage areas of about 100 mi^2. Resulting basins were screened for specific conditions that were not representative of most stream basins in

Table 1. Median of the Average, Standard Deviation, and Skew of the Common (Base 10) Logarithms and Geometric Mean of Event-Mean Concentrations of TP for Different Land Cover Categories

Land cover category	Number of sites	Geometric mean (mg/L)	Average*	Standard deviation*	Skew*	Source
Agricultural (Washington State)	4	0.193	−0.7135	0.2117	0.3221	*20*
Developed (Nationwide)	95	0.159	−0.7988	0.2953	0.3296	*21*
Forest (Washington State)	4	0.02	−1.693	0.2765	1.901	*20*
Highways (Oregon)	4	0.214	−0.6687	0.2888	−0.5693	*22*

*Statistics are for the common (base 10) logarithms of EPA parameter p00665, phosphorus water, unfiltered, milligrams per liter (mg/L).

the state. Basins with atypical proportions of canals, reservoirs, lakes, or ponds were screened out so that the basins would represent free-flowing stream conditions. Basins that passed this screening test were submitted to the StreamStats batch-processing tool to obtain the basin characteristics, which included imperviousness, land cover percentages, and the lengths of minor roads, major roads, and state highways within the delineated drainage areas. The manual method, which is comprised of zooming to a point of interest, clicking on a stream cell, and requesting a basin characteristics report, was used to get the data used in the example loading analysis (*13*). StreamStats users most commonly delineate a basin and obtain basin properties and flow statistics by using the manual method.

Stormwater-Quality Analysis Methods

Long-term annual yields of TP were simulated with version 1.02 of SELDM by using methods described by Granato and Jones (*3*). GIS layers of SELDM precipitation statistics created by Risley and Granato were used to generate storm events for these analyses (*4*). Yields were calculated by using long-term average event volumes, event durations, and time between event mid points of 0.693 in., 13.4 h, and 136 h, respectively. Roadway runoff coefficients were simulated by using SELDM statistics for completely impervious roadway areas (*11*). Runoff coefficients for other land cover areas were calculated by using non-roadway runoff coefficient statistics based on the assumption that total impervious percentages for developed land covers were 33.4%, agricultural land covers were 0.5%, and undeveloped land covers (primarily forest) were 0% (*11*). The area of state highways was subtracted from the developed land cover area to estimate total loads for each category.

The simulated populations of annual runoff yields for the different land covers depend on the input concentration statistic values and the runoff coefficients. Yields in pounds of TP per acre from each land cover are applied to the different areas obtained from StreamStats to calculate long-term average annual loads. Stormwater runoff quality from undeveloped and agricultural areas was simulated by using concentrations from forest and agricultural runoff from the Washington State Department of Ecology database (*20*). Stormwater runoff quality from developed non-highway areas was simulated by using BMP inflow concentration statistics from the September 2016 version of the International BMP database (*21*). Highway runoff quality was simulated by using statistics from four highway runoff sites in Oregon with TP data (*22*). The median of statistics for the average, standard deviation, and skew of the common logarithm of TP from sites in each data set were used for the annual-yield simulations (Table 1). The geometric mean TP concentration of the forest runoff is about an order of magnitude less than the geometric means of the other land uses, which are similar to each other. The standard deviations of the TP concentrations are similar for all the different land uses.

The effect of stormwater control measure BMPs were also simulated to examine potential improvements with treatment for highways and other developed land covers (*11,12*). The BMP performance statistics selected by Granato and Jones were used in these simulations (*3*). They simulated a BMP based on the median performance statistics from seven BMP categories to represent a generic BMP. They used values of 0.116, 0.548, 0.657, and 1.233 for the minimum, lower bound of the most probable value (LBMPV), upper bound of the most probable value (UBMPV), and maximum flow reduction ratios, respectively. They used a rank correlation coefficient (rho) of 0.21 between simulated inflow volumes and flow reduction ratios. Granato and Jones used values of 0.105, 0.226, 0.561, and 2.158 for the minimum, LBMPV, UBMPV, and maximum TP reduction ratios, respectively (*3*). They used a rho value of −0.52 between simulated inflow concentrations and concentration reduction ratios. They used a minimum irreducible concentration of 0.02 mg/L.

Geographic Analysis of Land Cover and Road Miles in Stream Basins within the Willamette Valley, Oregon

Multiple iterations of GIS analysis were conducted to balance the allowable percentage deviation from the drainage area of interest against the number of basins that qualify for use. Early iterations resulted in too few delineated basins within the prescribed bands of allowable drainage areas. For example, only 11 basins were delineated with a drainage area

Table 2. Basin Characteristics of 25 Selected Watersheds in the Willamette Valley, Oregon, with Delineated Drainage Areas from 80 mi^2 to 120 mi^2

Land Cover	Minimum	Maximum	Average	Median	Standard deviation
Entire basin area (mi^2)	84	120	101	101	11
Impervious areas (%)	0.05	6.4	0.72	0.14	1.28
Agricultural areas (%)	0.0	65	15	6.0	22
Developed areas (%)	0.0	8.0	2.3	1.0	2.4
Undeveloped areas/ forest (%)	16	100	82	91	25
Non-state minor roads (mi)	107	359	195	189	65
Non-state major roads (mi)	1.5	111	27	22	24
Non-state roads (total mi)	112	470	222	222	77
State highways (mi)	0.0	37	10	7.7	11

of 100 mi^2 and an allowable tolerance of 0.01%. Intuitively, one might expect that the number of delineated basins for all the drainage area sizes would be much larger, but several factors influence the delineation process. Physiographically, drainage area increases exponentially with increasing channel length. For example, Granato did an analysis of drainage area and main channel length at 845 streamgages nationwide and found that drainage area increased at a rate of 0.426 times the main channel length to the 1.74 power (*11*). Individual streamgages, however, varied from this regression line because additional drainage area is accumulated in blocks as the main channel crosses topographic crenulations and tributary areas. Furthermore, the National Hydrography Dataset used in StreamStats is discretized to a 30-m grid, which accentuates the step changes in drainage area with increasing channel length. For example, an increase of one grid cell (0.186 mi) in length would increase the drainage area by 0.04 mi^2, which is an increase of 0.8% at the 5-mi^2 scale. At the 100-mi^2 scale, each additional grid cell could be expected to increase drainage area by 0.14%. Consequently, the allowable percentage deviation from the drainage area of interest was not held as a consistent percentage between small and large drainage areas during later iterations of the GIS analysis. For the last GIS analysis iteration, bands of 20% deviation were allowed (80–120 mi^2), which resulted in 90 delineated basins within the Willamette Valley. A large percentage of these 90 basins were multiple basins nested within the same watershed. To avoid basins that covered the same geographic area, these basins were removed, leaving 25 basins for this analysis. These delineated basins have a wide range of basin characteristics (Table 2).

For these selected basins, the estimated percentage of developed areas comprised of all state and non-state roadways was 55%, and this estimated total road area represents 92% of the total impervious area (TIA). The percentage of impervious area comprised of roadways in this analysis is much greater than the literature would suggest. For example, Slonecker and Tilley did an analysis of 11 urbanized watersheds in five areas of the country and found that, on average, roads represented 28% of the impervious areas (*23*). They found that buildings, parking lots, driveways, sidewalks, and other anthropogenic impervious surfaces represented 29%, 25%, 9%, 3%, and 3% of the TIA, respectively. Similarly, Roy and Shuster examined a stream basin with 20% TIA and estimated that 28% of TIA was buildings, 25% was driveways, 23% was streets, 12% was parking areas, 6% was sidewalks, and 6% was other anthropogenic impervious surfaces (*24*). The large percentages of developed and impervious areas comprised of roads in the current study reflect the comparatively low percentages of developed and impervious areas in the Willamette Valley (averages of 2.3% and 0.72%, respectively). In basins with limited development, there are comparatively large stretches of roads connecting small clusters of developed area.

The amount of impervious area in a watershed is positively correlated with the total length of roads, although there is a large amount of variability for watersheds that are comprised of less than 1% impervious area (Figure 1). On average, within the 25 selected basins, state roads account for 4% of all road miles and are less abundant than major roads (12%), which are in turn less abundant than minor roads (84%). The number of stream crossings follows the same pattern, with minor roads crossings being most abundant, and state-road crossings least abundant (Figure 2).

Example Stormwater-Quality Analysis of Yields and Loads of TP from Highways and other Land Covers in the Willamette Basin, Oregon

The first step in this example analysis was to go to the StreamStats application, navigate to the area of interest and select a point on the Willamette River for analysis; in this case the stream cell above USGS streamgage 14198000 Willamette River at Wilsonville, Oregon (45.29791, −122.75033) was used for analysis (*13*). StreamStats was used to delineate the basin above this point and obtain the basin properties report. This streamgage is located above the city of Portland, but below Salem, Albany, Corvallis, Eugene, and other small cities. Although this basin includes many highly developed areas, it is predominately undeveloped and

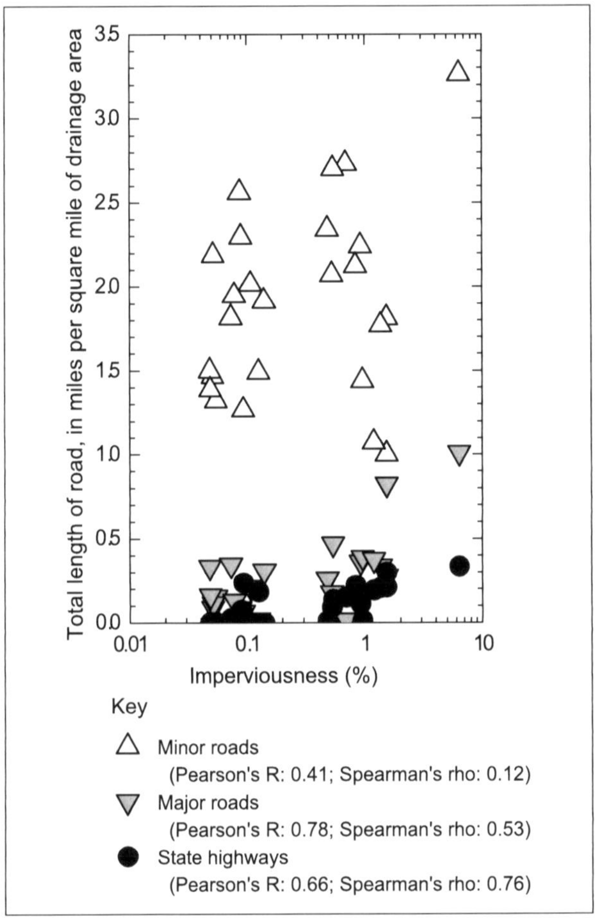

Figure 1. Relations between percent imperviousness and total length of roads for 25 selected Willamette Valley, Oregon, watersheds with drainage areas from 80 mi^2 to 120 mi^2.

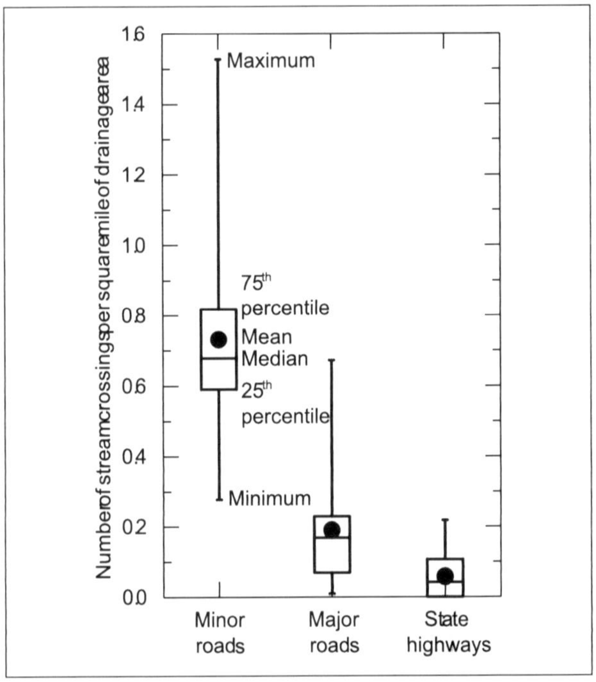

Figure 2. Figure plot showing the number of stream crossings per square mile of drainage area for state highways, major roads, and minor roads in 25 selected Willamette Valley, Oregon, watersheds with drainage areas from 80 mi^2 to 120 mi^2.

agricultural with a TIA of only 1.77% (Table 3). Differences in land cover and impervious percentages between this delineated basin (Table 3) and the delineated 100-mi^2 basins (Table 2) are caused by the concentration of agriculture and development in the valley floor. Although the percentages of developed area and imperviousness area in this basin (Table 3) are more than twice the values for the smaller subbasin averages (Table 2), the basin is mostly a rural area with a large road network connecting the developed areas.

In this basin, the long-term SELDM analyses demonstrate the effect of the selected concentration statistics and imperviousness for each land cover type on the TP yields. The geometric mean TP concentration estimated for highway runoff (0.214 mg/L) was 1.11, 1.35, and 10.7 times the geometric means of agriculture, developed lands, and forest respectively (Table 1). The median of statistics from the 95 selected non-highway sites in the International BMP database was used for analysis of runoff quality from non-state roadways and developed areas. However, geometric mean TP concentrations ranged from 0.029 to 3.09 mg/L among sites in the International BMP database. The geometric mean concentration from the Oregon highway data set was greater than 66% of the values for non-highway sites in the BMP database. Agricultural data from Washington State were used to calculate statistics for this analysis (Table 1), but the literature indicates that agricultural EMCs for TP may be much higher than the Washington data would suggest, with mean or median TP concentrations on the order of 0.14 to 1.56 mg/L (*25*). As such, TP yield calculations that were done by using the agricultural values in Table 1 may overestimate the contribution of highway runoff. The differences in concentrations are compounded by the differences in imperviousness and therefore the runoff coefficient statistics used for the simulations (*11*). Highway and roadway yields are for the impervious area only, whereas the yields from other land covers are attenuated by the relatively low impervious percentages used in these simulations (Figure 3).

In these simulations, the annual highway runoff yields were higher than runoff yields from developed, agricultural and undeveloped/forest land covers (Figure 3). The highway runoff yields were higher than other land covers even with generic BMPs in place. This is because of the relatively large highway concentrations and the high runoff coefficients associated with the impervious pavement. If the developed land concentration statistics (Table 1) are applied to the impervious non-highway roadway areas, then the long-term average highway yield (Table 4) is 1.30 times the average

Table 3. Land Cover Statistics for the Entire Willamette Basin above USGS Streamgage 14198000 Willamette River at Wilsonville, Oregon Developed by using StreamStats

Land cover	Area (mi^2)	Percentage of area
Entire basin area	8,440	100
Agricultural areas	1,688	20
Developed areas (including roadways)	422	5
Undeveloped/forest areas	6,330	75
Sum of areas	8,440	100
TIA	149	1.77
Non-state roads (19,010 mi)	102.5	1.21
State highways (1,220 mi)	9.7	0.11
Sum of all roads	112.2	1.32

roadway yield without BMP treatment, and 1.33 times the roadway yield with the BMP treatment. If the developed land concentration statistics are applied to the developed areas with an impervious percentage of 33.4%, then the highway and roadway yields are 3.66 and 2.81 times the developed area yields on average. Although the geometric mean concentration for agriculture was higher than the developed land concentration (Table 1), the long-term average runoff yield for the developed land cover was 2.42 times the yield from the agricultural land cover (Table 4) because of the simulated difference in imperviousness (33.4 versus 0.5%). The long-term average runoff yield from the forested land cover was about an order of magnitude less than the agricultural yield because the forest runoff concentrations were lower than the agricultural concentrations, and the impervious percentage of forest areas was simulated as zero.

Although these simulations indicate that state highway yields are substantially larger than yields from other land covers, the state highway contributions to long-term annual runoff loads are a small fraction of the total annual runoff loads. Without any BMPs, the simulated long-term basin-wide average annual runoff load from agricultural areas, developed areas, undeveloped areas, non-highway roads in undeveloped areas, and state highways was 6.26×10^5 lb/year. Runoff from state highways contributed 2.1% of this long-term average annual runoff load (Table 4). Agricultural areas, developed areas, undeveloped areas, and non-highway roads in undeveloped areas contribute 40%, 25%, 16%, and 17% of the total load, respectively. Simulation results indicate that the use of generic BMPs would reduce the TP loading from state highways by about half. Alternatively, this load reduction (6.64×10^3 lb/year) could also be achieved by adopting land management practices that result in a 2.64% decrease in annual runoff loads from agricultural lands, or a 4.33% decrease in annual runoff loads from developed lands. In comparison, application of the generic BMP performance to the developed areas alone results in a load reduction (8.10×10^4 lb/year), which is more than an order of magnitude more than the highway reduction. Furthermore, application of BMP to developed clusters may be more effective than construction and maintenance of BMPs distributed along linear highway systems (*6*).

Another approach to evaluate basin-wide runoff load reductions is to use the calculated loads to evaluate land cover specific reductions needed to achieve specific targets (Table 5). For example, if the basin-wide reduction target is 5%, one approach would be to reduce the average annual runoff loads from agricultural lands by 12% or to reduce the average annual runoff loads from developed lands by 20%. Reductions in state highway loads alone cannot be used to meet this target because the 5% reduction is 2.4 times the total highway load. Adding resource-intensive structural BMPs to state highways or developed areas results in the highest yield reductions, but not the largest load reductions (*6*). Other land cover treatment options, such as widespread use of agricultural best practices, do not need to be as efficient as structural BMPs to produce similar reductions in annual runoff loads (Table 5).

Limitations of this Analysis

This analysis was designed to produce planning level estimates of basin properties, runoff yields, and runoff loads to foster discussions about the relative contributions of different land covers and potential effectiveness of different management measures. Planning level estimates include substantial uncertainties that are commonly on the scale of one or more orders of magnitude (*11*). As such, more refined runoff quality statistics may be needed to estimate yields and loads for actual TMDL analyses in Oregon and other areas. As indicated in the discussion, TP concentration statistics for non-highway land uses were calculated with available data and may not be representative of runoff from areas in the Willamette River Basin. Highway concentration statistics were calculated by using four highway runoff sites in Oregon, which may not be representative of typical highway runoff in the Willamette River Basin. BMP treatment statistics represent the effects of a generic BMP rather than individual BMP designs, which may be optimized for the hydrology and water quality in this area (*3,12*). Pavement yields for highway and non-highway roads may substantially overestimate roadway contributions to streams because large sections of the road network drain to the local land surface rather than to connected waterbodies; this is evidenced by the low density of stream crossings (Figure 2). Furthermore, the yields and loads discussed in this paper are for stormwater runoff only; they do not include loadings from wastewater treatment plants or baseflow loading from fertilizer applications, agricultural underdrains, septic-system effluent, other commercial or industrial sources, or natural geologic sources.

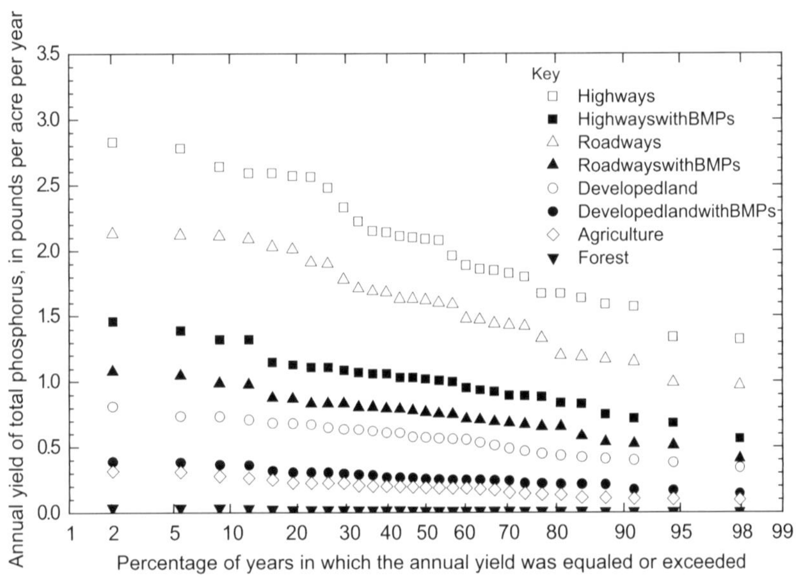

Figure 3. Probability plot showing annual stormwater runoff yields of TP, in pounds per acre per year, for a 29-year simulation period.

Table 4. Average Annual Stormwater Runoff Yields and Loads of TP from Different Land Cover Areas in the Entire Willamette Basin above USGS Streamgage 14198000 Willamette River at Wilsonville, Oregon

Land cover	Area (acres)	Average yield (Pounds of TP per acre per year)	Average annual load (Pounds of TP per year)
Agricultural areas*	1.07×10^6	0.235	2.51×10^5
Developed areas**	2.70×10^5	0.569	1.54×10^5
Developed areas** with BMPs	2.70×10^5	0.269	7.26×10^4
Undeveloped/forest areas*	3.99×10^6	0.0249	9.94×10^4
Non-highway roads***	6.77×10^4	1.60	1.08×10^5
Non-highway roads*** with BMPs	6.77×10^4	0.758	5.13×10^4
State highways	6.21×10^3	2.08	1.29×10^4
State highways with BMPs	6.21×10^3	1.01	6.27×10^3

*Areas are reduced by the basin-wide average road area percentage (1.32%).
**Areas include roadway but not state highway areas.
***Area includes non-highway roads in undeveloped and agricultural areas.

Table 5. Percentage Reduction of Average Annual Loads of TP Needed to Reach Specific Reduction Targets by Land Cover for the Entire Willamette Basin above USGS Streamgage 14198000 Willamette River at Wilsonville, Oregon. A Reduction Greater than 100% Indicates that Elimination of this Source Alone Cannot Be Used to Reach the Reduction Target

	Target reduction value of basin-wide annual TP load (%)				
Land cover	1	2	5	10	20
Agriculture*	2.5	5.0	12	25	50
Developed lands**	4.1	8.1	20	41	82
Undeveloped areas/Forest*	8.6	12	29	56	115
Non-highway roads***	5.8	12	29	58	116
State highways	48	97	240	480	970

*Areas are reduced by the basin-wide average road area percentage (1.32%).
**Areas include roadway but not state highway areas.
***Area includes non-highway roads in undeveloped and agricultural areas.

Conclusions

State DOTs need ready access to quantitative information about the percentages of different land cover categories and the number of roadways and highways above any given highway stream crossing. This information is needed to assess runoff quality issues including TMDLs. To this end, the USGS integrated land cover and road-length data into the web-based StreamStats tool in cooperation with the Oregon DOT and the FHWA. StreamStats is a web-based application that incorporates a GIS that enables users to derive basin characteristics, delineate drainage areas, and estimate streamflow statistics upstream from any given point along a stream. In this study, land cover and roadway length values from StreamStats and TP concentration statistics from the literature were used with SELDM to simulate long-term runoff yields and loads from different land covers above USGS streamgage 14198000 Willamette River at Wilsonville, Oregon. SELDM was used to simulate yields and loads from state highways, minor and major non-state roads, and developed land covers with and without use of stormwater BMPs. This paper demonstrates practice-ready methods to provide planning level load estimates.

A geographic analysis of 25 subbasins in the Willamette Valley of Oregon, with drainage areas between 80 mi^2 and 120 mi^2, was done with the StreamStats batch-processing tool to evaluate land cover categories in the basin. Results indicate that these basins are primarily forest and other undeveloped lands (median value of 91%), and tend to have little impervious area (median value of 0.14%). Minor roads are more abundant in these watersheds than major roads, which are in turn more abundant than state highways (median length values of 189 mi, 22.0 mi and 7.7 mi, respectively). Analysis of the number of stream crossings by road type follows the same pattern as the roadway length analysis. The amount of impervious area for a given watershed is positively correlated with the total miles of road, although the correlation is relatively weak for watersheds, with less than 1% imperviousness.

Results of the stormwater loading analysis with SELDM indicate that state highways produce the highest TP load per acre (the yield), followed by roadway pavement, developed, agricultural, and forest land covers. However, highways produce the lowest TP loading (2%) of the four land cover categories evaluated, because state highways constitute a relatively small percentage of the watershed. Simulated agricultural and developed land covers, which are 41% and 24% of the long-term average annual loads, account for most average annual TP loads in the simulated basin.

The potential effectiveness of BMPs was also evaluated in these simulations. Simulations indicate that, on average, loads from state highways could be reduced by 50% if typical BMPs are used throughout the basin. Targeted BMP treatment of highway or non-highway road runoff results in the largest reduction in runoff yields, but these areas are a small part of the basin, and highway and roadway loads are a small part of basin-wide runoff loads. This analysis also indicates that relatively modest runoff reductions from land management practices in large agricultural areas or clustered developed areas could result in load reductions equal to installing structural BMPs to treat all highway and roadway runoff in the basin.

Acknowledgments

This research was funded by the FHWA Office of Project Development and Environmental Review, the Oregon Department of Transportation, and the USGS Cooperative Matching Funds Program.

References

1. Abbasi, S. A., and K. Antti. *NCHRP Synthesis of Highway Practice 444: Pollutant Load Reductions for Total Maximum Daily Loads for Highways*. Transportation Research Board, Washington, D.C., 2013.
2. Granato, G. E., and S. C. Jones. Stochastic Empirical Loading and Dilution Model for Analysis of Flows, Concentrations, and Loads of Highway Runoff Constituents. *Transportation Research Record: Journal of the Transportation Research Board*, 2014. 2436: 139–147.
3. Granato, G. E., and S. C. Jones. Estimating Total Maximum Daily Loads with the Stochastic Empirical Loading and Dilution Model. *Transportation Research Record: Journal of the Transportation Research Board*, 2017. 2638: 16.
4. Risley, J. C., and G. E. Granato. *Assessing Potential Effects of Highway Runoff on Receiving-water Quality at Selected Sites in Oregon with the Stochastic Empirical Loading and Dilution Model* (SELDM). Scientific Investigations Report 2014–5099. U.S. Geological Survey, 2014. http://dx.doi.org/10.3133/sir20145099.
5. Environmental Protection Agency. *National Summary of Impaired Waters and TMDL Information*. United States Environmental Protection Agency, 2017. https://iaspub.epa.gov. Accessed July 5, 2017.
6. Taylor, S., M. Barrett, M. Leisenring, S. Sahu, D. Pankani, A. Poresky, A. Questad, E. Strecker, N. Weinstein, and M. Venner. *NCHRP Report 792: Long-Term Performance and Life-Cycle Costs of Stormwater Best Management Practices*. Transportation Research Board, Washington, D.C., 2014. http://dx.doi.org/10.17226/22275.
7. Granato, G. E., and P. A. Cazenas. *Highway-Runoff Database (HRDB Version 1.0)—A Data Warehouse and Preprocessor for the Stochastic Empirical Loading and Dilution Model*. FHWA-HEP-09-004. FHWA, U.S. Department of Transportation, 2009. https://doi.org/10.5066/F74Q7T5X.
8. Smith, K. P., and G. E. Granato. *Quality of Stormwater Runoff Discharged from Massachusetts Highways, 2005–07*. Scientific Investigations Report 2009–5269. U.S. Geological Survey, 2010. https://pubs.usgs.gov/sir/2009/5269/.
9. Wagner, C., S. Fitzgerald, and M. Lauffer. Characterization of Stormwater Runoff from North Carolina Bridges and Determination of Effects on Selected Receiving Waters. *Transportation Research Record: Journal of the Transportation Research Board*, 2012. 2309: 178–199.
10. American Association of State Highway and Transportation Officials. *Stormwater Management Community of Practice. State-of-the-Practice Report: TMDLs*. AASHTO, Washington,

D.C., 2010 https://environment.transportation.org/pdf/communities_of_practice/tmdls.pdf. Accessed July 5, 2017.

11. Granato, G. E. Stochastic Empirical Loading and Dilution Model (SELDM) Version 1.0.0. *Techniques and Methods*, book 4, chap. 3. U.S. Geological Survey, 2013. https://pubs.usgs.gov/tm/04/c03/.
12. Granato, G. E. *Statistics for Stochastic Modeling of Volume Reduction, Hydrograph Extension, and Water-Quality Treatment by Structural Stormwater Runoff Best Management Practices (BMPs)*. Scientific Investigations Report 2014–5037. U.S. Geological Survey, 2014. https://doi.org/10.3133/sir20145037.
13. Ries, K. G., III, J. G. Guthrie, A. H. Rea, P. A. Steeves, and D. W. Stewart. *StreamStats—A Water Resources Web Application*. Fact Sheet 2008-3067. U.S. Geological Survey, 2008. https://pubs.usgs.gov/fs/2008/3067/.
14. Cooper, R. M. *Estimation of Peak Discharges for Rural, Unregulated Streams in Western Oregon*. Scientific Investigations Report 2005-5116. U.S. Geological Survey, 2005. https://pubs.usgs.gov/sir/2005/5116/.
15. Risley, J. C., A. J. Stonewall, and T. L. Haluska. *Estimating Flow-Duration and Low-Flow Frequency Statistics for Unregulated Streams in Oregon*. Scientific Investigations Report 2008-5126. U.S. Geological Survey, 2008. https://pubs.usgs.gov/sir/2008/5126/.
16. Jin, S., L. Yang, P. Danielson, C. Homer, J. Fry, and G. Xian. A Comprehensive Change Detection Method for Updating the National Land Cover Database to Circa 2011. *Remote Sensing of Environment*, Vol. 132, 2013, pp. 159–175. https://doi.org/10.1016/j.rse.2013.01.012.
17. U.S. Geological Survey. *National Land Cover Database 2011 (NLCD 2011)*. U.S. Geological Survey, 2011. http://www.mrlc.gov/nlcd2011.php. Accessed July 28, 2017.
18. Homer, C. G., J. Dewitz, L. Yang, S. Jin, P. Danielson, G. Z. Xian, J. Coulston, N. Herold, J. Wickham, and K. Megown. Completion of the 2011 National Land Cover Database for the Conterminous United States—Representing a Decade of Land Cover Change Information. *Photogrammetric Engineering and Remote Sensing*, Vol. 81, 2015, pp. 345–354. http://www.ingentaconnect.com/content/asprs/pers/2015/00000081/00000005/art00002.
19. Oregon Department of Transportation. *Data and Maps*. https://www.oregon.gov/odot/td/tdata/pages/gis/odotgis.aspx. Accessed July 28, 2017.
20. Herrera Environmental Consultants, Inc. *Toxics in Surface Runoff to Puget Sound: Phase 3 Data and Load Estimates*. Publication 11-03-010. Washington State Department of Ecology, 2011. https://fortress.wa.gov/ecy/publications/summarypages/1103010.html.
21. Water Environment and Reuse Foundation. *International BMP Database Version 2016-09*. http://www.bmpdatabase.org. Accessed October 21, 2016.
22. Granato, G. E. *Highway Runoff DataBase (HRDB) Version 1.0.0b*. U.S. Geological Survey 2018. https://doi.org/10.5066/F74Q7T5X. Accessed March 13, 2018.
23. Slonecker, E. T., and J. S. Tilley. An Evaluation of the Individual Components and Accuracies Associated with the Determination of Impervious Area. *GIScience and Remote Sensing*, Vol. 41, No. 2, 2004, pp. 165–184.
24. Roy, A. H., and Shuster, W. D., 2009. Assessing Impervious Surface Connectivity and Applications for Watershed Management. *Journal of the American Water Resources Association*, Vol. 45, No. 1, pp. 198–209. https://doi.org/10.1111/j.1752-1688.2008.00271.x.
25. Lin, J. P. *Review of Published Export Coefficient and Event Mean Concentration (EMC) Data*. ERDC TN-WRAP-04-3. U.S. Army Engineer Research and Development Center, 2004.

The Standing Committee on Stormwater (AFB65) peer-reviewed this paper (18-05240).

Article

TRR
JOURNAL OF THE TRANSPORTATION RESEARCH BOARD

Highway Cross-Slope Measurement using Mobile LiDAR

Transportation Research Record
2018, Vol. 2672(39) 88–97

DOI: 10.1177/0361198118756371
journals.sagepub.com/home/trr
SAGE

Alireza Shams[1], Wayne A. Sarasua[1], Afshin Famili[1], William J. Davis[2], Jennifer H. Ogle[1], Leonildo Cassule[1], and Adika Mammadrahimli[1]

Abstract

Ensuring adequate pavement cross-slope on highways can improve driver safety by reducing the potential for ponding to occur or vehicles to hydroplane. Mobile laser scanning (MLS) systems provide a rapid, continuous, and cost-effective means of collecting accurate 3D coordinate data along a corridor in the form of a point cloud. This study provides an evaluation of MLS systems in terms of the accuracy and precision of collected cross-slope data and documentation of procedures needed to calibrate, collect, and process this data. Mobile light detection and ranging (LiDAR) data were collected by five different vendors on three roadway sections. The results indicate the difference between ground control adjusted and unadjusted LiDAR derived cross-slopes, and field surveying measurements less than 0.19% at a 95% confidence level. The unadjusted LiDAR data incorporated corrections from an integrated inertial measurement unit and high-accuracy real-time kinematic GPS, however it was not post-processed adjusted with ground control points. This level of accuracy meets suggested cross-slope accuracies for mobile measurements (±0.2%) and demonstrates that mobile LiDAR is a reliable method for cross-slope verification. Performing cross-slope verification can ensure existing pavement meets minimum cross-slope requirements, and conversely is useful in identifying roadway sections that do not meet minimum standards, which is more desirable than through crash reconnaissance where hydroplaning was evident. Adoption of MLS would enable the South Carolina Department of Transportation (SCDOT) to address cross-slope issues through efficient and accurate data collection methods.

Highway pavement cross-slope is a crucial cross-sectional design element as this provides positive lateral drainage from the roadway surface, and helps to minimize the occurrence of ponding, thus improving driver safety by reducing potential for hydroplaning.

SCDOT minimum cross-slope design criteria apply to tangent alignments. On high-speed roadways, the normal crown cross-slope is 0.25 in./ft (2.08%) on tangent sections with some exceptions depending on the number of lanes (*1*). Accommodating other horizontal design features, for example, super-elevation for circular and spiral curves, requires transitioning from a normal cross-slope.

It is important for roadways to meet minimum pavement cross-slope design criteria, and that maximum criteria are not exceeded. Cross-slopes that are too steep can cause vehicles to drift, skid laterally when braking, and become unstable when crossing over the normal crown to change lanes. Table 1 shows potential adverse impacts to safety and operations if minimum and maximum design criteria are not met.

One of the primary objectives for conducting this research was to investigate efficient methods for identifying highway sections that do not meet minimum criteria for pavement cross-slope. Current approaches for locating cross-slope sections identified for improvement include roadway ponding, cross-slope verification (particularly after rehabilitation projects) using conventional surveying techniques, crash analysis, and tort litigation. In cases of bodily injury or fatalities related to hydroplaning crashes, when site investigations determined prevailing pavement cross-slope did not meet minimum design criteria, SCDOT has been found at-fault in tort claims brought against the Department. Application of conventional survey methods to determine locations of pavement cross-slope problems system-wide, for all practical purposes, is cost prohibitive. Mobile laser scanning (MLS) may provide an efficient and practical solution to addressing this challenge. Accurate pavement cross-slope data is crucial for implementing repaving and rehabilitation programs and

[1]Clemson University, Clemson, SC
[2]The Citadel, Charleston, SC

Corresponding Author:
Address correspondence to Alireza Shams: ashamse@g.clemson.edu

Table 1. Potential Adverse Safety Impact of Deviation from Design Criteria (2)

Safety and operational issues	Freeway	Expressway	Rural two-lane	Urban arterial
Run-off-road crashes	×	×	×	
Slick pavement	×	×	×	×
Water ponding on the pavement surface	×	×	×	×
Water spreading onto the traveled lanes				×
Loss of control when crossing over a high cross-slope break	×	×	×	

Note: Freeway = high-speed, multi-lane divided highway with interchange access only (rural or urban); Expressway = high-speed, multi-lane divided arterial with interchange access only (rural or urban); Rural two-lane = high-speed, undivided rural highway (arterial, collector, or local); Urban arterial = urban arterial with speeds 45 mph or less.

projects that can provide targeted corrective action to addressing cross-slope problems.

In this study the researchers conducted a survey of state highway agencies across the U.S., which determined that although 70% collect some type of cross-slope data, only 23% of respondents did so to determine cross-slope compliance and almost none did so system-wide. Most of the states only performed cross-slope verification on Interstate and primary routes, as states lack necessary resources to conduct the surveying work needed. Conventional surveying for cross-slope verification can only be conducted at sample locations and may not be representative of segments between the samples. SCDOT's emphasis on ensuring that adequate pavement cross-slopes are maintained through verification is predicated upon two principles: deployment of a safe and efficient method for collecting cross-slope data; and system-wide adoption so an accurate and comprehensive network-based cross-slope database can be maintained.

A variety of techniques can be used for acquiring roadway cross-slope data including contractor as-built plans if available, photogrammetry using high-resolution stereo images, conventional surveying, attitudinal GPS, remote sensing data such as USGS digital elevation models (DEMs), and measuring with an inertial device such as a digital gyroscope or an accelerometer (*3, 4*). Factors such as accuracy, safety, cost, and time of performance play important roles in the selection of one method over another (*4*). Conventional surveying methods provide accurate results at sampled locations; however, this approach is very time-consuming (especially for short intervals) and poses safety risks to personnel due to the close proximity to traffic (*3*). Stereo photogrammetry is an accurate method for collecting topographic data but processing time and the need for extensive ground control to produce reasonable cross-slope accuracy, plus collecting high-resolution aerial imagery, is an expensive option (*3*). A vehicle-mounted inertial device can collect data at highway speeds but can only obtain measurements for one travel lane at a time. Multiple lanes would require several passes to determine cross-slopes for the entire roadway. MLS is capable of collecting an entire cross-section, with an exception at steep side slopes, at highway speeds in a single pass (*5*).

The strengths of MLS include continuous and comprehensive data collection, high-resolution capability, reduced number of field visits, elimination of roadside work hazards for survey crews, and multiple end users and opportunities to share for various applications (*6*). The weaknesses of MLS include expensive up-front cost, line-of-sight requirements, adjustment for vehicles scanned within the traffic stream, and need to automate the classification of large numbers of points (*6*). Further, very accurate ground control points are needed to adjust and calibrate MLS data for applications that require a high level of accuracy.

This research evaluates the use of MLS for collecting accurate cross-slope to ensure that adequate cross-slope and proper drainage exist on highways. The light detection and ranging (LiDAR) data were collected on three roadway test sections, including representative urban and rural restricted roadway locations, and rural parkways. MLS data from five vendors was used in conducting this evaluation. MLS is evaluated in terms of the accuracy of the collected cross-slope data, as well as procedures to calibrate, collect, and process the data. Conventional surveying methods were also used for comparison purposes.

Literature Review

The literature review focused on mobile methods for collecting cross-slope data and the relative accuracies of the collected data. Inertial devices as a sole cross-slope data collection device is not covered because, while they can be extremely accurate, they can only collect a single lane of data with one pass. The use of MLS to collect cross-slope data requires an integrated inertial measurement unit (IMU) for location adjustments and to compensate for the roll of the vehicle.

Baffour discussed the need for roadway geometry in transportation projects (*4*). Although some geometry information may be extracted from existing road plans, but some of the current characteristics may not match with the original design due to undocumented changes. The paper discussed the use of multi-antenna configurations that are synchronized with a single global positioning system (GPS) receiver to determine the 3D orientation of the moving vehicle. After designing the antenna platform, all the data collected were compared with standard data collected by conventional surveying. The cross-slopes were collected at 50 ft intervals, and the accuracy was at 0.01%. The results showed attitudinal GPS has promise as a tool for collecting this data (*4*). A drawback of attitudinal GPS is that, similar to an inertial device, only data from one lane can be collected and, thus, multiple passes would be required for multi-lane roads.

Souleyrette et al. attempted to collect grade and cross-slope from LiDAR data on tangent highway sections (*3*). Measurements were compared against grade and cross-slope collected using an automatic level for 10 test sections along Iowa Highway 1. The physical boundaries of shoulders and lanes were determined by visual inspection from 6 in. resolution orthophotos, 12 in. orthophotos by Iowa Department of Transportation (DOT), and triangular irregular network (TIN) from LiDAR. Multi-linear regression analysis was conducted to fit the plane to the LiDAR data corresponding to each analysis section. Vendor accuracy was 0.98 ft root-mean-square error (RMSE) and vertical accuracy of 0.49 ft. Whereas the grade was successfully calculated within 0.5% for most sections, and 0.87% for all sections, the cross-slope data were much less accurate. Cross-slope estimated from LiDAR deviated from field measurements by 0.72% to 1.65%. The results indicated that the cross-slope could not be practically estimated using a LiDAR surface model.

Jaakkola et al. discussed the necessity for laser-based mobile mapping for transportation study due to the large amount of data produced (*5*). Data were collected by the Finnish Geodetic Institute (FGI) Roamer Mobile Mapping System (MMS), relating to points on the painted markings on the road, found the curb stones from the height of the image and modeled the pavement as a TIN. As they processed the raster image, which is more efficient than point cloud, it was possible to locate most curbstones, parking spaces, and a zebra crossing with mean accuracies of about 80% or better.

Zhang and Frey attempted to model road grade using LiDAR to estimate vehicle emissions (*7*). It was problematic to measure road grade directly from portable emissions monitoring systems (PEMS) and the available GPS data have not been proven to be reliable for road grade estimation. The LiDAR-based method was used to model the road grade, and fit a plane using regression techniques on interstate highways I-40 and I-540, as well as major arterials. RSME was utilized to assess the precision of LiDAR data used in this work, in the range from 7.7 to 25 cm, which was found to be smaller than changes in elevation that were significant regarding emissions. LiDAR data were shown to be reliable and accurate for road grade estimation for vehicle emission modeling.

Tsai et al. proposed a mobile cross-slope measurement method using mobile LiDAR technology (*8*). The proposed method instruments included: a mobile LiDAR system, a high-resolution video camera, an accurate positioning system composed of a GPS, an IMU, and a distance measurement instrument. The accuracy and repeatability of the proposed method were critically validated through testing in a controlled environment. Results showed the proposed method achieved desirable accuracy with a maximum difference of 0.28% cross-slope (0.17°) and an average difference of less than 0.13% cross-slope (0.08°) from the digital auto level measurement. Repeatability results showed standard deviations within 0.05% (0.03°) at fifteen benchmarked locations in three runs. The acceptable accuracy is typically 0.2% (or 0.1°) during construction quality control. The case study on I-285 demonstrated the proposed method could efficiently conduct the network-level analysis. The GIS-based cross-slope measurement map of the 3-mile section of studied roadway can be derived in under 2 person-hours with use of the collected raw LiDAR data.

Holgado-Barco et al. (*9*) attempted to extract road geometric parameters through the automatic processing of mobile LiDAR system point clouds. Their methodology was carried out in several different steps: data capture; segmentation to simplify the point cloud to extract the road platform; applying principal component analysis (PCA)-based on orthogonal regression to fit the best plane on points; and extracting vertical and cross-section geometric parameters and analysis. The study method proposed an alternative automated development of the as-built plan. The experimental results validated the method within relative accuracies under 3.5%.

Study Area and Data Collection

This research evaluated the use of MLS from five vendors to obtain accurate cross-slope data. Three roadway test sections were used in performing the research evaluation including: a four-lane parkway without any curb cuts (driveways) in Anderson, SC; a section of urban restricted access highway in Spartanburg, SC; and a rural restricted access highway just west of Easley, SC.

Study Section 1: East–West Parkway (Using Adjusted Point Cloud)

The first study section is a 3-mile corridor along East–West Parkway (EW Pkwy) in Anderson, SC shown in Figure 1. The study section originates at US-76 (Clemson Boulevard) and terminates at the SC-81 (East Greenville Street). EW Pkwy is a limited access four-lane two-way mostly divided highway. It has a variety of geometric design elements including 15 vertical curves, 7 horizontal curves (all super-elevated), one bridge, two intersections, traversable and non-traversable medians, two lanes per direction with an additional turning lane at intersections, and sections with adjacent bike lanes and separate bike paths.

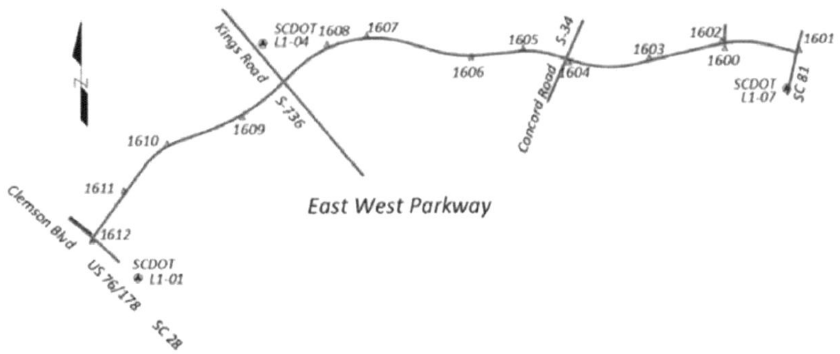

Figure 1. GCPs and check points along the 3-mile study area section 1.

Figure 2. GCPs and panel points along the study area section 2.

MLS combines precise ranging, with high-accuracy GPS and an integrated IMU to obtain a very dense point cloud. The resulting point cloud can be useful for many applications such as asset data collection (lane widths, presence of median, etc.) or navigation, but may not be accurate enough for surveying or some engineering applications, such as precise quantity take-offs. To improve accuracy for this research, a ground control survey was conducted that identified primary and secondary geodetic control point (GCP) locations throughout the corridor. At least two primary GCPs were used by vendors as base station locations for GPS differential correction and all the GCPs (both primary and secondary) were used for post-processing adjustment. Figure 1 shows the GCP locations along the study corridor.

The corridor was also surveyed to locate 100 ft stations along white edge lines. These locations were marked with PK surveying nails. Eight of these locations were selected along the corridor as cross-slope test sections. The test sections were selected to ensure diverse roadway cross-slope characteristics including differing lane geometry, normal crown, and superelevated sections. PK surveying nails were also added to the yellow centerline markings. Reflective pavement marking tape was used to ensure that PK nail locations could be identified in the LiDAR data using the intensity attribute.

Study Section 2: Intestate 85 Business Loop (using Adjusted Point Cloud)

The second study section is a 3.4-mile corridor along Interstate 85 business loop (I-85 BL) in Spartanburg, SC, shown in Figure 2. The study section originates at I-585 and terminates at I-85. I-85 BL is a restricted access four-lane two-way divided freeway. Researchers measured cross-slopes at selected locations prior to the test. These locations correspond with panel points P78, P91, P98, P103, P126, and P127 (note that P103, P126, and P127 are on ramps). All panel points are marked with a painted chevron, yellow reflective pavement marking tape, and a PK nail. Detailed surveying of horizontal/vertical elements was not conducted within the travel way of this study section, however, primary and secondary GCPs were established along paved shoulders. The GCPs were used for GPS differential correction and for post-process adjustment.

Study Section 3: US-123 (using Unadjusted Point Cloud)

The third study section is a 1-mile corridor along US-123 just west of Easley, SC. This section of US-123 is a restricted access four-lane two-way divided highway. The survey crew measured cross-slopes at selected locations prior to the test. These locations correspond with different traffic signs located at six pre-designated stations along the corridor. The LiDAR measurements were combined with high-accuracy GPS and IMU measurements to create a point cloud. However, on US-123 the point cloud was not adjusted through post-processing with GCPs. It is not uncommon to use unadjusted mobile LiDAR point clouds for applications that do not require the highest level of accuracy such as statewide asset management or autonomous vehicle applications.

Table 2. Vendor Data Collection Specifications for Test Sections 1 and 2

	Vendor A	Vendor B	Vendor C	Vendor D
Brand	Riegl	Teledyne Optech	Teledyne Optech	Leica
Model	VMX 450	M1	SG1	9012
Single laser or dual	Dual	Dual	Dual	Single
Measurement rate	1100 kHz	500 kHz/sensor	600 kHz (each laser)	1000 kHz

Field Surveying using Auto Leveling

Conventional surveying (auto leveling combined with taping and total station measurements) was used to develop ground-truth cross-slopes for all three test sections. Each of the cross-section stations were leveled using two different instrument setups to ensure accuracy and adjust for random error. The cross-slope along each section was computed for each lane from the elevation difference between lane lines, along with horizontal distances in between, which was measured by tape or total station.

LiDAR Data Collection

LiDAR data for test sections 1 and 2 were collected by two vendors on June 30, 2016 and two other vendors on August 30, 2016. Data for test section 3 were collected in 2015. The section 1 and 2 vendors and their stated equipment specifications are provided in Table 2. On section 3, the vendors LiDAR system was a Riegl VMX 450. Vendors were allowed to calibrate their systems both before and after data collection runs. A primary benefit of a MLS is that point cloud data can be collected for multiple travel lanes with a single pass. For this study, vendors were asked to collect data by direction by driving in the right lane. Only a single pass was allowed for each direction. Vendors were asked to follow a lead vehicle that drove at the posted speed limit. For section 1, traffic control was provided by two trailing SCDOT vehicles driving side by side so that no cars could pass the vendor data collection vehicles; however, for practical purposes, there was no traffic control for the opposing travel direction. There was no traffic control for section 2 or section 3.

Extracting Cross-Slope from Point Cloud

There were two potential methods to define the cross-section line at each test section as follows: in cases where the location of the PK nails on two ends of the test section were distinctly identified, a reference line was drawn between the two points; or the LiDAR image of the pavement marking tape pointing to the PK nails was used to create the reference line. Using the reference line from either method, a 4-in. buffer of points was clipped in an automated fashion using ArcGIS. Two separate mesh grid surfaces were fitted to the LiDAR derived points using nearest neighbor interpolation within the buffer area. One mesh grid included continuous values of easting, northing, and elevation, fitted to the LiDAR points (Figure 3). The second mesh grid included the easting, northing and intensity of the points.

Using the reference line, a continuous cross-section is extracted including elevation and intensity. The yellow and white pavement markings have higher intensity values and are easily identifiable (Figure 4). The cross-slope is calculated from the rise and run between the lane lines. These LiDAR derived cross-slopes are directly comparable with the field survey cross-slopes.

Comparison of LiDAR and Conventional Survey Data

The use of LiDAR to extract pavement cross-slope dimensions on three study sections was compared against cross-slope measurements collected using conventional surveying for eight specific roadway stations along EW Pkwy Anderson, SC, six stations on I-85 BL, and at six sign locations on US-123. The mobile LiDAR data collected by the vendors were provided as dense point clouds and evaluated using a number of comparative methods. Reference lines within each roadway study location were created between two distinct surveyed points established with PK nails and reflective pavement marking tape. Elevation and intensity of points along the reference lines were extracted from the mesh grid fitted to LiDAR point clouds within 4 in. thickness across each station of interest. Owing to the difference of reflectivity of the materials, which resulted in different intensities in the point cloud, the edge of the pavement, lane lines, and centerline were readily extracted from LiDAR data by matching intensity and elevation results. The pavement cross-slope for each travel lane was calculated by dividing the difference in elevations by the distance between two pavement markings. Pavement cross-slopes were directly measured in the field for each test section using automatic leveling. Field measurements were used as reference data for comparison against vendor collected LiDAR derived data.

A cross-slope comparison for different test sections at three different study areas are shown in Tables 3, 4, and 5, respectively. The comparison is based on each traveling lane and the vendor names have been removed and are shown in random order.

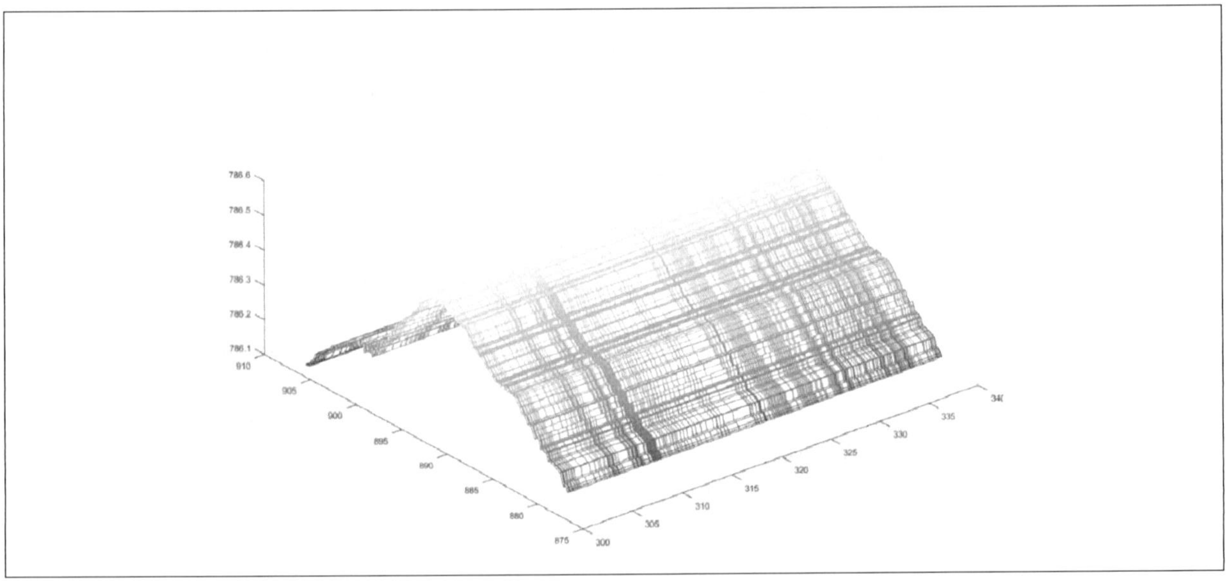

Figure 3. Mesh grid fitted to points within buffer area.

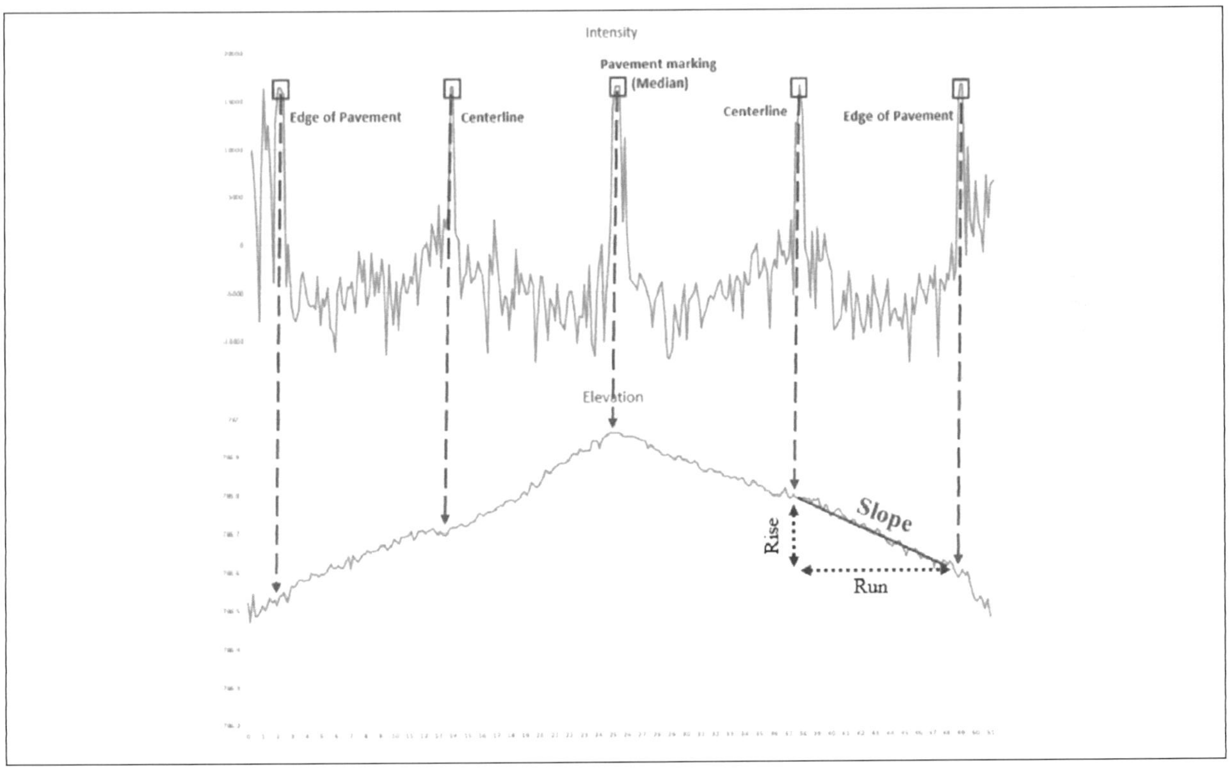

Figure 4. Pavement marking extraction and corresponding elevations.

Evaluation of Results

In evaluating cross-sectional data at reference station locations, cross-slope estimates from adjusted LiDAR differed from field surveyed measurements ranging from 0% to 0.98% with an average of 0.19% for all vendors, as shown in Table 6. The comparison between unadjusted LiDAR data and field surveying varies from 0% to 0.24%. With regards to SHRP2 guide specification a slope tolerance value of ± 0.2% of the design value would be acceptable for final measurement after project completion (*10*).

Table 3. Cross-Slope Comparison between Surveyed Data and LiDAR Derived Cross-Slope: Section 1

				Difference from surveyed data			
Station	Lane	Lane width(HD)	Surveyed data	Vendor A	Vendor B	Vendor C	Vendor D
110+00	EB Outer	12.02	1.75%	0.25%	0.30%	0.34%	0.11%
	EB Inner	12.18	1.97%	0.00%	0.22%	0.71%	0.11%
	WB Outer	12.04	1.83%	0.07%	0.10%	0.24%	0.22%
	WB Inner	11.74	2.22%	0.14%	0.00%	0.55%	0.22%
124+00	EB Outer	11.72	4.61%	0.23%	0.18%	0.07%	0.08%
	EB Inner	12.93	5.14%	0.30%	0.55%	0.40%	0.54%
	Turning	14.41	4.82%	*	0.42%	0.66%	0.80%
	WB Outer	11.7	4.79%	0.20%	0.90%	0.24%	0.35%
	WB Inner	12.04	4.32%	0.02%	0.47%	0.04%	0.02%
128+00	EB Outer	11.72	2.39%	0.24%	0.02%	0.10%	0.09%
	EB Inner	12.19	2.26%	0.10%	0.11%	0.15%	0.37%
	Turning	12	1.58%	0.26%	0.19%	0.23%	0.37%
	WB Outer	12	0.46%	0.24%	0.16%	0.02%	0.00%
	WB Inner	12	0.04%	0.03%	0.20%	0.05%	0.00%
149+00	EB Outer	11.6	0.86%	0.26%	0.01%	0.03%	0.56%
	EB Inner	11.64	0.69%	*	0.10%	0.01%	0.21%
	WB Outer	11.77	2.63%	0.22%	0.15%	0.12%	0.19%
	WB Inner	11.96	2.80%	0.05%	0.39%	0.12%	0.19%
203+00	EB Outer	11.94	3.81%	0.09%	0.22%	0.02%	0.00%
	EB Inner	11.83	4.65%	0.08%	0.02%	0.04%	0.23%
	WB Outer	11.57	3.59%	0.07%	0.50%	0.09%	0.07%
	WB Inner	11.86	4.60%	0.06%	0.46%	0.00%	0.19%
208+00	EB Outer	11.62	2.32%	0.28%	0.08%	0.07%	0.05%
	EB Inner	11.88	2.48%	0.17%	0.06%	0.06%	0.02%
	Turning	11.19	2.01%	0.30%	0.01%	0.06%	0.02%
	WB Outer	11.9	1.09%	0.06%	0.34%	0.15%	0.12%
	WB Inner	11.42	0.00%	0.24%	0.12%	0.00%	0.00%
227+00	EB Outer	11.73	2.39%	0.00%	0.29%	0.03%	0.19%
	EB Inner	12.13	2.14%	0.03%	0.37%	0.00%	0.19%
	WB Outer	11.81	1.91%	0.98%	*	*	0.46%
	WB Inner	11.95	1.88%	0.04%	0.32%	0.01%	0.05%
232+00	EB Outer	11.7	2.48%	0.00%	0.04%	0.07%	0.10%
	EB Inner	11.75	2.77%	0.12%	0.50%	0.03%	0.01%
	WB Outer	11.48	2.79%	0.02%	0.13%	0.05%	0.05%
	WB Inner	11.92	1.97%	0.02%	0.57%	0.02%	0.00%

*Data were missing in point cloud.

The LiDAR derived point clouds on Sections 1 and 2 were adjusted using IMU measurements and through post-processing with ground control points, however, the Section 3 point cloud was adjusted only with the integrated IMU data. The one-sided *t*-test for both adjusted and unadjusted LiDAR indicates that the difference of the LiDAR derived slopes and field surveying was less than 0.19% (Table 6) at a 95% confidence level. Cross-slope calculations are based on relative elevation of points along reference lines. Study results indicate that regardless of whether data are adjusted or unadjusted through post-processing with ground control points, cross-slopes can be estimated accurately, within acceptable tolerance, using LiDAR surface model data.

Cross-Slope Sensitivity Analysis

The typical range for cross-slopes along urban arterials is 1.5% to 3% (*11*); the lower portion of this range is appropriate where drainage flow is across a single lane and higher values are appropriate where flow is across several lanes (*11*). On high-speed roadways, SCDOT recommends that the normal cross-slope be 2.08% on tangent sections with some exceptions depending on the number of lanes (*1*). Inherent characteristics of paving operations leads to deviations from design cross-slope values. As discussed previously, these deviations can potentially compromise safety. Identifying roadway sections that do not meet minimum criteria requires accurate cross-slope measurements. To quantify the safety

Table 4. Cross-Slope Comparison between Surveyed Data and LiDAR Derived Cross-Slope: Section 2

Station	Lane	Lane width (HD)	Surveyed data	Difference from surveyed data		
				Vendor A	Vendor B	Vendor C
P78	WB outer lane	12.04	3.26%	*	0.12%	0.08%
	WB inner lane	11.62	1.40%	*	0.18%	0.02%
	EB inner lane	11.87	1.31%	0.42%	0.15%	0.31%
	EB outer lane	12.09	1.45%	0.24%	0.11%	0.06%
P91	WB outer lane	12.01	3.41%	0.12%	0.19%	0.07%
	WB inner lane	11.82	1.27%	0.07%	0.23%	0.12%
	EB inner lane	11.72	1.71%	0.03%	0.19%	0.03%
	EB outer lane	12.07	1.91%	0.02%	0.16%	0.13%
P98	WB outer lane	12.04	1.96%	0.00%	0.00%	0.04%
	WB inner lane	11.62	1.03%	0.42%	0.25%	0.34%
	EB inner lane	11.87	1.60%	0.01%	0.19%	0.01%
	EB outer lane	12.07	2.50%	0.03%	0.12%	0.05%
P103	WB outer lane	11.77	6.69%	0.63%	0.73%	0.70%
	WB inner lane	11.51	7.54%	0.54%	0.56%	0.57%
P126	WB outer lane	11.97	3.97%	*	0.14%	0.12%
	WB inner lane	12.09	4.47%	*	0.33%	0.24%
P127	WB outer lane	11.43	1.40%	0.48%	*	0.04%
	WB inner lane	12.24	1.12%	0.67%	0.80%	0.12%

*Data were missing in point cloud.

Table 5. Cross-Slope Comparison between Surveyed Data and LiDAR Derived Cross-Slope: Section 3

Station	Lane	Lane width (HD)	Surveyed Data	Vendor E	Difference from surveyed data
34+31	EB outer lane	11.98	1.50%	1.30%	0.20%
	EB inner lane	12.00	1.92%	2.08%	0.16%
38+52	EB outer lane	12.00	1.75%	1.91%	0.16%
	EB inner lane	11.96	0.92%	1.08%	0.16%
44+20	EB outer lane	11.98	2.00%	2.17%	0.17%
	EB inner lane	12.00	1.16%	1.33%	0.17%
44+68	EB outer lane	12.00	2.16%	2.25%	0.09%
	EB inner lane	11.95	1.25%	1.42%	0.17%
45+92	EB outer lane	12.00	1.92%	2.00%	0.08%
	EB inner lane	11.97	0.92%	1.16%	0.24%
57+39	EB outer lane	11.96	8.08%	8.08%	0.00%
	EB inner lane	11.97	6.58%	6.41%	0.17%

effects of MLS cross-slope measurement errors, the researchers conducted a cross-slope sensitivity analysis on hydroplaning potential.

When rain falls on a sloped pavement, the path that runoff takes to the pavement edge is called the drainage path, and the water depth that accumulates on the pavement can be calculated from the following equations (*12*):

$$L_f = L_x((1+{S_g}/{S_x})^2)^{0.5} \qquad (1)$$

$$\mathrm{WD}_0 = 0.00338\mathrm{TXD}^{0.11}L_f^{\ 0.43}I^{0.59}S_x^{\ -0.42} - \mathrm{TXD} \qquad (2)$$

$$\mathrm{WD} = (\mathrm{WD}_0 + \mathrm{TXD})\times\sqrt{1+({S_g}/{S_x})^2} - \mathrm{TXD} \qquad (3)$$

where:

S_x = cross-slope (ft/ft);
S_g = longitudinal grade (ft/ft);
L_x = pavement width (ft) from crown of the pavement;
L_f = length of flow path;
WD = water depth above the top of the surface asperities (in.);
TXD = texture depth (in.); and
I = intensity of rainfall (in./h).

Table 6. Summary of Cross-Slope Comparison

Section 1, East–West Parkway

	EB outer lane	EB inner lane	Turning lane	WB inner lane	WB outer lane
Min	0%	0%	0.01%	0%	0%
Max	0.56%	0.71%	0.80%	0.57%	0.98%
Mean	0.14%	0.19%	0.30%	0.14%	0.22%
Median	0.09%	0.11%	0.26%	0.05%	0.15%
One-sided *t*-test	Margin of error	*n*	*p*-value		Significant
	0.18%	136	< 0.05		Yes

Section 2, I-85 Business Loop

	EB outer lane	EB inner lane	WB inner lane	WB outer lane
Min	0.02%	0.01%	0.02%	0.00%
Max	0.24%	0.42%	0.80%	0.73%
Mean	0.1%	0.15%	0.34%	0.23%
Median	0.11%	0.15%	0.29%	0.12%
One-sided *t*-test	Margin of error	*n*	*p*-value	Significant
	0.19%	49	< 0.05	Yes

Section 3, US-123

			EB outer lane		EB inner lane
Min			0.16%		0.00%
Max			0.24%		0.20%
Mean			0.18%		0.12%
Median			0.17%		0.13%
One-sided *t*-test	Margin of error	*n*		*p*-value	Significant
	0.18%	12		< 0.05	Yes

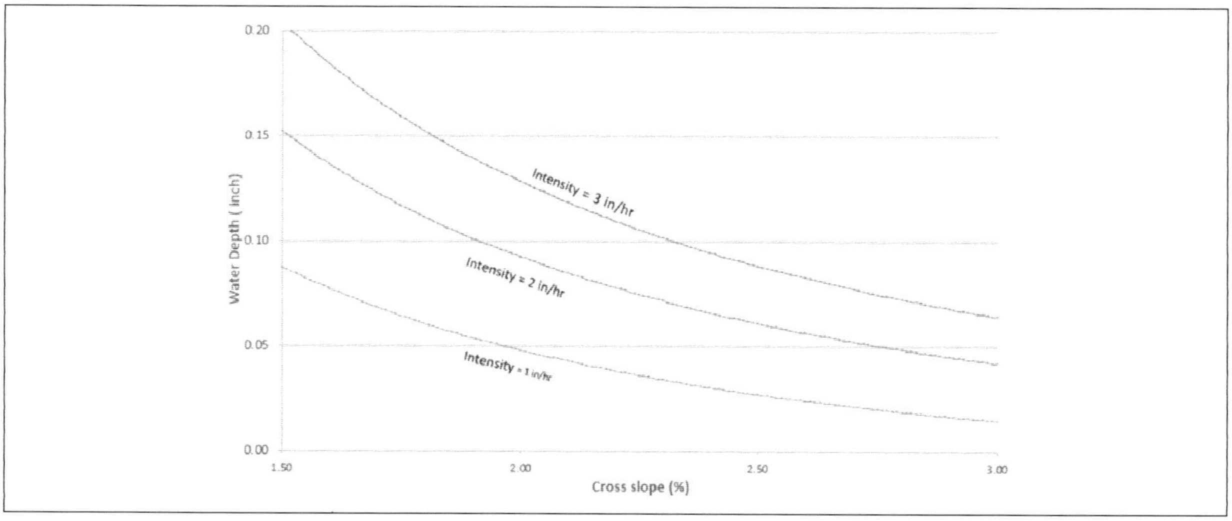

Figure 5. Cross-slope sensitivity analysis on pavement water depth.

On wet pavement, when tires lose contact with the pavement due to water film depth, hydroplaning is likely to occur (*12*). A water depth of 0.15 in. can lead to hydroplaning for a passenger vehicle traveling at highway design speeds (*13*). To determine how the difference in cross-slope values impact the water depth, the following assumption has been made: $S_g = 4.5\%$, TXD = 0.04 (50 percentile) (12). Using the above equations, the impact of changes in cross-slope on water depth accumulation by rainfall intensity were calculated and the results are shown in Figure 5.

Driving visibility is reduced when rainfall intensity exceeds 2 in./h and becomes poor when intensity exceeds 3 in./h (*14*). It is expected that vehicle operators will refrain from driving or drive very slowly during such heavy rainfall periods (*12*). The SCDOT uses a maximum construction tolerance of ± 0.348% (*1*). For a highway section with a typical cross-slope of 2.08%, an allowable minimum cross-slope would be 1.73%. Using the SHRP 2 suggested slope acceptable measurement error ± 0.2% (*10*), which is greater than the average MLS measurement error of ± 0.19% found in this research, a cross-slope of 1.53% can potentially be considered acceptable when incorporating a ± 0.2% error. According to Figure 5, a cross-slope of 1.53% corresponds to a water depth of 0.147 inches which has a low potential for hydroplaning for vehicles traveling at highway speeds for rain fall intensities less than 2 in./h. For longitudinal grade over than 4.5% the MLS needs supplemented sample survey data. This suggests that typical MLS measurement error is acceptable for cross-slope verification purposes.

Conclusion

The use of mobile LiDAR to extract the cross-slope has been evaluated on 20 stations including 203 travel lanes. The results of this research have proved the feasibility of automated data collection vehicles in comparison with human collection methods to collect data efficiently, accurately, and reliably. The results of *t*-test statistical analysis indicated the average deviation between LiDAR data and field surveying measurements was less than the minimum acceptable accuracy value (± 0.2% specified by SCDOT and SHRP 2) at a 95% confidence level. Both adjusted and unadjusted LiDAR data met the SCDOT standard.

Common survey data collection methods are time-consuming and require data collectors to be located on the road, which poses a safety issue, however, methods such as MLS are available to capture accurate cross-slope, grades, location, and a variety of other geometric design characteristics. These applications increase productivity, minimize road crew exposure and, create robust information products that serve multiple uses such as flood mapping, hydroplaning, and road inventory.

Acknowledgments

The authors thank SCDOT and the Federal Highway Administration for sponsoring this research.

References

1. SCDOT. *South Carolina Highway Design Manual*. South Carolina Department of Transportation, 2003.
2. Stein, W. J., and T. R. Neuman. *Mitigation Strategies for Design Exceptions*. FHWA-SA-07-011, FHWA, U.S. Department of Transportation, 2007.
3. Souleyrette, R, S. Hallmark, S. Pattnaik, M. O'Brien, and D. Veneziano. *Grade and Cross Slope Estimation from LIDAR-based Surface Models*. U.S. Department of Transportation Research and Special Programs Administration, 2003.
4. Baffour, R. A. Collecting Roadway Cross slope Data Using Multi-Anenna–Single Receiver GPS Configuration. In *ASCE Proceedings of the International Conference on Applications of Advanced Technology in Transpotation Engineering*, 2002.
5. Jaakkola, A, J. Hyyppä, H. Hyyppä, and A. Kukko. Retrieval Algorithms for Road Surface Modelling using Laser-Based Mobile Mapping. *Journal of Sensors*, Vol. 8, No. 9, 2008, pp. 5238–5249.
6. Olsen, M. J., J. D. Raugust, and V. Roe. *Use of Advanced Geospatial Data, Tools, Technologies, and Information in Department of Transportation Projects*. Transportation Research Board, Washington, D.C., 2013.
7. Zhang, F. K., and H. C. Frey. Road Grade Estimation for On-Road Vehicle Emissions Modeling Using Light Detection and Ranging Data. *Journal of the Air and Waste Management Association*, Vol. 56, No. 6, 2012, pp. 777–788.
8. Tsai, Y., A. Chengbo, W. Zhaohua, and E. Pitts. Mobile Cross-Slope Measurement Method Using Lidar Technology. *Transportation Research Record: Journal of the Transportation Research Board*, 2013. 2367: 53–59.
9. Holgado-Barco, A., D. Gonzalez-Aguilera, P. Arias-Sanches, and J. Martines-Sanches. An Automated Approach to Vertical Road Characterisation using Mobile LiDAR Systems: Longitudinal Profiles and Cross-Sections. *Journal of Photogrammetry and Remote Sensing*, Vol. 96, 2014, pp. 28–37.
10. Hunt, J. E., A. Vandervalk, and D. Snyder. *SHRP2 Report S2-S03-RW-01: Roadway Measurement System Evaluation*. Transportation Research Board of the National Academies, Washington, D.C., 2013.
11. AASHTO. *A Policy on Geometric Design of Highways and Streets*. Washington D.C: American Association of State Highway and Transportation Officials, 2004.
12. Guven, O., and J. G. Melville. *Pavement Cross Slope Design*. Auburn: Auburn University, 1999.
13. Mraz, A., and A. Nazef. Innovative Techniques with a Multipurpose Survey Vehicle for Automated Analysis of Cross Slope Data. *Transportation Research Record: Journal of the Transportation Research Board*, 2008. 2068: 32–38.
14. Gallaway, B. M, D. L. Ivey, G. Hayes, W. B. Ledbetter, R. M. Olson, D. L. Woods, and R. F. Schiller. *Pavement and Geometric Design Criteria for Minimizing Hydroplaning*. College station, TX: Texas Transportation Institute, 1979.

The Standing Committee on Geospatial Data Acquisition Technologies (AFB80) peer-reviewed this paper (18-03882).

Article

TRR
JOURNAL OF THE TRANSPORTATION RESEARCH BOARD

Transportation Research Record
2018, Vol. 2672(39) 98–106

DOI: 10.1177/0361198118758685
journals.sagepub.com/home/trr
SAGE

Automated Extraction of Horizontal Curve Attributes using LiDAR Data

Suliman Gargoum[1], Karim El-Basyouny[1], and Joseph Sabbagh[1]

Abstract

Horizontal curves are designed to provide a safe and smooth transition between straight segments on a highway network. Although curves are often designed to meet very stringent standards, imperfections during construction and high operating speeds mean that they are still prone to collisions. Therefore, it is essential that attributes of curves are surveyed to ensure they meet design requirements. Moreover, knowledge of the locations of horizontal curves and their attributes is also required to provide drivers with accurate information in advanced curve-warning systems, which are expected to enhance safety. Unfortunately, conventional techniques to obtain information about horizontal alignments are extremely tedious and, in some cases, impractical. This paper proposes a method by which horizontal curves can be automatically detected and their attributes automatically measured on scans of the highways obtained using light detection and ranging (LiDAR) technology. The proposed method is tested on two different highway segments at the Province of Alberta, Canada, where LiDAR data were collected. Moreover, testing was also conducted using virtual highways with curves with known attributes generated in AutoCAD Civil 3D. The results show that the code is successful in detecting all curves on a highway segment; moreover, the attributes of those curves were estimated with a high degree of accuracy (average difference <3%).

Horizontal alignments on highways are designed to ensure that drivers traveling at the road's speed limit can safely negotiate the curve. Curves with sharp radii or inadequate superelevations could result in drivers losing control of their vehicles, increasing the risk of overturning or sliding off the highway. In fact, statistics show that approximately one-quarter of fatalities on highways occur on horizontal curves (*1*). As a result, design guidelines dictate that curve radius and other attributes on a curve meet specific requirements. The minimum curve radius is estimated as a function of the road's design speed, the superelevation of the road, and the friction factor, which is used as a proxy for the amount of friction the pavement provides. If the radius on a curve drops below the minimum radius estimated in design equations, the likelihood of a driver safely negotiating the curve reduces significantly.

Despite all efforts to construct road alignments in accordance with design standards, budget limitations, and site imperfections mean that deviations from design plans can occur. Therefore, it is essential to survey horizontal curves and their geometric attributes after construction to ensure they meet design requirements. Obtaining such information in an efficient manner helps authorities take timely corrective action in cases in which design limitations exist. Obtaining information about attributes of horizontal curves is also essential for the development of operating speeds models. These models are often used to predict operating speeds on roads based on a road's geometry. The prediction is then used to set a safe speed limit that is consistent with the road's geometry. In the realm of connected vehicle technology and vehicle-to-infrastructure (V2I) communications, curve-warning systems also require that curve information is readily available and is efficiently communicated to drivers in real time, therefore the importance of having an efficient method to obtain such information. The knowledge of curve geometry is also essential for autonomous driving at locations where lane-marking information might not be sufficient to guide the vehicle through a curve.

Regardless of the reason, surveying the attributes of horizontal alignments on roads is an integral step to efficient management of road infrastructure. Unfortunately, existing techniques such as the chord offset method and compass method used to measure horizontal alignment information are time-consuming, labor-intensive, and in some cases inaccurate (*2*). As a result, transportation agencies have been looking for more efficient methods to obtain such information. Global Positioning System (GPS) data have been commonly used as an alternative to conventional curve estimation techniques. Many agencies collect GPS data points along

[1]Department of Civil and Environmental Engineering, University of Alberta, Edmonton, AB, Canada

Corresponding Author:
Address correspondence to Suliman Gargoum: gargoum@ualberta.ca

Figure 1. LiDAR point cloud.

highways and import them into geographic information system (GIS) tools such as curve estimators to identify attributes of curves (*3*). Although using GPS has helped overcome the burden associated with site visits and manual data collection, the low density of points collected in GPS surveys and the low accuracy sometimes result in poor estimates of the horizontal alignment geometry. In other words, the scattered nature of points collected in GPS surveys without an inertial measurement unit (IMU), caused by inaccuracy in the position information, means attributes of curves measured off those points could be highly inaccurate. Moreover, the GPS data only being collected along a single path makes estimating curve geometry at different approaches on a roadway or different lanes on an approach a challenging task.

In this paper, a method to automatically detect and measure attributes of horizontal curves on highways using mobile light detection and ranging (LiDAR) data is proposed. LiDAR data consist of closely spaced points with known coordinates, resulting in a virtual image of the scanned highway segment, as seen in Figure 1. In addition to overcoming the limitations associated with using GPS, the high point density of LiDAR datasets helps use the same dataset to assess sight distance on curves and measure the allowable horizontal offset. In fact, even outside the context of detecting horizontal curves and measuring their attributes, LiDAR datasets have been of great value to many applications in the field of transportation, ranging from automated sign detection (*4–7*) and sight distance assessments (*8, 9*) to measuring cross-slopes and side-slopes of roads (*10*).

The proposed algorithm uses changes in azimuth of vectors aligned along the road's axis to estimate the locations of horizontal curves. Once the points of curvature (PC) and points of tangency (PT) are defined for a particular curve, the algorithm then uses linear regression of the tangents connecting the curve to locate the point of intersection (PI). The PC and PT are then re-estimated. The final step involves locating the center of the curve. Once this is identified, the curve's radius, deflection angle, length, and chord length are all estimated.

The proposed algorithm was tested on two mobile LiDAR datasets. Moreover, to verify the accuracy of the proposed algorithm, it was also tested on simulated data generated using AutoCAD Civil 3D.

Previous Work

Several recent studies have proposed techniques to detect and measure attributes of horizontal curves using GPS data and GIS platforms. Findley et al. (*11*) compared estimating attributes of curves using Curvature Extension (CE) and Curve Finder (CF). Curve Estimator is a plugin added into ArcGIS, which estimates the radius and length of a horizontal curve given the end-points of the curve. CF is a tool that uses GIS polylines to determine curve length and radius based on GPS data. Unlike CE, CF has the ability to detect curves while also estimating their radii and lengths. The two tools were tested on two two-lane roads and a multilane interstate highway in North Carolina. CF was able to detect 67% of curves along the three routes. For 65% of the curves, CF estimated a radius within 20% of that obtained using CE (used as a benchmark). Length estimate was within 20% for only 31% of the curves. The authors attribute the inability to detect some curves to their large radii. The authors also emphasized the importance of collecting highly dense points to obtain a higher level of accuracy.

Rasdorf et al. (*3*) also evaluated several GIS platforms and techniques used for extracting curve information from GIS line points. The paper compared the accuracy of three publicly available GIS applications (i.e., CF, Curve Estimator, and Curve Calculator). The study concluded that Curve Estimator is most effective when used for individual curve analysis, while Curve Calculator was only accurate if a substantial number of GIS points were available. For network assessments, the study recommended using CF.

Andrášik and Bíl (*12*) proposed a method to classify road segments as tangential or circular using digital vector data. The technique involved (i) geometric generalization, (ii) learning based on the classification procedure and explanatory variables, and (iii) postprocessing. For geometric generalization, the Douglas–Peucker algorithm was used to reduce the redundancy of the data points. The learning procedure involved classifying points into curved or circular, based on six explanatory variables that relate to road geometry. These variables include the cumulative angle between consecutive points, the density of the points, and how well a set of points fit to a curve of a specific radius. In the postprocessing stage, the authors defined a set of heuristic conditions that points must meet to maintain initial classification. The proposed method was tested on road segments in the Czech Republic. The procedure was able to accurately detect three of four curves on one of the segments. The authors report a 2.08% error for radius estimation.

Xu and Wei (*13*), proposed a technique that involved the identification of curves by estimating the difference in bearing angle between consecutive vertices along a highway. The authors define a threshold for the change in bearing that indicates the start of a curve. Once the PC and PT locations of the curve are identified, the length of the curve is measured. The length is then used to calculate other attributes of the curve, including radius and deflection angle. The authors report success rates of up to 95% for detection of curves. It is also reported that radius estimation was within 5% of the true radius. The authors do acknowledge, however, that their procedure was not capable of detecting spiral curves or measuring their attributes.

Li et al. (*14*) also used difference in bearing angle to classify segments into curved and non-curved. They also attempted further classification of curves into simple or compound curves based on the tangent length between two curves. The length of a curve is calculated using the PC/PT locations and compared to the actual length. If the error is >2.5%, then the curve is compound. The proposed algorithm was tested on 10 county roads in Wisconsin and was 96.7% accurate in detecting curves. Moreover, 97% accuracy was achieved in classification of curves into simple and compound curves. However, the algorithm was unable to distinguish between spiral and circular curves.

Ai and Tsai (*15*) took a slightly different approach in their curve-detection attempts. The authors put the data through an iterative process to identify straight, curved, and tangent sections. The curved sections were identified using Kasa circle fitting (KCF), which takes several neighboring points and fits them onto a circle. An iterative process repeats the KCF, increasing the number of neighboring points until a large enough error is obtained. Once a curve is obtained, further filtering is applied in which, if a straight segment is falsely detected, it is expected that it has fewer GPS points and a large radius. Therefore, thresholds for the number of points and radius length were defined. Tangent sections were defined by looking at consecutive sections. Validated straight and curved sections from the previous step are connected and if consecutive sections are both straight, then the section in between is a straight section; if consecutive sections contain a curved section then the section in between is a tangent section. Also, a radius value is given to each "tangent" by interpolating the radius values of the prior and subsequent sections. This algorithm is tested using both simulated data and field data. For the simulated data, an average radius error of 1.48% ± 1.2% is obtained, while for the field data, the error ranged from 4.83% ± 1.8%.

Imran et al. (*16*) proposed using GPS data to detect curves and measure their attributes. First, data points are separated into linear and nonlinear sections based on the change in angle between consecutive points. A PC/PT estimate is obtained and 75% of the tangent points are used to fit a line for tangent segments (75% so as to avoid points that fall close to the curve). The Jacobian matrix method is then used to fit circles onto curved segments, and any points between the circle and tangent segments are considered transition spirals. For the circular segment, the code takes the first, middle, and last points of the middle 50% of points to estimate the coordinates of the center of the circle and the radius. After multiple iterations, the code outputs the length, center, and radius of a curve, as well as the length of spiral. The proposed method was tested on a highway segment that included nine curves. For ground truth assessment, a digitized map was developed using aerial survey data. The fitted radii were found to have an average error of 1.55% from the actual values. The error is attributed to the layout survey data being different from the true as-built data. For curve length, an average difference of 16.7% was reported.

Holgado-Barco et al. (*17*) carried out one of a few studies that attempted horizontal alignment extraction from LiDAR. The semi-automatic method proposed in the paper involves segmenting, parameterizing, and filtering the point cloud. In the segmentation stage, points tracing the road's trajectory (lane markings) are filtered out. These points are then classified into curved and straight segments based on changes in azimuth and curvature between consecutive points. Data are classified into three segment types: straight segments (if zero curvature is detected); circular arcs (in case of constant curvature); and clothoids (if curvature varies across consecutive points). After detecting the curves, radii and the transition lengths – if any – are estimated. The proposed algorithm was tested on simulated data and data collected on a Spanish highway. Comparing results obtained from the algorithm to those estimated by an experienced topographer, errors in length of up to 3.8 m (2.0%) for circular arcs and 5.6 m (0.4%) for circular radius were reported. When testing on the simulated segment, the errors were 0.3 m (0.1%) and 0.9 m (1.1%) for arc radius and length, respectively.

As evident from the review, attempts to extract curve attributes using GIS platforms have recently grown in popularity. However, the majority of existing work is limited to using GPS data to estimate such attributes. Moreover, previous studies focus on estimating curve radii alone, without much interest in other attributes. Finally, some previous studies use commercial software such as ArcGIS curve estimators to obtain horizontal curve data. In an attempt to address some of these limitations, this paper develops an algorithm to automatically detect and survey horizontal curves on highways using mobile LiDAR data.

Proposed Method

The procedure proposed in this study is divided into multiple stages.

Trajectory Vector Definition

The first step of the extraction process involves defining trajectory vectors that trace the road's axis. The vectors are estimated using points (or position vectors) in the LiDAR point cloud that trace the path of the data collection truck. These points are obtained by filtering the LiDAR point cloud based on the scanner angle. The points are automatically filtered out of the LiDAR point cloud as part of the MATLAB script. After reading in the LAS file, the MATLAB script is written to filter out points parallel to the road's axis that lie in the nadir view of the scanner (i.e., points of scanner angle 0 that lie directly below the scanner are extracted and used to represent the points tracing the road's axis).

Although these points are a very good representation of the truck's trajectory and consequently the road's axis, occasional deviation in points exists because of changes in the truck's yaw angle. Therefore, the moving average technique was used to estimate the new position vector based on the positions of a set of three position vectors. The new position vectors were then used to define the end-points of each trajectory vector.

Let P_1 be the position vector defining the start-point of the trajectory vector (V_i). Similarly, let P_2 represent the position vector representing the end-point of the trajectory vector. In that case, the trajectory vector ($\widehat{V}_i$) at a point i along the road can be defined as follows:

$$\widehat{V}_i = \widehat{P}_2 - \widehat{P}_1 \quad (1)$$

where $\widehat{P}_1 = \begin{bmatrix} x_1 \\ y_1 \\ z_1 \end{bmatrix}$ and $\widehat{P}_2 = \begin{bmatrix} x_2 \\ y_2 \\ z_2 \end{bmatrix}$, therefore $\widehat{V}_i$ can be rewritten as follows:

$$\widehat{V}_i = \begin{bmatrix} x_2 \\ y_2 \\ z_2 \end{bmatrix} - \begin{bmatrix} x_1 \\ y_1 \\ z_1 \end{bmatrix} = \begin{bmatrix} x_2 - x_1 \\ y_2 - y_1 \\ z_2 - z_1 \end{bmatrix} \quad (2)$$

Multiple trajectory vectors ($i = 1$ to n) are defined between each pair of consecutive position vectors. The number of trajectory vectors defined depends on the length of the LiDAR segment analyzed and the length of each trajectory vector, which is predefined by the user. In this paper, 20 m long trajectory vectors were used (i.e., the set of position vectors representing P_1 and P_2 are chosen to be 20 m apart). This parameter can be altered by the user depending on the curve sharpness expected on a particular segment (i.e., sharpness of curves that exist on the portion of a highway that is to be analyzed). For detection of smooth curves, higher spacing is recommended; for sharp curves even a spacing of <5 m may be sufficient.

Curve Detection

For curve detection, the algorithm loops through all the trajectory vectors and compares the difference in the azimuth between consecutive vectors. If the change in azimuth exceeds a predefined threshold (0.5° in this paper), this point is assumed to mark the start or the end of a curve. To ensure that the threshold for detecting a curve is effective for all curves on a particular segment, the difference in azimuth between consecutive points is normalized. This is done by dividing change in azimuth for each pair of consecutive trajectory vectors by the maximum change in azimuth detected on the entire segment. Let αV_1 and αV_2 represent the azimuth of two consecutive trajectory vectors V_1 and V_2 respectively, in this case $\Delta\alpha$, which is the change in azimuth, can be calculated as follows:

$$\Delta\alpha = \alpha V_1 - \alpha V_2 \quad (3)$$

The normalized change in azimuth ($\Delta\alpha_{norm}$) can be computed as follows:

$$\Delta\alpha_{norm} = \Delta\alpha / \Delta\alpha_{max} \quad (4)$$

where $\Delta\alpha_{max}$ denotes the highest change in azimuth detected between any two consecutive vectors on the analyzed segment.

To distinguish the beginning of a curve from its end, the code checks the change in azimuth of the previous pair of points; if this does not exceed the threshold, the point is a start-point, otherwise it is the end of the curve. Furthermore, to ensure that the curve does indeed start at a particular point where a change in azimuth is detected, the code requires that the change is persistent (i.e., the change in azimuth is sustained for a particular length). This is done to minimize the possibility of detecting short changes in azimuth caused by data collection inaccuracies as horizontal curves (i.e., minimize false-positives).

Feature Identification

Once the whereabouts of the curves are identified, its geometric elements can be measured; however, before that is done it is essential to accurately identify the defining point of the curve, including the PI, PC, and PT.

Point of Intersection. To estimate PI, the algorithm uses a set of points before reaching the candidate start-point of the curve and another set of points after passing the candidate end-point of the curve, as indicated in Figure 2*a*. Linear regression equations are then developed for the two tangents. The algorithm then finds the PI of the curve by intersecting those equations. The deflection angle is also estimated by calculating the difference in azimuth between the lines.

Point of Curvature and Tangency. Although the code identifies candidate locations for the start and end of each curve, these points are not always accurate. The true PC and PT could be a few points before or ahead of the location identified based on change in azimuth. Therefore, more accurate estimation for the beginning and end of the curve are achieved by minimizing the difference in northings estimated using the regression equation and those obtained from the LiDAR data which trace the road's axis. The code starts at PI and moves toward both PC and PT until the difference in northings drops below a specific threshold (0.3 m in this paper), as illustrated in

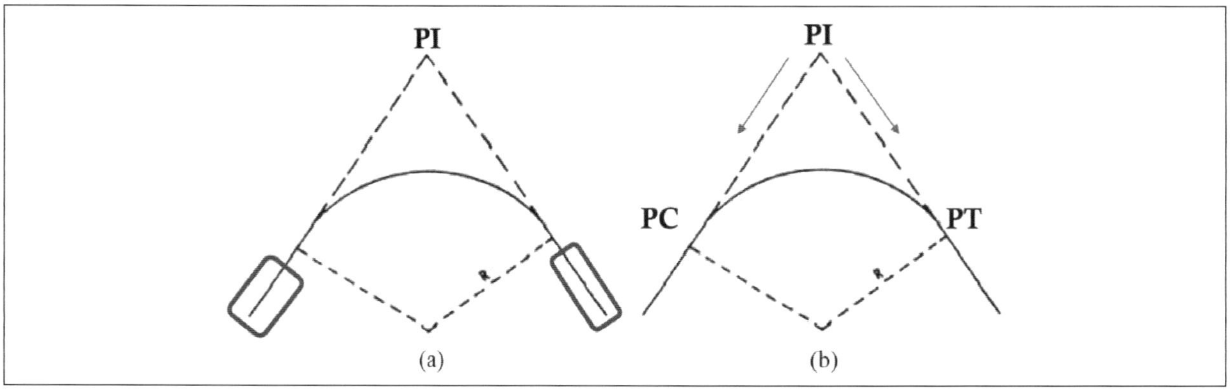

Figure 2. PI, PC, and PT estimation. (*a*) Developing regression equations to locate PI. (*b*) Locating accurate PC and PT.

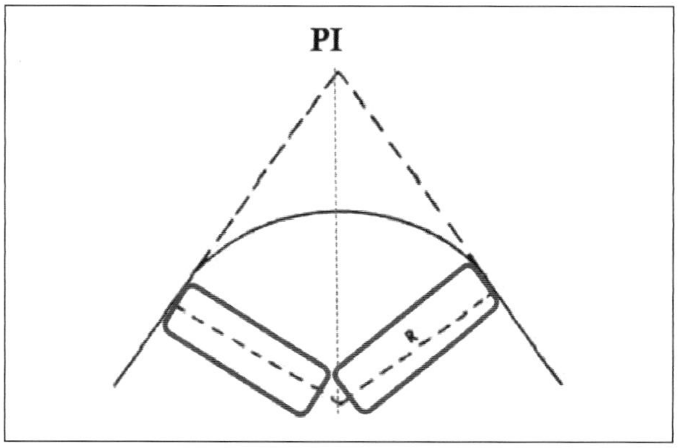

Figure 3. Locating the curve's origin and measuring its radius.

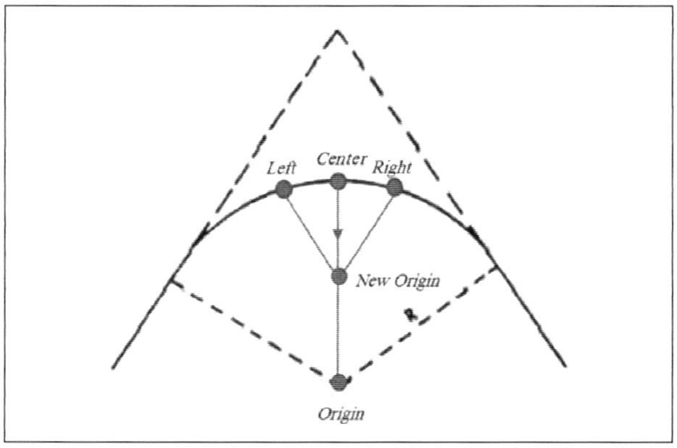

Figure 4. Left and right points with 25% of the center of the curve.

Figure 2*b*. Once that point is reached, the new locations of PC and PT are defined.

Curve Tangent, Radius, and Chord Length. The tangent of the curve (T) is estimated by calculating the Euclidean distance between PI to both PC and PT. It is expected that, for a symmetric curve, the estimates for T obtained from either side of the curve are equal.

To identify the radius of the curve, the normal line for each of the two tangent lines previously developed using regression is estimated. The lines are estimated such that they pass through PC and PT of the curve. The intersection point of the two lines, which represents the origin of the curve, is then located as evident in Figure 3. After locating the origin of the curve, the radius is measured by finding the Euclidean distance between PC/PT and the origin (i.e., the intersection point). Moreover, the chord length is also measured by finding the Euclidean distance between PC and PT.

Spiral Transition Detection

In an attempt to identify whether the curve includes spirals or not, a separate procedure is proposed. In this procedure, the first step is to locate the center point along the curve's arch between PC and PT. Once this point is located, a point to the right and left of this point, which fall on the curve and are within 25% of the distance between the center point and the end-points, are located, as shown in Figure 4.

After identifying the three points *left*, *right*, and *center*, the code calculates the distance between each of the three points and points along the line which connect the center of the curve and its origin, as illustrated in Figure 4. The code loops through multiple points along that line until it reaches a point at which the distance to the left, right, and center are almost equal. The idea here is that if a spiral transition exists, the three distances (three radii) measured will only be equal at a single point before reaching the previously identified origin. The reason here is that the previously identified origin is the center of the curve including the spiral, whereas the new point where the three distances are equal would be the origin of the circular portion of the curve and, therefore, the radius measured at this origin would be that of the circular curve.

The code is written such that it minimizes the difference between the three radii. Once the difference meets an acceptable threshold defined by the user, the code outputs the coordinates of the proposed origin of the circular curve and provides an estimate for the circular radii if a spiral exists.

Figure 5. Point cloud data at test highways. (a) Highway 36. (b) Highway 32.

The entire procedure was coded in Microsoft Visual Basic version 7.1.

Case Study

The proposed algorithm was tested on data collected in mobile LiDAR scans of existing highways and data simulated in AutoCAD Civil 3D. The next few paragraphs provide information on the LiDAR data collection system and the two test segments.

LiDAR Data

The LiDAR data used in this study were collected by Alberta Transportation (AT) using third-party services. The data were collected using a RIEGL VMX-450 scanning system, which is mounted on a pavement surface profiling vehicle (PSP-7000). The RIEGL VMX-450 system is equipped with two VQ-450 scanners and IMU/Global Navigation Satellite System units to collect the LiDAR data. The laser scanners are symmetrically configured on the left and right sides, pointing toward the rear of the vehicle at a heading angle of approximately 145°. The VQ-450 scanner has a scan rate of up to 1.1 million points per second and a scan speed of 400 lines per second, a precision of 5 mm, and an accuracy of 8 mm (*18*).

The scanning system collects 3D point cloud data as the data collection truck travels through the highway corridor of interest. Surveys are conducted within normal traffic flow while traveling at the posted speeds of up to 100 km/h. Provincial surveys conducted at 90 km/h result in LiDAR point densities on the road surface of 150–1000 points/m^2 (*19*).

Test Segments

The two highway segments on which the algorithm was tested are shown in Figure 5. Curve detection is based on changes in azimuth, which are typically high for sharp curves and low for mild curves. Therefore, to ensure that the algorithm is accurate enough in detecting even the mildest of curves, testing was conducted on segments in which existing curves had a very large radius (i.e., curves were not sharp).

Highway 36. A 4 km segment on Highway 36 was used in this study. The segment is a two-lane undivided rural road with a speed limit of 100 km/h, located in the southern region of Alberta, southeast of the city of Lethbridge. The roadside area is relatively clear, with very low vegetation.

Highway 32. A 4 km segment on Highway 32 was used in this study. The segment is also a two-lane undivided rural road with a 100 km/h speed limit. The segment is located in Central Alberta, northwest of the City of Edmonton. Unlike Highway 36, vegetation is relatively high on this portion of Highway 32.

Simulated Data

In addition to the two LiDAR highway segments, the algorithm was also tested on curved segments of known geometric features developed in AutoCAD Civil 3D. This was done to verify the accuracy of the proposed algorithm since these curves are more reliable than as-built information provided for the LiDAR segments, were not up to date, and, in some cases, did not represent existing conditions. Moreover, some information about the curves was missing from the as-built data and had to be calculated based on other attributes. Two different curves were drawn in Civil 3D; one was a simple circular curve (Simulated Curve I) and the other was a curve with spiral transitions at either end (Simulated Curve II).

Results and Discussion

Tables 1 and 2 show the results of running the algorithm on the segments, listing the information about all curve attributes, including radii, deflection angle, arc length, and tangent length for each of the test segments. For the LiDAR segments, Table 2 lists information obtained from the as-built roads provided by AT, and information estimated using

Table 1. Test Results on Simulated Curves

	Simulated simple curve				Simulated spiral curve			
	Actual	Detected	Difference (m)	% Difference	Actual	Detected	Difference (m)	% Difference
Radius (m)	2000	1976	24	1.20	1000	1009.2	9.2	0.92
Deflection angle (degrees)	90	89.12	0.88	0.98	90	88.89	1.11	1.23
Length (m)	3141.5	3130.99	10.51	0.33	1570.75	1603.82	33.07	2.10
Tangent length (m)	2000	1985	15	0.75	1000	1004.93	4.93	0.49

Table 2. Test Results on LiDAR-Detected Curves

	Highway 36				Highway 32			
	As-built	Detected	Difference (m)	% Difference	As-built	Detected	Difference (m)	% Difference
Radius (m)	N/A	1955.3	N/A	N/A	1164.3	1218.3	54.02	4.6
Deflection Angle (degrees)	N/A	24.28	N/A	N/A	52.1	51.7	0.34	0.7
Length (m)	N/A	838.8	N/A	N/A	1134.5	1126.2	8.27	0.7
Tangent Length (m)	N/A	398.5	N/A	N/A	579.2	571.2	8	1.4
Chord Length (m)	N/A	810.4	N/A	N/A	1021.7	1064.9	43.2	4.2

the proposed procedure. It is worth noting that as-built information was only available for Highway 32. Similarly, for the Civil 3D segments, Table 1 lists the actual attributes used to draw the curves in Civil 3D and the information obtained from the proposed procedure. For all three curves, the tables also show the absolute difference in the two measurements (actual versus experimental) for each of the attributes. Moreover, the percentage difference is also computed and shown in the tables.

For the actual detection, the results show that the algorithm was successful in automatically detecting all the curves on the four test segments. In fact, the algorithm was also able to detect a second curve toward the end of the Highway 36 segment; however, the attributes of this curve were not measured (not listed in Table 2) since only a portion of the curve was captured on the analyzed LiDAR segment.

The results also reflect the high accuracy achieved when estimating curve attributes using the proposed algorithm. For the three curves for which reference measurements were available, the average percentage difference between measurements obtained from the code and other measurements was <2.25% for all attributes.

For Simulated Curve I, the percentage difference between the measured and the actual attributes of the curve range from 0.33% to 1.2% for each of the attributes. For Simulated Curve II, the differences range from 0.92% to 2.1%. It is worth noting here that the values obtained from the algorithm and recorded in Table 1 are for the curve including spiral curve transitions (i.e., they are not attributes of the circular portion of the curve).

The result accuracy was also high for the LiDAR highway where as-built data was available. On Highway 32, the percentage difference between the results obtained using the proposed algorithm and the information read off the as-built drawings ranged between 0.7% and 4.6%. The highest difference of 4.6% (54 m) was recorded when estimating the radius of the curve. In contrast, the smallest difference (0.7%) was for the arc length 8.27 m and the deflection angle of the curve 0.34°. Recall that a 20 m position vector spacing was used, therefore a difference of 54 m could be a matter of the code being inaccurate by one or two points when locating the PC and PT of the curve.

In the case of Highway 36, although no as-built data were available to assess the accuracy of the results, further analysis on the curve data revealed that the curve was not symmetric and that a spiral transition existed on one end of the curve. The code was able to detect this difference when estimating the tangent (T) of the curve. Two different estimates were obtained either side of the curve (on all the other curves the estimates either side were very close). To verify the existence of a spiral transition on one end of the curve, the azimuth change diagram shown in Figure 6 was plotted. It is noted on the figure that at one end of the curve (2200 m) the change in the azimuth is linear, whereas on the other end it is nonlinear (2950 m). This proves the existence of a spiral transition at the nonlinear end and not the other.

As demonstrated by the results, the proposed algorithm is highly accurate in detecting simple curves and estimating their attributes; however, its ability to distinguish spiral transitions from circular curves depends on the spacing used to estimate the vectors in the initial stage. As already mentioned, testing was conducted on segments in which existing curves had a very large radius (i.e., curves were not very sharp). In order for the algorithm to be able to detect curves of very large radius, sensitivity analysis revealed that a position vector spacing of 20 m had to be used. This ensures that the change in azimuth

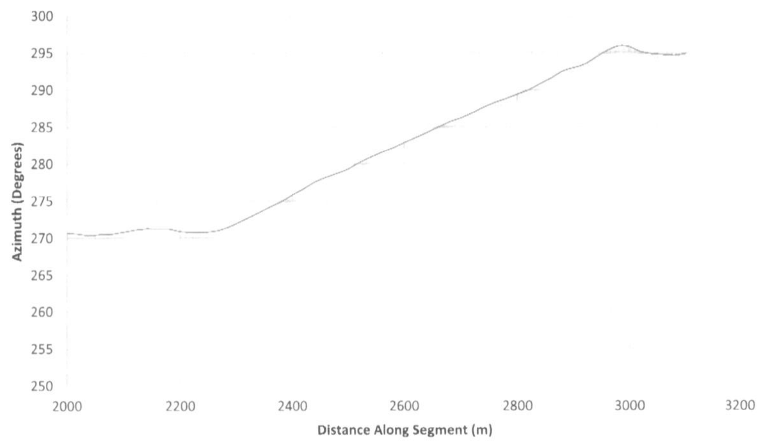

Figure 6. Change in Azimuth along Highway 36 (no change denotes straight segment; linear change denotes curved segment; nonlinear change denotes spiral transition).

between consecutive vectors is sharp enough for the curve to be detected. While this is essential to the detection process, it does not help when the aim is to detect spiral transitions. Since spiral transitions are often relatively short compared to the total length of the curve's arc, a lower point spacing is required to distinguish the radius of the circular portion of a curve from the radius of the curve including the spirals.

Conclusions and Future Research

This paper proposes an algorithm that can be used to automatically detect and measure the attributes of horizontal curves on highways that are remotely sensed using LiDAR technology. The proposed algorithm involves assessing changes in the azimuth between consecutive trajectory vectors aligned to the road's axis to detect the presence of horizontal curves on a highway segment. The PC, PT, and PI for each curve are then identified using regression analysis and the origin of the curve is located to measure its radius. The code also measures the length of the curve and the length of its chord. The algorithm was tested using LiDAR data collected on two highway segments in the Province of Alberta, as well as two simulated curves developed in Civil 3D. Testing revealed that the code was successful in detecting all curves on a highway segment. Moreover, the attributes of those curves were estimated with a high degree of accuracy. The proposed algorithm can be used to survey curves in a more efficient manner. This helps reduce the burden associated with conventional surveying tools. The applications of the developed algorithm range from providing connected vehicle applications and autonomous vehicles with real-time information on curve geometry to using the extracted information in speed-prediction models and horizontal-offset assessments to improve safety on curves. Although the algorithm is effective in estimating attributes of simple horizontal curves, more work is required to distinguish simple curves from curves with spiral transitions. More work is also required to assess the performance of the algorithm on reverse curves. Future work could also consider supplementing the algorithm with a portion that would estimate horizontal sight offset on existing curves.

Acknowledgments

The authors would like to thank Alberta Transportation for sponsoring and providing data for this study. Acknowledgment is also extended to Alberta Innovates and Alberta Advanced Education.

References

1. Federal Highway Administration. *RDs and Curves 2005–2008*. FHWA, FHWA Office of Safety, Washington, D.C., 2010.
2. Easa, S. M. Simple Numerical Method for Solving Horizontal Circular Curves. *Journal of Surveying Engineering*, Vol. 120, No. 1, 1994, pp. 44–48.
3. Rasdorf, W., D. J. Findley, C. V. Zegeer, C. A. Sundstrom, and J. E. Hummer. Evaluation of GIS Applications for Horizontal Curve Data Collection. *Journal of Computing in Civil Engineering*, Vol. 26, No. 2, 2011, pp. 191–203.
4. Ai, C., and Y. C. J. Tsai. Critical Assessment of an Enhanced Traffic Sign Detection Method Using Mobile LiDAR and INS Technologies. *Journal of Transportation Engineering*, Vol. 141, No. 5, 2014, p. 04014096.
5. Gargoum, S., K. El-Basyouny, J. Sabbagh, and K. Froese. Automated Highway Sign Extraction Using LiDAR Data. *Transportation Research Record: Journal of the Transportation Research Board*, 2017. 2643: 1–8.
6. Riveiro, B., L. Díaz-Vilariño, B. Conde-Carnero, M. Soilán, and P. Arias. Automatic Segmentation and Shape-Based Classification of Retro-Reflective Traffic Signs from Mobile LiDAR Data. *IEEE Journal of Selected Topics in Applied Earth Observations and Remote Sensing*, Vol. 9, No. 1, 2016, pp. 295–303.

7. Soilán, M., B. Riveiro, J. Martínez-Sánchez, and P. Arias. Traffic Sign Detection in MLS Acquired Point Clouds for Geometric and Image-Based Semantic Inventory. *ISPRS Journal of Photogrammetry and Remote Sensing*, Vol. 114, 2016, pp. 92–101.
8. Gargoum, S., K. El-Basyouny, and J. Sabbagh. Assessing Stopping and Passing Sight Distance on Highways Using Mobile LiDAR Data. *Journal of Computing in Civil Engineering*, 2018.
9. Castro, M., J. A. Anta, L. Iglesias, and J. A. Sánchez. GIS-Based System for Sight Distance Analysis of Highways. *Journal of Computing in Civil Engineering*, Vol. 28, No. 3, 2013, p. 04014005.
10. Tsai, Y., C. Ai, Z. Wang, and E. Pitts. Mobile Cross-Slope Measurement Method Using LiDAR Technology. *Transportation Research Record: Journal of the Transportation Research Board*, 2013. 2367: 53–59.
11. Findley, D. J., C. V. Zegeer, C. A. Sundstrom, J. E. Hummer, W. Rasdorf, and T. J. Fowler. Finding and Measuring Horizontal Curves in a Large Highway Network: A GIS Approach. *Public Works Management & Policy*, Vol. 17, No. 2, 2012, pp. 189–211.
12. Andrášik, R., and M. Bíl. Efficient Road Geometry Identification from Digital Vector Data. *Journal of Geographical Systems*, Vol. 18, No. 3, 2016, pp. 249–264.
13. Xu, H., and D. Wei. Improved Identification and Calculation of Horizontal Curves with Geographic Information System Road Layers. *Transportation Research Record: Journal of the Transportation Research Board*, 2016. 2595: 50–58.
14. Li, Z., M. Chitturi, A. Bill, and D. Noyce. Automated Identification and Extraction of Horizontal Curve Information from Geographic Information System Roadway Maps. *Transportation Research Record: Journal of the Transportation Research Board*, 2012. 2291: 80–92.
15. Ai, C., and Y. Tsai. Automatic Horizontal Curve Identification and Measurement Method Using GPS Data. *Journal of Transportation Engineering*, Vol. 141, No. 2, 2014, p. 04014078.
16. Imran, M., Y. Hassan, and D. Patterson. GPS–GIS-Based Procedure for Tracking Vehicle Path on Horizontal Alignments. *Computer-Aided Civil and Infrastructure Engineering*, Vol. 21, No. 5, 2006, pp. 383–394.
17. Holgado-Barco, A., D. González-Aguilera, P. Arias-Sanchez, and J. Martinez-Sanchez. Semiautomatic Extraction of Road Horizontal Alignment from a Mobile LiDAR System. *Computer-Aided Civil and Infrastructure Engineering*, Vol. 30, No. 3, 2015, pp. 217–228.
18. RIEGL. *RIEGL VMX-450 Data Sheet*. RIEGL Laser Measurement Systems. http://www.riegl.com/uploads/tx_pxpriegldownloads/DataSheet_VMX-450_2015-03-19.pdf. Accessed April 2017.
19. Steel, P., D. Mesher, and J. Adamson. Development of a Road Safety Audit Network Screening Tool. In *Twenty-Fourth Canadian Multidisciplinary Road Safety Conference*, Vancouver, 1-4 June 2014.

The Standing Committee on Geospatial Data Acquisition Technologies (AFB80) peer-reviewed this paper (18-04619).

Article

TRR
JOURNAL OF THE TRANSPORTATION RESEARCH BOARD

Transportation Research Record
2018, Vol. 2672(39) 107–117

DOI: 10.1177/0361198118783170
journals.sagepub.com/home/trr
SAGE

CFD-DEM Two-Way Coupled Numerical Simulation of Bridge Local Scour Behavior under Clear-Water Conditions

Junhong Li[1] and Junliang Tao[1]

Abstract

Bridge local scour involves extremely complex interactions among the flow, the sediments, and the pier. Our understanding of the underlying mechanism, especially at the microscale, is far from complete. The newly emerged multiphase numerical simulation overcomes the technical limitations imposed on physical experiments. To advance knowledge of the sediment dynamics under turbulent junction flows, this paper employed a CFD-DEM two-way coupled numerical model to simulate the local scour behavior of uniform spherical mass particles around an oblong pier. In this coupled system, the flow phase was described as a continuum by a detached eddy simulation (DES) model, and motion of the discrete particles was governed by Newton's laws of motion (DEM). The flow–sediment interactions were directly considered through the momentum exchange between the two phases. The two-way coupled model was shown to successfully capture microscopic sediment dynamics similar to those observed in the literature. Sediment grains in the bed-load layer were shown to be eroded in the form of sliding and saltation. The erosional forces acting on two representative target particles, together with their correspondent motions/trajectories, provide valuable insights into the mechanism that dictates the scour initiation.

Bridge local scour is a particular form of surface erosion that occurs in the vicinity of bridge foundations (i.e., bridge piers or abutments) as a result of the impact of hydrodynamic forces. It involves extremely complex interactions between the water flow, the sediments, and the in-stream structure. The problem of bridge local scour has been widely studied during the past few decades because of the complexity and significance of this problem. However, an understanding of the local scour mechanism is still far from complete, despite the tremendous efforts being made on the topic. This may be partially attributed to the lack of high-quality experimental data on bridge scour, owing to the technical difficulties in making reliable measurements at the microscale. Most of the previous experimental studies have concentrated on investigating the scour process at the macroscale, such as the evolution of the scour hole and the sediment flux with time. Experimental surveys in the literature regarding the erosion and subsequent transportation mechanism for sediment grains from around bridge foundations are limited because it is very difficult, if not impossible, to adequately obtain direct measurements. Meanwhile, as numerical models are free from the restrictions imposed on a physical experiment, they have received increasing attention from the erosion research community as high-performance computer technology and the developed numerical schemes have advanced tremendously during the past two decades or so.

As a special case of flow-induced erosion problems, bridge scour has been widely investigated using single-phase computational fluid dynamics (CFD) models at different advanced levels (*1–3*). Previous numerical studies in the literature have greatly advanced our knowledge of the physics of turbulent junction flows. However, the value of single-phase models is limited when it comes to studying the microscopic scour mechanism at the particle scale. In single-phase numerical models, only the flow physics is solved, and the sediment scour process is simulated indirectly by incorporating a certain empirical sediment transport model (*4*) into the CFD model. However, existing sediment transport models have certain drawbacks. Firstly, as most, if not all, of the sediment transport formulas are based on the excess shear stress (or velocity) theory, the accuracy of the erosion process prediction using single-phase models is compromised. The excess shear stress theory emphasizes the time-averaged boundary shear stress, and is therefore widely recognized to be incapable of accounting for the sediment dynamics in turbulent flows by itself (*5*). It is generally agreed that wall-generated turbulent bursts (i.e., so-called sweep and ejection events) are primarily responsible for sediment entrainment

[1]Department of Civil Engineering, University of Akron, Akron, OH

Corresponding Author:
Address correspondence to Junliang Tao: jtao2@uakron.edu

(*6–8*). Moreover, most sediment transport models are empirical, or at best semi-empirical, and were derived from experimental studies on sediment entrainment, in which the determination of incipient motion of the particles was subjective, by using either a visually based or a reference-based method. Finally, the single-phase models assume a clear-water condition, which is not always the case in reality. With the clear-water assumption, single-phase numerical models intrinsically fail to account for the flow–sediment interaction, especially in a flow with a high concentration of sediment particles. Therefore, although single-phase models with certain sediment transport models can predict the scour process with acceptable accuracy for many industrial applications, they provide limited value for advancing our current understanding of the scour mechanism.

By comparison, the recently developed two-phase numerical models treat the flow and solids as independent phases and thus avoid the use of sediment transport models. Among the two-phase models, the two-fluid model (TFM) and the discrete particle method (DPM) are the two most popular methods for simulating granular flows. While the TFM considers the flow and solid as two interpenetrating continuous media and solves their interactions in an Eulerian–Eulerian scheme, DPM computes the flow field in an Eulerian framework and directly resolves the dynamics of individual solid particles in a Lagrangian scheme. The advantage of the DPM models over the TFM mainly lies in the capability of the DPM for investigating the behavior of particular bed material grains rather than model the sediment transport based on the concept of concentration.

DPM models broadly fall into two categories: resolved and unresolved (*9*). Resolved DPM, such as the immersed boundary method (IBM), prescribes the fluid boundary conditions as the no-slip boundary of each discrete solid particle. This requires a sophisticated grid technique to treat the flow–solid boundary and extremely high computation cost. On the other hand, the unresolved DPM (referred to as the CFD-DEM coupled model) introduces the solid void fraction term and uses drag closure equations to calculate flow–solid interactions. In previous studies, a DPM coupled model has been demonstrated to predict the flow–solid interactions with considerable accuracy and to outperform the conventional single-phase models in terms of simulation accuracy (*10*). More importantly, a DPM model is capable of providing a more holistic and fundamental understanding of the microscopic erosion mechanism. Since their early development, DPM coupled models (both resolved and unresolved) have been extensively applied to study gas–solid fluidization phenomena (*11–14*), and have been employed to study the underlying physical mechanism of surface erosion problems by some researchers (*15–18*). For example, Ji et al. employed a resolved DPM coupled model (i.e., a combination of direct numerical simulation, the combined finite-discrete element method, and IBM) to investigate the sediment entrainment in a turbulent channel flow, and their study revealed a strong correlation between sediment entrainment and sweep events (*18*). To the best knowledge of the authors, DPM coupled models have been rarely employed to study scour under turbulent junction flows, which is a particular form of bed surface erosion. This is possibly because of the extremely high demand for computational resources and the challenge involved in code implementation (i.e., complex geometry and the necessity to resolve small-scale turbulence). A few studies in the literature have been identified on evaluating bridge local scour using DPM techniques (*19*, *20*). Nevertheless, the Lagrangian models in these studies only considered one-way coupling and involved a small number of sediment particles. One-way coupling assumes that the presence of particles produces negligible local modifications of the flow field. Such an assumption is valid only when sediment particles have small sizes and the particle concentration is low. In addition, the small particle numbers in existing studies (typically only a couple of particle layers) would fail to account for the role that the momentum exchange between surface and subsurface flows plays in the scour mechanism. This aspect is rarely considered in existing bridge scour studies. However, the role of subsurface flow physics is widely recognized in studies of wave-induced erosion around coastal structures. The wave-induced surface pressure variation was sometimes found to generate such a large excess pore pressure within the bed that the submarine sediments were momentarily liquefied and therefore eroded away readily. In a bridge local scour scenario, the vortex-induced surface pressure variation is likewise expected to induce vertical pore pressure gradient within the bed, which will compromise the effective self-weight of the particles. Therefore, the momentum exchange across the bed surface is an important addition to our current knowledge of the scour mechanism and should be accounted for in the model.

As a pioneering study, this paper employed a CFD-DEM two-way coupled numerical model (i.e., an unresolved DPM) to simulate the local scour behavior of uniform spherical mass grains around an oblong pier. Two-way coupled simulation is particularly difficult to implement because it involves correct modeling of both the turbulence structure and the sediment dynamics, as well as taking into account the effect of flow on particle motion and the local modification of the flow field induced by the presence of a particle. Interactions between particles (i.e., collisions and frictional sliding) should also be considered in the numerical scheme. In this coupled system, the continuum-scale flow was solved by the detached eddy simulation (DES) approach, whereas the disperse particles were treated as a collection of discrete particles, the motion of which was governed by Newton's laws of motion (DEM). The flow–sediment interactions were directly considered through the momentum exchange between the two phases. The sediment bed depth was over 12 times the particle diameter so that the effect of the surface–subsurface mass and momentum exchange on scouring was taken into account. The sediment dynamics in the bed-load, as well as evolution of the vortex structure, will be studied at the microscale. Time history of the hydrodynamic forces exerted on representative particles, together with their correspondent coordinates will be presented to shed light on the mechanism that dictates the scour initiation.

Numerical Methods

In the coupled numerical model of this study, the fluid flow was described as a continuum phase in Eulerian frame and the disperse particles were treated as a collection of discrete particles with Lagrangian formulation. In an Eulerian specification, positions for the meshed flow domain were deterministic and the flow variables were studied at fixed points. In a Lagrangian specification, on the other hand, movement of an individual solid particle was "tracked" through space and time, and the computed variables (e.g., momentum, forces, etc.) were associated with each particle. The continuous flow was strongly coupled with the discrete solid particles through mass, momentum, and energy exchanges. The formulations that governed the behavior of fluid and solid particles in the coupled numerical model are briefly described below. The continuous fluid phase was modeled by solving the modified Navier–Stokes (NS) governing equations for incompressible fluid by including the effect of particles. The modified NS equations included a mass balance equation and a momentum conservation equation, as shown in Equations 1 and 2.

$$\frac{\partial(\alpha_f \rho_f)}{\partial t} + \nabla \cdot (\alpha_f \rho_f \mathbf{v}_f) = 0 \tag{1}$$

In the mass balance equation above, α_f is the local volume fraction of the fluid, representing the effects from particles; ρ_f is the fluid density; $\mathbf{v}_f$ is the fluid velocity and t is the time. Similarly, the momentum conservation equation is written as Equation 2:

$$\begin{aligned}\frac{\partial(\alpha_f \rho_f \mathbf{v}_f)}{\partial t} + \nabla \cdot (\alpha_f \rho_f \mathbf{v}_f \cdot \mathbf{v}_f) = \\ -\alpha_f \nabla p + \nabla \cdot (\alpha_f \tau_f) + \alpha_f \rho_f \mathbf{g} - \mathbf{f}_p\end{aligned} \tag{2}$$

where p is the hydrodynamic pressure, $\boldsymbol{\tau}_f$ is the fluid stress, $\mathbf{g}$ is the gravitational acceleration, and $\mathbf{f}_p$ represents the momentum transfer between the fluid and the particles.

In this study, the flow field was solved by the Sparlart–Allmaras (SA)-based DES model (*21*), which treated the boundary layers using the Unsteady Reynolds-averaged Navier–Stokes (URANS) model with an SA closure and resolved the outer detached eddies in the remaining domain using the large eddy simulation (LES) model of Smagorinsky, 1963 (*22*). The transition from URANS to LES was controlled by a modified distance function: $\tilde{d} \equiv \min(d, C_{DES}\Delta)$, where d is the URANS turbulent length scale; C_{DES} is a constant having an empirical value of 0.65; and Δ is the largest dimension of the grid cell, where $\Delta \equiv \max(\Delta x, \Delta y, \Delta z)$. This modified distance function caused the DES model to behave as a URANS SA model when $d \ll \Delta$ and behave as a Smagorinsky LES model when $\Delta \ll d$. Because of space limitations, the closure methods for the DES numerical model are not discussed here. Please refer to a previous publication of the authors for more details (*23*).

The translational and rotational motions of particle i were governed by Newton's second law of motion, as shown in Equations 3 and 4, respectively.

$$m_{p,i} \frac{\partial \mathbf{v}_{p,i}}{\partial t} = -V_i \nabla p + \sum_{j=1}^{n_i^c} \mathbf{f}_{ij}^c + \mathbf{f}_i^f + m_{p,i}\mathbf{g} \tag{3}$$

In the equation above, $m_{p,i}$ is the mass of particle i, $\mathbf{v}_{p,i}$ denotes the translational velocity of particle i, $-V_i \nabla p$ is pressure gradient force, V_i is volume of the particle, and $\mathbf{f}_{ij}^c$ is the contact force acting on particle i by particle j or the boundary wall. The particle–particle or particle–wall contact force was computed by the Hertz–Mindlin contact model (*24*), which is a nonlinear model taking into account the elastic deformation of spherical solids and the changing contact radius. The term n_i^c is the number of contacts for particle i, and $\mathbf{f}_i^f$ represents the fluid–particle interaction force acting on particle i. The fluid–particle interaction forces mainly include drag force as well as buoyance force, and shear lift force was neglected in this model. In this study, the Gidaspow (*25*) drag model was employed, which has been demonstrated to result in good agreement with experimental observations (*26*).

$$I_{p,i} \frac{\partial \boldsymbol{\omega}_{p,i}}{\partial t} = \sum_{j=1}^{n_i^c} M_{ij} + M_{f,i} \tag{4}$$

In Equation 4, $I_{p,i}$ is the moment of inertia for particle i, $\omega_{p,i}$ denotes the angular velocity of particle i, and M_{ij} is the torque acting on particle i by particle j or the boundary wall. The term $M_{f,i}$ represents the torque acting on particle i by the continuous flow phase. In this work, a rolling friction coefficient of 0.001 was applied to the Hertz–Mindlin contact model to account for the rolling resistance.

Test Conditions and Model Setup

Computational experiments in this study were carried out in a rectangular channel with a length and width of 1,240 mm and 240 mm, respectively. The clear-water depth (H) of the open channel flow was 50 mm. An oblong-shaped pier consisting of two half-circles and one rectangle was mounted on the fixed bed surface at a distance of 700 mm downstream of the inlet boundary, where the turbulence was fully developed. The pier width was 20 mm and the aspect ratio of pier cross-section was 1:4. The long axis of the pier was aligned with the flow direction. A recess was designed around the

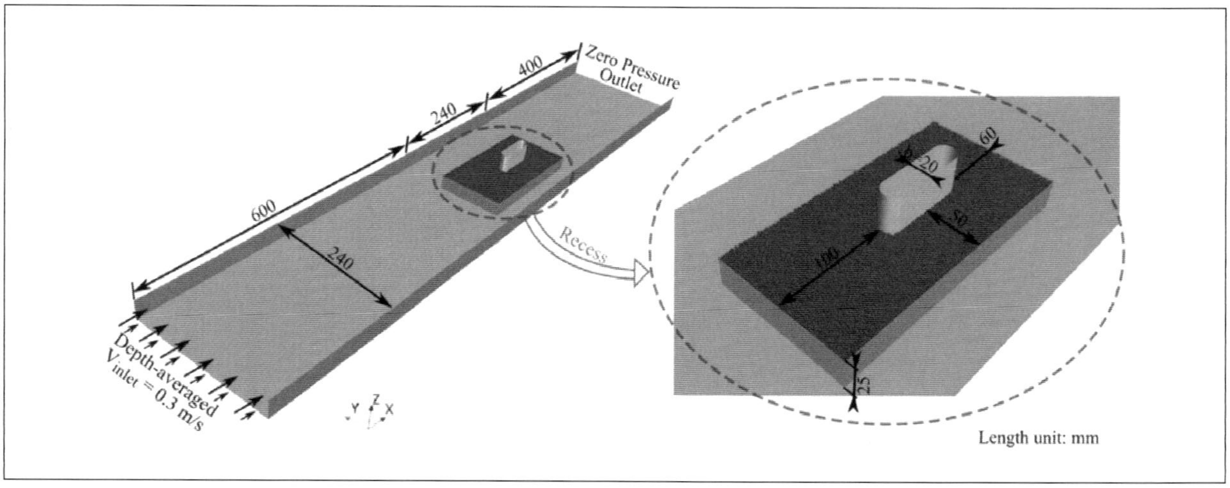

Figure 1. Illustration of test conditions.

pier to retain inertial sediment grains. All spherical particles had a uniform diameter (*d*) of 2 mm and a density of 2,650 kg/m^3. The initial sediment top surface was leveled with the fixed planer bed beyond the recess region to ensure that water entered the sediment bed smoothly, and also to minimize the number of particles needed in the coupled model. As shown in Figure 1, the recess was extended to 5 times the pier width (b = 20 mm) upstream of the pier's leading nose, 3 times downstream of the pier's wake nose and 2.5 times in the transverse direction away from the pier sidewalls; this was done to minimize the recess wall effects. The velocity profile at the inlet boundary followed the log-wake law and entered the granular bed smoothly. The depth-averaged inlet velocity was 0.3 m/s to ensure a clear-water condition, which indicated that only sediment grains in the vicinity of the pier would be entrained by the intensified turbulence and accelerated flows. The outlet boundary was set at zero pressure, and the flume top surface was set as a slip surface. The remaining boundaries of the flow domain were modeled as nonslip walls. The Reynolds number was 6,000 according to the above model setup.

In the coupled system, there is a requirement on the DEM time step in order to obtain valid results. In DEM modeling, movement of a particle is affected not only by its immediate neighboring particles but also by particles far away from it through the propagation of disturbance waves. Basically, the DEM time-step size should be sufficiently small to enable the contacts to last for several time steps (e.g., 10 time steps) to avoid problems on energy conservation owing to numerical integration (*27*). In order to avoid the disturbance from particles outside its immediate neighborhood during a single time step, the DEM time step is typically determined as a minimum of three time scales: (1) 20% of the time that it takes the Rayleigh wave to propagate across the particle surface to the opposite pole (*28*); (2) 10% of the duration of impact between two particles, assuming Hertz contact theory given by the Timoshenko equation (*29*), in order to resolve particle collisions correctly; and (3) 10% of the time that it takes for the particle to move the full length of the particle radius in order to prevent missing contacts between DEM particles. Based on these time-step-limiting criteria, the DEM time step was set at 1E-5 sec in this study. Ideally, the CFD time step should be equal to the one used in the DEM in order to exchange the interaction forces immediately after each step of calculation. However, doing so requires excessive computational effort. In this study, the CFD time step was set at 1E-4 sec to reduce the computational cost. The flow and particles were coupled at the end of each CFD time step.

In a two-way coupled model, the flow domain generally needs to be discretized with a grid size larger than the particle diameter. This step is taken to avoid having the whole cell volume occupied by a single particle and thus reporting unrealistic source values back to the continuous phase. At the same time, however, the flow grids need to be fine enough to capture the small-scale dynamics of the turbulent juction flow with high accuracy. Such contradictory requirements posed by the resolution of the small-scale turbulence and the important geometric features versus the necessity to maintain smoothless in particle solid fraction distribution often results in inaccuracy or instability of the solution. To overcome this challege, the software STAR-CCM+ was used in this study, as it provides a two-grid procedure for two-way coupling that clusters groups of contiguous original fine cells together to create a second coarser grid (*30*). With the two-grid method, the coupled model was used to calculate the flow–sediment interactions using the second coarser cells (about 2.5 times the particle diameter). Then the solution from the interaction calculation was distributed evenly across the component fine cells by volume fraction contribution,

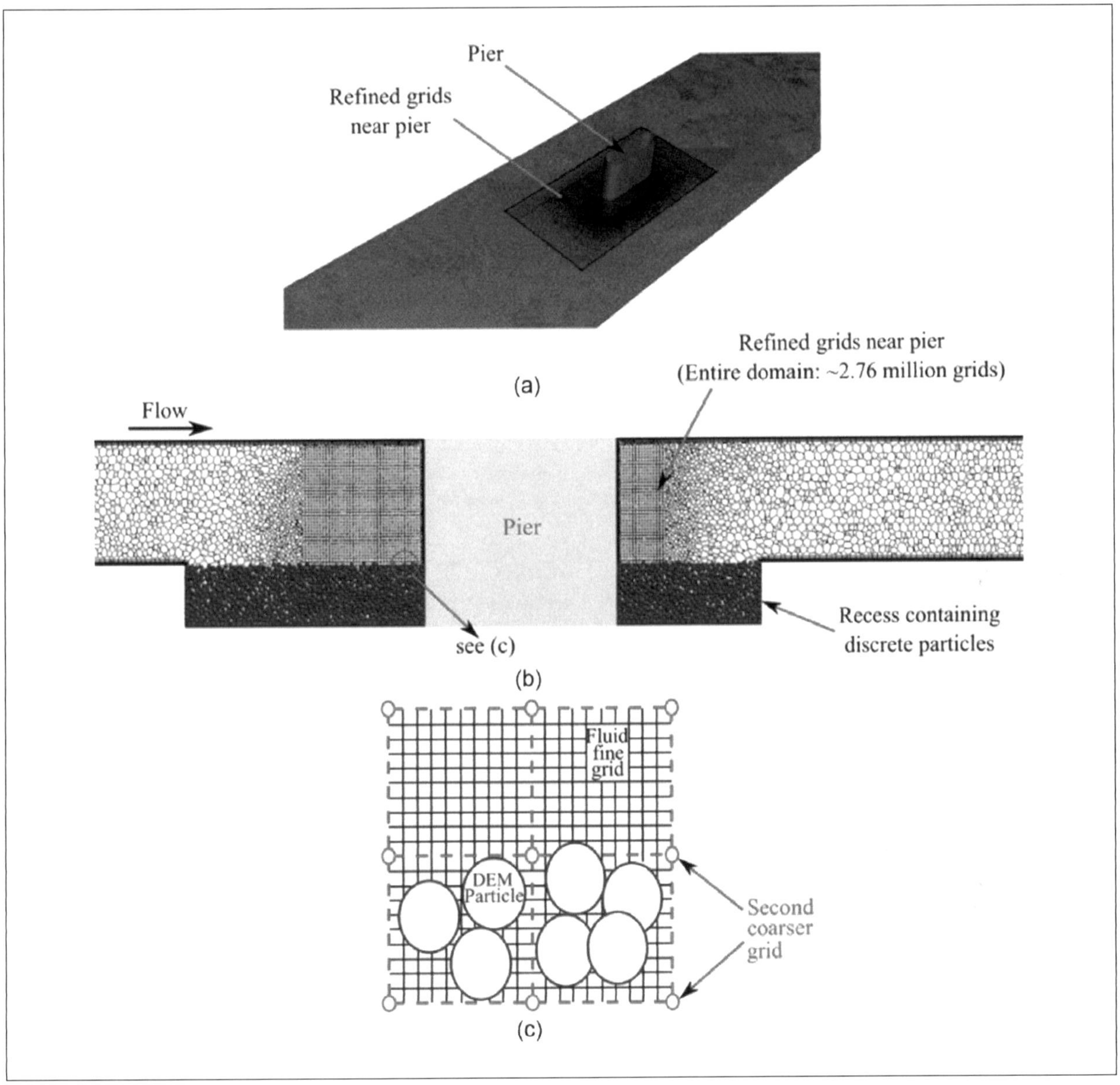

Figure 2. (*a*) 3D visualization of flow domain mesh; (*b*) flow mesh and discrete particles at the symmetry plane; (*c*) illustration of the two-grid method.

momentum source terms, and other transferred quantities. Details of the flow physics were thus resolved based on the fine meshes, which are independent of particle size. This two-grid method concept is illustrated in Figure 2(*c*). In the simulation case for this study, the total number of grids in the flow domain was about 2.76 million, and over 70% of all computational grids were concentrated in the area of interest (i.e., in the vicinity of the pier), as shown in Figure 2, (*a*) and (*b*). The total number of particles in this study was around 90,000, and it took 1 day for 20 parallel cores to simulate a time duration of 2 seconds. The simulation parameters are summarized in Table 1.

Simulation Results and Discussion

In this section results from the computational experiment are presented to elucidate the flow field and particle motion around an oblong pier. According to the Shield's diagram (*31*), the depth-averaged flow velocity was only about 60% of the critical velocity of the uniform particles (d = 2 mm) used in this study. Thus, only grains in the vicinity of the oblong pier were expected to be entrained by the locally rich dynamics of flow. At the start of the simulation, all particles in the flume recess were at repose. Initially, the motion of all particles was frozen deliberately in order to avoid undesired particle motion

Table 1. Summary of Simulation Parameters

Spherical particles	
Number of particles	~90,000
Diameter	2 mm
Density	2,650 Kg/m^3
Poisson's ratio	0.3
Young's modulus	1 Mpa
Particle–particle and particle–wall interaction	
Friction coefficient	0.5
Normal restitution coefficient	0.85
Tangential restitution coefficient	0.3
Rolling friction coefficient	0.001
Fluid	
Density	997 Kg/m^3
Dynamic viscosity	8.90E-4 Pa•s
Computation parameters	
Time step for DEM	1E-5 s
Time step for CFD	1E-4 s

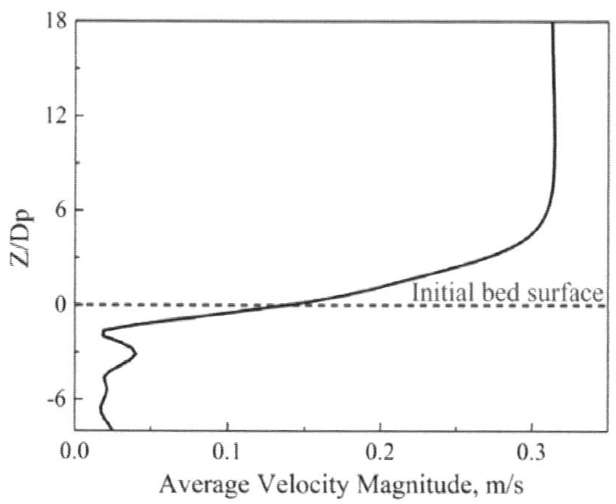

Figure 3. Vertical profile of time-averaged velocity upstream of the horseshoe region.

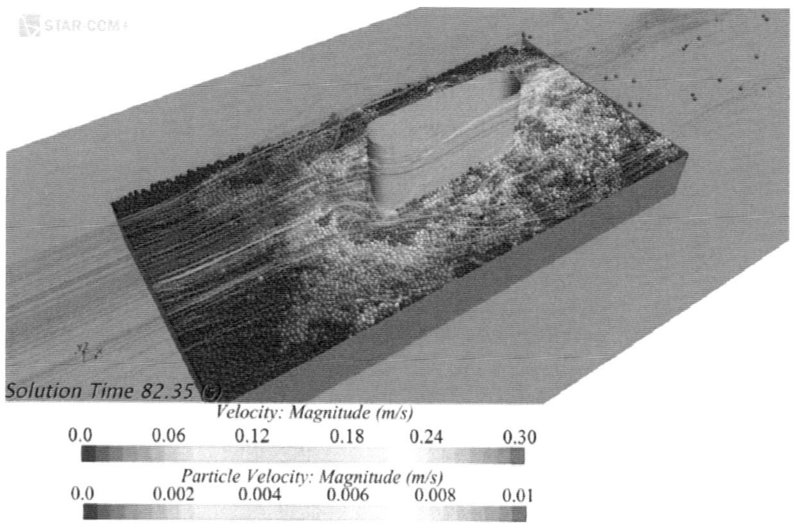

Figure 4. 3D visualization of instantaneous flow streamline and particle transport.

being triggered by computational instability at the initial iterations of the simulation. After a steady initial flow solution was obtained, the particles were allowed to freely interact with the flow. Figure 3 reports the time-averaged velocity profiles in regions farther upstream of the horseshoe zone, where influence from the pier on the flow field was negligible. The brown dashed line in the figure represents the initial top surface of the sediment bed. The *Z* coordinate was normalized by the particle diameter (*Dp*). Figure 3 indicates a flow penetration of about 2 particle diameters into the porous bed, which conforms to the transition layer thickness revealed by existing studies (*32*). Below the transition layer, however, the velocity remained around 0.02 m/sec and did not reduce to the Darcy scale. This may be attributed to the method used in a CFD-DEM coupled model to solve the flow–particle interaction. The resolved DPM model (e.g., lattice Boltzmann method coupled with the DEM model and with the IBM) is likely to perform better in the prediction of the flow field within the porous bed. Fortunately, an inaccuracy at a point so far below the transition layer is expected to have a minimal influence on the simulation of the sediment entrainment and transport.

As expected for a clear-water condition, the depth-averaged inlet velocity (0.3 m/sec) was insufficient to erode the sediment grains at locations far upstream of the pier. Figure 4

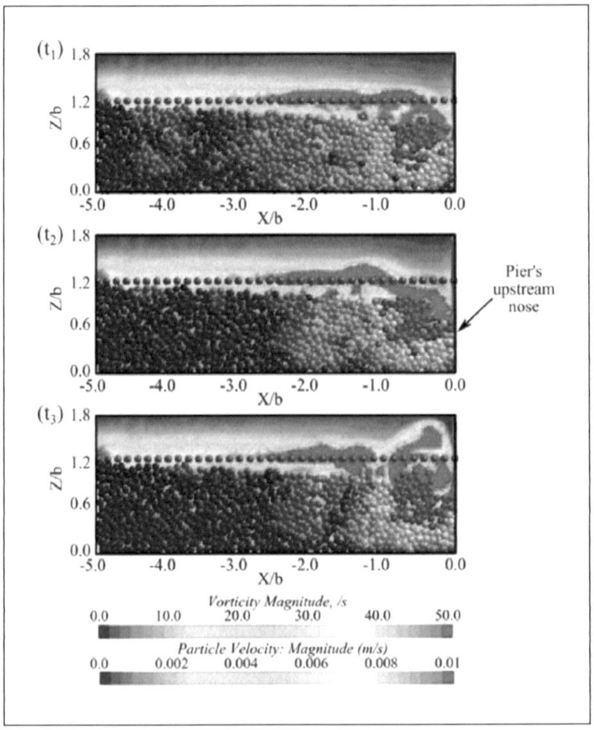

Figure 5. Instantaneous flow vorticity contour and particle motion at symmetry plane.

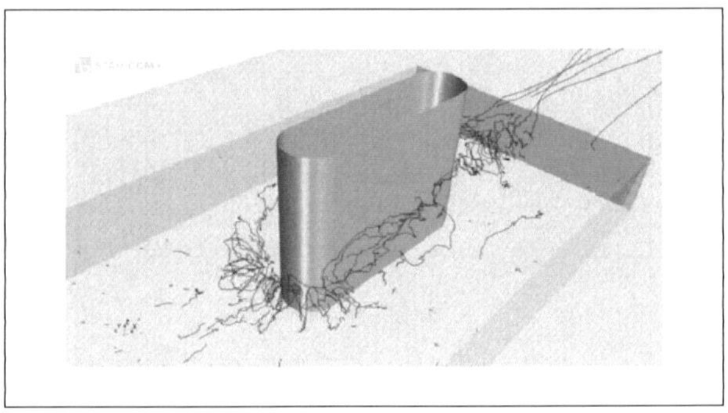

Figure 6. Trajectory of selected particles showing slide and saltation in motion.

shows the instantaneous flow streamline and particle transport around the oblong pier. The red color in the figure indicates larger particle or flow velocities. Only particles within the horseshoe region upstream of the pier and in the wake region were shown in motion. Moreover, the morphology of the "active" particles in the upstream horseshoe region was shown to align with the stretch of the vortical structure, which would bend downstream toward the direction of flow. Particles shown in red in the horseshoe region were either being transported downstream or were being trapped by the near-wall primary vortices.

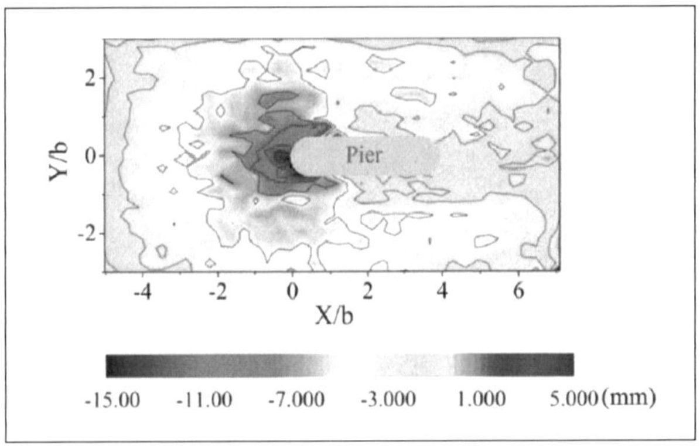

Figure 7. Scoured bed contour at the final stage.

Figure 5 shows a sequence of instantaneous sediment dynamics at the vertical symmetry plane. The background contour map shows the solution of the flow vorticity, while the color of the particles indicates the magnitude of the particle velocity. Both the X axis and Z axis were normalized with the pier width (b). A shallow scour hole was already formed upstream of the pier at the instants before those shown in Figure 5. The primary vortex was shown to be intensified within the scour hole and tended to change continuously as the scour hole gradually developed. As can be seen from the motion sequence in Figure 5, the vortical structure gradually evolved into the scour hole and brought particles up to the surface. At time t_3 in Figure 5, certain particles were abruptly ejected into a suspension state by the horseshoe vortex instabilities. Those suspended grains in the bed-load layer were either readily eroded away or remained within the scour hole under the vortex constraint. Figure 5 also shows that flow separation near the wall boundary occurred at about 2.5 times the pier diameter upstream of the pier, which is close to the observation in the experimental study by Dargahi (*33*). From previous studies on the bed erosion mechanism, wall-generated turbulent bursts (i.e., ejection and sweep events) are generally recognized as the primary reason for the entrainment and transport of sediment particles (*8*, *18*, *34*, *35*). In the bridge local scour scenario, however, the primary reason for sediment entrainment is different. Closer to the pier surface, the pressure greatly increases as a result of obstruction of the pier to the channel flow. Thus, the pressure gradients as well as the vortex structure are expected to play crucial roles in the scour mechanism.

As the flow condition was well below the threshold of motion for the particles in this study, all the particle motion was induced by the instantaneous hydrodynamic forces. According to studies in the literature (*36*, *37*), two well-known modes of sediment transport have been observed under clear-water conditions, namely sliding and saltation. The essential characteristics of these two transport modes are illustrated in Figure 6, which presents representative trajectories of some

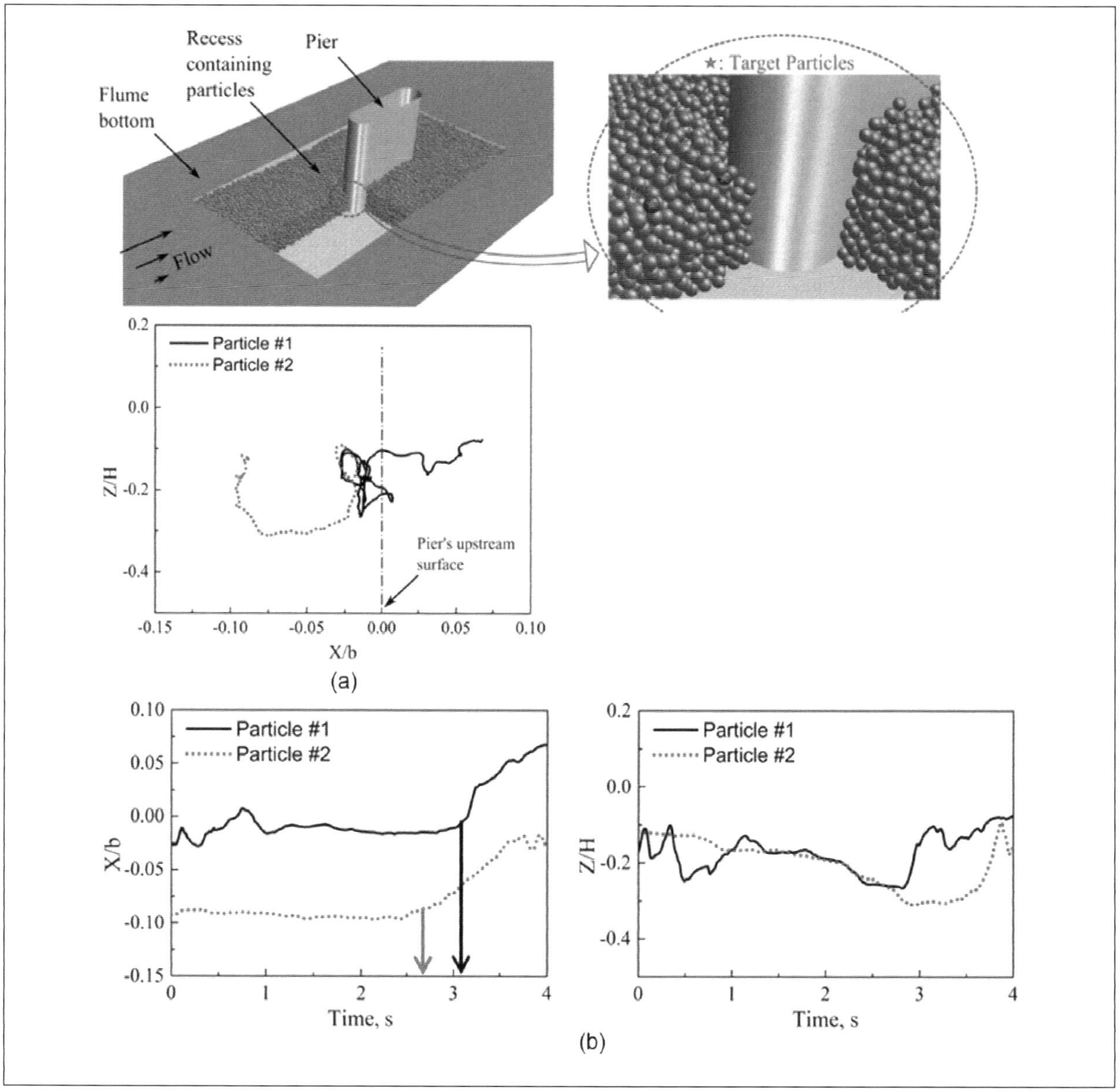

Figure 8. (*a*) Trajectory on the *XZ* plane; (*b*) time histories of coordinates for two representative particles.

selected particles. Grains were shown to be trapped by the horseshoe vortex upstream of the pier for most of the time. Once the grains were beyond the constraint of the vortical structure, however, they were quickly transported downstream along the legs of the horseshoe vortex. All eroded particles were transported as a bed-load and had intermittent contact with the bed surface (i.e., saltation). In the wake region, the particles in motion seemed to first become trapped in the vertical vortex cores, which had the maximum vorticity and minimum pressure, before they were transported farther downstream.

The two-way coupled model in this study was not designed to simulate the scour process at the pier. The aim of the computational experiment was to investigate the sediment dynamics and the mechanism governing the scour initiation. Nonetheless, a contour of the scoured bed at the equilibrium state is provided in Figure 7. When plotting the contour map, any DEM particles in a suspension state were assumed to be eroded away at a later time. As expected, Figure 7 shows that the scour hole was primarily developed within the horseshoe region upstream of the pier under clear-water conditions. The maximum scour depth was about 19

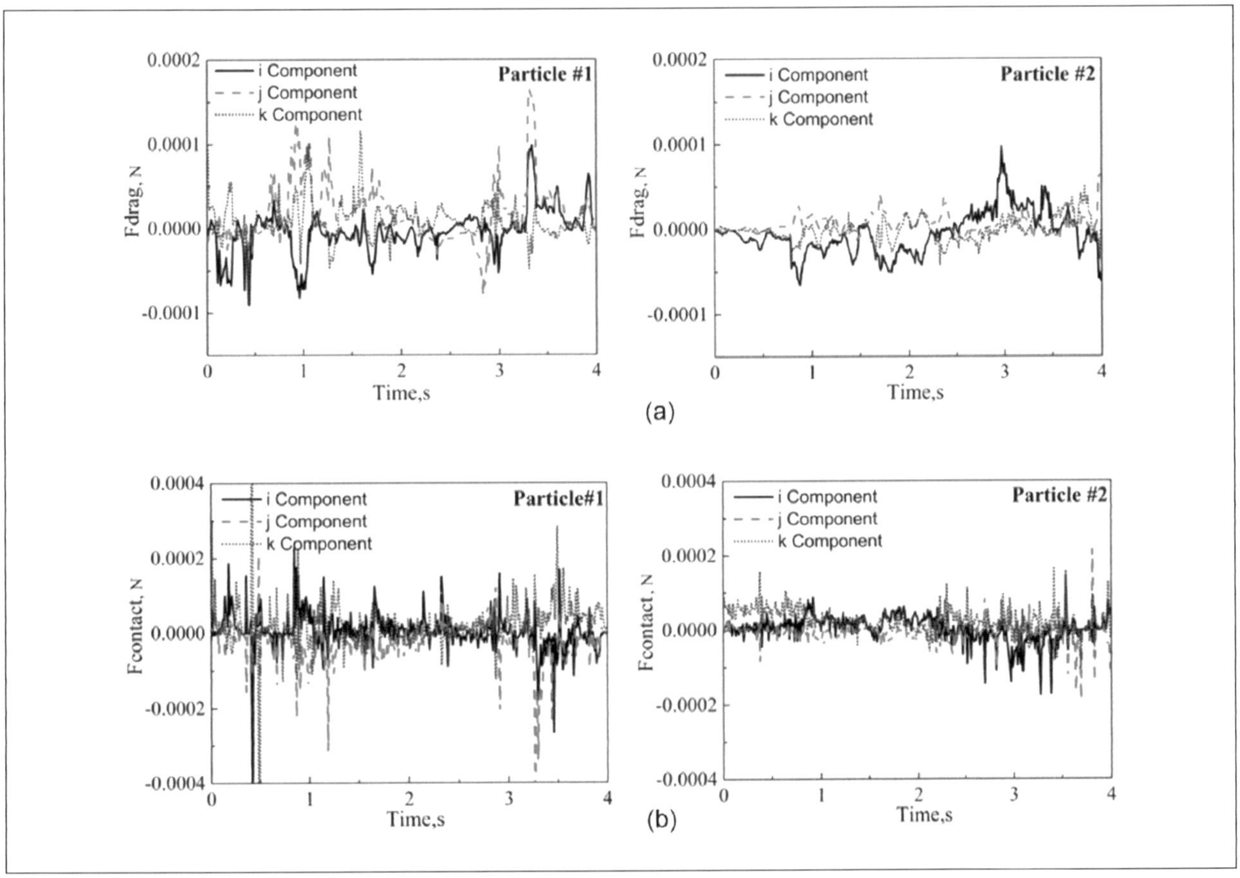

Figure 9. Time histories of (*a*) drag force components and (*b*) contact force components acting on two representative particles.

mm for the simulation case in this study, which was close to the pier width ($b = 20$ mm). This is consistent with the empirical prediction of the popular Melville–Chiew scour equation for narrow piers (*38*).

In order to further reveal the mechanism of scour initiation, time histories of the hydrodynamic forces exerted on two representative particles (location marked at the top of Figure 8) were studied with their correspondent coordinates to reveal the underlying physical mechanism that dictates the scour initiation. The results are presented in Figures 8 and 9. At the start of the time series, Particle 1 was already constrained within the primary vortex for some time; Particle 2 was at rest at the peripheral edge of the evolving scour hole. Figure 8(*a*) shows the trajectories of the two particles on the *XZ* plane. It was observed that Particle 1 was confined within the horseshoe vortex core just upstream of the pier's leading face for a certain time before being eroded downstream along the pier sides; Particle 2 gradually slid into the scour hole and was constrained within the scour hole at the end of the time series reported. The time at entrainment for both particles can be determined from Figure 8(*b*) as around 3.1 s and 2.7 s, respectively. Figure 9 indicates the time histories of three drag and contact force components for both particles. From Figure 9(*a*), a sudden increase in the drag force component (the *i* component, represented by a black line) was observed for both particles around the time of entrainment. It is also interesting to observe from Figure 9(*b*) that after the time of entrainment for Particle 1, the contact force component in the vertical upward direction (represented by a blue line) displayed a violent fluctuation. This indicated that Particle 1 was transported downstream with a mode of saltation. Furthermore, a previous numerical study conducted by Ji et al. (*16*) has revealed a strong correlation between the suddenly decreasing streamwise velocity and the increasing vertical velocity at collision, which indicates the saltation of particles during the transport process.

Summary and Conclusions

This study carried out a pioneering computational experiment of the local scour behavior at an oblong pier under clear-water conditions using a CFD-DEM two-way coupled numerical model. Results from the simulation test showed that the two-phase model has great potential in simulating the scour process. Under clear-water conditions, only sediment grains in conditions near the threshold of motion were eroded by the intensified turbulence in the vicinity of the pier. Scour was initiated upstream of the pier by the dynamic horseshoe vortex, where the maximum scour depth occurred. Unlike a bed erosion scenario in which wall-generated turbulent bursts (i.e., ejection and sweep events) are generally recognized as the primary reason for entrainment, the pressure gradients and primary horseshoe vortex are expected to play a crucial role in the mechanism of bridge local scour. During the scour process, the primary vortex brought grains on the surface layer into a state of suspension; some grains were then readily eroded away along the stretch of horseshoe vortex, whereas others were confined within the scour hole under the vortex constraint. Simulation results showed that the coupled model was also able to capture the microscopic erosion behavior of grains at the bed–flow interface, which is of great value in revealing the microscopic scour mechanism. Trajectories of selected particles revealed that sediment grains were eroded in the form of sliding and saltation. Time histories of the hydrodynamic forces exerted on two representative particles were studied with their correspondent coordinates, and the results reveal the mechanism that dictates the scour initiation. A correlation was observed between the suddenly increased drag force component (the *i* component) and the particle entrainment. As the particles were transported downstream, an increased vertical contact force was observed as particles collided with packed bed particles at a certain incident angle.

Author Contributions

The authors confirm contribution to the paper as follows: study conception and design: Junhong Li, Junliang Tao; data collection: Junhong Li; analysis and interpretation of results: Junhong Li, Junliang Tao; draft manuscript preparation: Junhong Li, Junliang Tao. All authors reviewed the results and approved the final version of the manuscript.

References

1. Roulund, A., B. M. Sumer, J. Fredsøe, and J. Michelsen. Numerical and Experimental Investigation of Flow and Scour Around a Circular Pile. *Journal of Fluid Mechanics*, Vol. 534, 2005, pp. 351–401.
2. Richardson, J., and V. Panchang. Three-Dimensional Simulation of Scour-Inducing Flow at Bridge Piers. *Journal of Hydraulic Engineering*, Vol. 124, No. 5, 1998, pp. 530–540.
3. Paik, J., C. Escauriaza, and F. Sotiropoulos. On the Bimodal Dynamics of the Turbulent Horseshoe Vortex System in a Wing-body Junction. *Physics of Fluids (1994-present)*, Vol. 19, No. 4, 2007, pp. -.
4. Engelund, F., and J. Fredsøe. A Sediment Transport Model for Straight Alluvial Channels. *Hydrology Research*, Vol. 7, No. 5, 1976, pp. 293–306.
5. Diplas, P., C. L. Dancey, A. O. Celik, M. Valyrakis, K. Greer, and T. Akar. The Role of Impulse on the Initiation of Particle Movement Under Turbulent Flow Conditions. *Science*, Vol. 322, No. 5902, 2008, pp. 717–720.
6. Jackson, R. G. Sedimentological and Fluid-dynamic Implications of the Turbulent Bursting Phenomenon in Geophysical Flows. *Journal of Fluid Mechanics*, Vol. 77, No. 3, 1976, pp. 531–560.
7. Sumer, B. M., and R. Deigaard. Particle Motions Near the Bottom in Turbulent Flow in An Open Channel. Part 2. *Journal of Fluid Mechanics*, Vol. 109, 1981, pp. 311–337.
8. Dwivedi, A., B. Melville, and A. Y. Shamseldin. Hydrodynamic Forces Generated on a Spherical Sediment Particle During Entrainment. *Journal of Hydraulic Engineering*, Vol. 136, No. 10, 2010, pp. 756–769.
9. Deb, S., and D. K. Tafti. A Novel Two-grid Formulation for Fluid–particle Systems Using the Discrete Element Method. *Powder Technology*, Vol. 246, 2013, pp. 601–616.
10. Cheng, L., and M. Y. Shen. A Three Dimensional k-ε-Ap Model for Water-sediment Movement. *International Journal of Sediment Research*, Vol. 25, No. 1, 2010, pp. 17–27.
11. Hoomans, B., J. Kuipers, W. Briels, and W. Van Swaaij. Discrete Particle Simulation of Bubble and Slug Formation in a Two-dimensional Gas-fluidised Bed: A Hard-sphere Approach. *Chemical Engineering Science*, Vol. 51, No. 1, 1996, pp. 99–118.
12. Xu, B., and A. Yu. Numerical Simulation of the Gas-solid Flow in a Fluidized Bed by Combining Discrete Particle Method with Computational Fluid Dynamics. *Chemical Engineering Science*, Vol. 52, No. 16, 1997, pp. 2785–2809.
13. Tsuji, Y., T. Kawaguchi, and T. Tanaka. Discrete Particle Simulation of Two-dimensional Fluidized Bed. *Powder Technology*, Vol. 77, No. 1, 1993, pp. 79–87.
14. Van der Hoef, M., M. van Sint Annaland, N. Deen, and J. Kuipers. Numerical Simulation of Dense Gas-solid Fluidized Beds: A Multiscale Modeling Strategy. *Annual Review of Fluid Mechanics*, Vol. 40, 2008, pp. 47–70.
15. Kidanemariam, A. G., C. Chan-Braun, T. Doychev, and M. Uhlmann. Direct Numerical Simulation of Horizontal Open Channel Flow with Finite-size, Heavy Particles at Low Solid Volume Fraction. *New Journal of Physics*, Vol. 15, No. 2, 2013, p. 025031.
16. Ji, C., A. Munjiza, E. Avital, D. Xu, and J. Williams. Saltation of Particles in Turbulent Channel Flow. *Physical Review E*, Vol. 89, No. 5, 2014, p. 052202.
17. Abdelhamid, Y., and U. El Shamy. Pore-scale Modeling of Surface Erosion in a Particle Bed. *International Journal for Numerical and Analytical Methods in Geomechanics*, Vol. 38, No. 2, 2014, pp. 142–166.
18. Ji, C., A. Munjiza, E. Avital, J. Ma, and J. Williams. Direct Numerical Simulation of Sediment Entrainment in Turbulent Channel Flow. *Physics of Fluids (1994-present)*, Vol. 25, No. 5, 2013, p. 056601.
19. Pasiok, R., and E. Stilger-Szydło. Sediment Particles and Turbulent Flow Simulation Around Bridge Piers. *Archives of Civil and Mechanical Engineering*, Vol. 10, No. 2, 2010, pp. 67–79.
20. Escauriaza, C., and F. Sotiropoulos. Lagrangian Model of Bed-load Transport in Turbulent Junction Flows. *Journal of Fluid Mechanics*, Vol. 666, 2011a, pp. 36–76.
21. Spalart, P. R., W. Jou, M. Strelets, and S. R. Allmaras. *Comments on the Feasibility of LES for Wings, and on a Hybrid RANS/LES Approach*, 1997. Advances in DNS/LES, https://www.researchgate.net/publication/236888805_Comments_on_the_Feasibility_of_LES_for_Wings_and_on_a_Hybrid_RANSLES_Approach. Accessed November 14, 2017.

22. Smagorinsky J. General Circulation Experiments with the Primitive Equations: I. The Basic Experiment. *Monthly Weather Review*. Vol. 91 (March), No. 3, pp. 99–164.
23. Li, J., J. Tao, and Y. Liu. DES Modeling of Erosional Forces around Streamlined Piers and Implications for Scour Countermeasures. *International Journal of Geomechanics*, 2016, p. 04016139.
24. Mindlin, R. D., and H. Deresiewicz. Elastic Spheres in Contact Under Varying Oblique Forces. *Journal of Applied Mechanics-Transaction of the ASME*, Vol. 20, No. 3, 1953, pp. 327–344.
25. Gidaspow, D. *Multiphase Flow and Fluidization: Continuum and Kinetic Theory Descriptions*. Academic Press, New York, 1994.
26. Du, W., X. Bao, J. Xu, and W. Wei. Computational Fluid Dynamics (CFD) Modeling of Spouted Bed: Assessment of Drag Coefficient Correlations. *Chemical Engineering Science*, Vol. 61, No. 5, 2006, pp. 1401–1420.
27. Jajcevic, D., E. Siegmann, C. Radeke, and J. G. Khinast. Large-scale CFD-DEM Simulations of Fluidized Granular Systems. *Chemical Engineering Science*, Vol. 98, 2013, pp. 298–310.
28. Johnson, K. L. *Contact Mechanics*. Cambridge University Press, Cambridge, 1987.
29. Timoshenko, S., and J. N. Goodier. *Theory of Elasticity*, Vol. 5. McGraw-Hill, New York, 1951, p. 500.
30. Siemens, P. S. *STAR-CCM+® Documentation Version 11.06*. 2016.
31. Shields, A. *Anwendung der Aehnlichkeitsmechanik und der Turbulenzforschung auf die Geschiebebewegung*. PhD thesis. The Technical University of Berlin Preussischen Versuchsanstalt für Wasserbau, 1936.
32. Goharzadeh, A., A. Khalili, and B. B. Jørgensen. Transition Layer Thickness at a Fluid-porous Interface. *Physics of Fluids (1994-present)*, Vol. 17, No. 5, 2005, p. 057102.
33. Dargahi, B. The Turbulent Flow Field Around a Circular Cylinder. *Experiments in Fluids*, Vol. 8, No. 1–2, 1989, pp. 1–12.
34. Hofland, B. *Rock and Roll: Turbulence-induced Damage to Granular Bed Protections*. Delft University of Technology, 2005.
35. Dwivedi, A., B. W. Melville, A. Y. Shamseldin, and T. K. Guha. Flow Structures and Hydrodynamic Force During Sediment Entrainment. *Water Resources Research*, Vol. 47, No. 1, 2011.
36. Schmeeckle, M. W., J. M. Nelson, J. Pitlick, and J. P. Bennett. Interparticle Collision of Natural Sediment Grains in Water. *Water Resources Research*, Vol. 37, No. 9, 2001, pp. 2377–2391.
37. Ancey, C., T. Böhm, M. Jodeau, and P. Frey. Statistical Description of Sediment Transport Experiments. *Physical Review E*, Vol. 74, No. 1, 2006, p. 011302.
38. Kandasamy, J., and B. Melville. Maximum Local Scour Depth at Bridge Piers and Abutments. *Journal of Hydraulic Research*, Vol. 36, No. 2, 1998, pp. 183–198.

The Standing Committee on Hydrology and Hydraulics (AFB60) peer-reviewed this paper (18-05939).

Article

TRR
JOURNAL OF THE TRANSPORTATION RESEARCH BOARD

Transportation Research Record
2018, Vol. 2672(39) 118–129

DOI: 10.1177/0361198118774745
journals.sagepub.com/home/trr
SAGE

Development of Transition between Free-Standing and Reduced-Deflection Portable Concrete Barriers

Mojdeh Asadollahi Pajouh[1], Robert W. Bielenberg[1], John D. Reid[2], Jennifer D. Schmidt[1], Ronald K. Faller[1], and Erik Emerson[3]

Abstract

Portable concrete barriers (PCBs) are often used in applications in which limited deflection is desired during vehicle impacts, such as bridge decks and work zones. In an earlier study, a reduced-deflection, stiffening system was configured for use with non-anchored, F-shape PCBs and was successfully crash tested under *Manual for Assessing Safety Hardware* (MASH) safety performance criteria. However, details and guidance for implementing this barrier system outside the length-of-need, including within transitions to other barrier systems, were not provided. The focus of this study was to develop a crashworthy transition design between the reduced-deflection, F-shape PCB system to free-standing, F-shape PCB segments using engineering analysis and LS-DYNA computer simulation. First, the continuous steel tubes in the reduced-deflection system were tapered down to the surface of the free-standing PCB segments to reduce the potential for vehicle snag. In addition, steel tube spacers were added at the base of the two joints upstream from the reduced-deflection system to increase the stiffness of adjacent free-standing PCBs. Simulations were performed to determine the critical impact points for use in a full-scale crash testing program. It was recommended that three full-scale crash tests be conducted, two tests with a 2270P pickup truck vehicle and one test with an 1100C passenger car, to evaluate the proposed design system with impacts at the recommended critical impact points.

Many State Departments of Transportation utilize the F-shape portable concrete barrier (PCB) that was developed by the Midwest Roadside Safety Facility (MwRSF) (*1–3*). However, this system deflected 80 in. under MASH TL-3 impact conditions, which can restrict its use in limited-sized construction zones. A PCB system with lower deflections was desired for use on construction projects associated with bridge decks as well as other limited-space work zones. The MwRSF researchers developed and crash tested a new, reduced-deflection system for the F-shape PCB that was not anchored to the roadway surface and was retrofitted to the existing PCB with minimal modifications (*4*). This system consisted of a cap plate bolted across the PCB joint and continuous steel tubes running along the sides of the barriers, as shown in Figure 1. The continuous steel tubes interlocked with the sheet metal of a vehicle making impact and provided increased vehicle stability by creating a more vertical barrier face.

Two full-scale crash tests, Test Nos. RDTCB-1 and RDTCB-2, were conducted on the reduced-deflection PCB design located with the back of the PCB system offset 24 in. laterally away from the edge of a simulated bridge deck. Test No. RDTCB-1 consisted of a 4,998-lb pickup truck striking the reduced-deflection PCB system at a speed of 63.6 mph and at an angle of 24.9°. The vehicle was safely and smoothly redirected in the test. The peak dynamic lateral deflection of the barrier system was 43.0 in., which represented a 46% reduction in deflection as compared with the free-standing, F-shape PCB crash testing under *Manual for Assessing Safety Hardware* (MASH) Test Level 3 (TL-3) (*5*). Following Test No. RDTCB-1, design modifications were investigated to further reduce dynamic system deflections. The modified system consisted of a cap plate bolted across the PCB joint, continuous steel tubes running along the sides of the barriers, and the additional attachment points. Test No. RDTCB-2, which was conducted to evaluate the modified barrier

[1]Midwest Roadside Safety Facility, University of Nebraska-Lincoln, Lincoln, NE
[2]Mechanical & Materials Engineering, Midwest Roadside Safety Facility, University of Nebraska-Lincoln, Lincoln, NE
[3]Wisconsin Department of Transportation, Bureau of Project Development, Madison, WI

Corresponding Author:
Address correspondence to Mojdeh Asadollahi Pajouh: mojdeh.asadollahipajouh@unl.edu

Figure 1. Reduced-deflection PCB—Test No. RDTCB-2 (*4*).

system, consisted of a 4,978-lb pickup truck striking the reduced-deflection PCB system at a speed of 64.8 mph and at an angle of 25.4°. The vehicle was safely and smoothly redirected in this test. The peak dynamic lateral deflection of the barrier system was 40.7 in., which represented a 49% reduction in deflection as compared with the free-standing, F-shape PCB under MASH TL-3 crash testing.

When implementing this reduced-deflection barrier system outside the length-of-need, including within transitions to other barrier systems, additional considerations must be taken into account. First, the deflections of the two systems (i.e., reduced-deflection and free-standing PCBs) vary by approximately 50%. Thus, a transition in joint stiffness was deemed necessary to prevent barrier pocketing and provide safe and stable vehicle redirection. Second, the horizontal steel tubes of the reduced-deflection system would induce potential for vehicle snag, climb, or instability. Therefore, the objective of this project was to develop a MASH TL-3 transition between the reduced-deflection PCB system and free-standing PCB segments. The transition design concepts were evaluated based on potential safety performance and ease of installation. LS-DYNA (*6*) simulations were used to evaluate proposed design concepts according to MASH TL-3 safety performance criteria. Furthermore, a simulation study was performed to determine critical impact points for full-scale crash testing. Transitions, including those between free-standing PCBs and reduced-deflection PCBs, should satisfy impact safety standards published in MASH (*5*). According to TL-3 of MASH, transitions must be subjected to two full-scale vehicle crash tests, as following: 1) test designation 3–20 with a 2,425-lb small car (designated 1100C) striking at a speed of 62 mph and angle of 25°; and 2) test designation 3–21 with a 5,000-lb pickup truck (designated 2270P) striking at a speed of 62 mph and angle of 25°.

Background

Transitions have been previously developed between PCB segments and permanent concrete barriers. The MwRSF researchers designed and successfully crash tested an approach transition between free-standing, F-shape PCBs and a permanent concrete median barrier (*7*). The researchers utilized a combination of free-standing and tied-down Kansas Temporary F-shape Barriers, nested thrie beam on each side of the barriers for the transition to the single-slope permanent barrier, and transition cap, as shown in Figure 2*a*. Two full-scale crash tests were performed on the system under MASH safety requirements. In crash Test No. TCBT-1, a 5,006-lb pickup truck struck the portable barriers 56⅞ in. upstream from the permanent barrier at a speed of 62.4 mph

Figure 2. PCB to permanent concrete barrier transition developed by (*a*) MwRSF (*7*) and (*b*) Texas A&M Transportation Institute (TTI) (*8*).

and an angle of 24.7°. Crash Test No. TCBT-2 was performed using a 4,990-lb pickup truck, which struck the temporary barriers 54 ft 6¾ in. upstream from the permanent barrier at a speed of 62.2 mph and an angle of 26.2°. Both tests were determined to be acceptable according to the MASH TL-3 evaluation criteria.

The Texas A&M Transportation Institute (TTI) developed a transition from the pinned-down F-shape temporary concrete barrier to a permanent concrete barrier (*8*). The transition was crash tested under MASH TL-3 criteria using a 32-in. tall, pinned-down, anchored F-shape temporary concrete barrier connected to a 42-in. tall, single-slope, concrete median barrier, as shown in Figure 2*b*. The transition consisted of a 12-gauge thrie beam connected to the traffic-side face of the barriers, a ⅛-in. thick steel cap attached to the top of the F-shape and single-slope barriers, and a ¼-in. thick steel plate fastened to the non-traffic-side faces of the barriers. The transition safely redirected the 2270P vehicle and performed acceptably to MASH criteria. Maximum dynamic deflection of the transition was 5.7 in.

Transition Design and Analysis

Design Criteria and Considerations

Before developing design concepts for a transition between the reduced-deflection PCB and free-standing PCB, the following design criteria were identified:

1. Design concepts should meet MASH TL-3 safety performance criteria.
2. The dynamic deflection of the transition system should range between deflections found for the reduced-deflection system (40.7 in.) and those found for the free-standing PCB system (79.6 in.).
3. The transition design should require minimal modification to the barrier and should not include modifications to the pin-and-loop connection.
4. The concept fabrication and installation should be easy.
5. The length of the transition and the required hardware should be minimized.

In addition, Euler angles, including roll, pitch, and yaw angles, were used to evaluate vehicle stability. Roll and pitch angles should not exceed 75° according to MASH (*5*). Occupant risk measures, which evaluate the degree of hazard to the occupants in the striking vehicle, included the longitudinal and lateral occupant impact velocities (OIVs) as well as longitudinal and lateral occupant ride-down accelerations (ORAs). According to MASH, longitudinal and lateral OIVs should fall below the maximum allowable value of 40.0 ft/s. MASH also states that longitudinal and lateral ORAs should fall below the maximum allowable value of 20.49 *g* (*5*). Other concerns including vehicle snag, vehicle climbing, and barrier deflection were also considered.

Discussion and Concept Selection

Initially, it was anticipated that the sharp and blunt ends of the HSS5 × 5 × $^{3}/_{16}$ continuous tubes running along the sides of the reduced-deflection PCB segments would increase the potential for vehicle snag. Therefore, it was recommended to taper the box beam ends along the reduced-deflection PCB segments at a 1:6 lateral-to-longitudinal rate. The researchers further brainstormed several concepts, including reduction of joint tolerance in the transition region, increasing barrier-to-ground friction, and increasing the barrier mass, to improve the transition performance from the reduced-deflection PCB to the free-standing PCB. Following the review of the design concepts, it was concluded that the concept of reducing joint tolerance by adding a joint spacer (e.g., a steel tube) would be a simple yet functional option as a transition in stiffness. The joint spacers would cause the barrier segments to engage each other more rapidly during impact to avoid "knee-shape" contact between the vehicle and the barrier segments.

LS-DYNA Model

LS-DYNA simulations were used to evaluate the proposed design concepts (i.e., tapering the box beam ends along

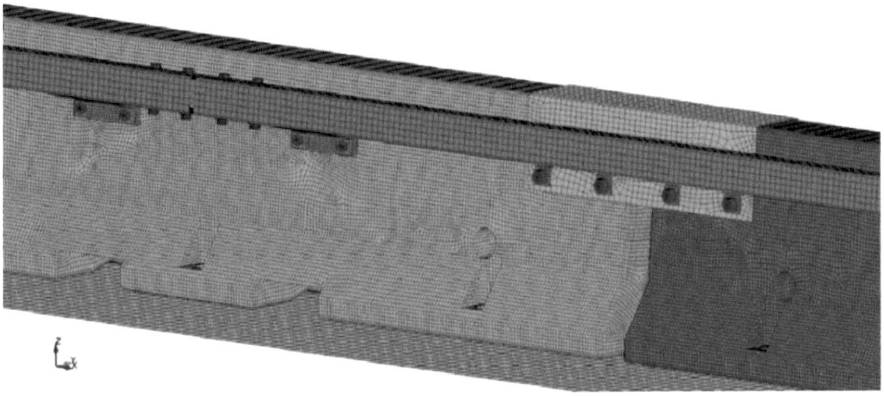

Figure 3. Test No. RDTCB-2 barrier model.

free-standing PCB segments and adding joint spacers) according to MASH TL-3 safety performance criteria.

Free-Standing PCB Model

The LS-DYNA model of the F-shape PCB consisted of the F-shape barrier, the end connection loops, and the connection pins. The main body of the F-shape barrier model was created using shell elements with a rigid material definition. The barrier segments were assigned a mass of 4,976 lb based on measurements taken from actual barrier segments. The use of the shell elements improved the overall contact of the barrier and the vehicle. The loops in the barrier model consisted of two sets of three rebar loops. The connection loops were modeled with a rigid material as previous testing of the barrier in various configurations showed little to no deformation of the connection loops. The connection pin was modeled with the MAT_PIECEWISE_LINEAR_PLASTICITY material in LS-DYNA with the appropriate properties for ASTM A36 steel. The barrier system model incorporated a total of 16 barrier segments for a total barrier length of 200 ft. The simulated F-shape barrier model was struck under the MASH TL-3 impact conditions for Test Designation No. 3–11. More details can be found in Bielenberg et al. (*4*). The simulation was then compared with the actual test on free-standing, F-shape barriers (Test No. TB-2) to ensure that it provided reasonable estimates of the barrier deflection before implementing deflection-limiting concepts (*4*).

Reduced-Deflection PCB Model

The PCB model geometry was modified to add holes for the reduced-deflection steel component connections (*4*). A simulation model of Test No. RDTCB-2, as shown in Figure 3, was constructed to serve as a baseline for comparison of proposed design modifications. The steel cap across the barrier joint and the steel tubes on the side of the barrier were added using shell elements. The tubes were welded to the cap using constrained nodal rigid bodies to create a simplified weld. The connection hardware, including the bolts, nuts, and splice plates were modeled explicitly in the model. Bolt preload was achieved using the INITIAL_STRESS_SECTION command in LS-DYNA. A Chevrolet Silverado Version 2 (v2) pickup truck originally developed by NCAC (*9*) struck the existing reduced-deflection model at a speed of 64.8 mph and at an angle of 25.4°. Comparison of the simulation results with full-scale crash Test No. RDTCB-2 indicated that the model provided good correlation with the full-scale test results and was appropriate for use in evaluation of the transition, as shown in Table 1. The LS-DYNA simulation tended to predict slightly larger lateral and longitudinal ORAs as compared with the crash testing results, which may be caused by lack of failure in the wheel, tire, and suspension model assembly.

Simulation of Proposed Transition Design Modifications

Tapered Tubes along Reduced-Deflection PCB Segments

The tubes along the reduced-deflection PCB segments were tapered down to the PCB surface upstream from the joint between the free-standing and reduced-deflection PCB systems at a 1:6 lateral-to-longitudinal rate over a 30-in. length, as shown in Figure 4*a*. Computer simulations of the barrier with only the tapered tubes and no transition in the barrier stiffness were evaluated to the MASH safety performance criteria. Several impact points upstream from the joint between the reduced-deflection PCB system and the free-standing PCB (Joint No. 1) were simulated using the 2270P Silverado model.

Table 1. Comparison of Reduced-Deflection PCB System Test and Simulation

Design No. RDTCB-2	Evaluation parameters							
	Maximum dynamic deflection (in.)	Maximum roll angle (°)	Maximum pitch angle (°)	Maximum yaw angle (°)	Long. ORA (g)	Lat. ORA (g)	Long. OIV (ft/s)	Lat. OIV (ft/s)
Physical test	40.7	11.8	6.3	35.1	5.6	8.5	12.5	21.3
Simulation	39.7	9.9	8.5	32.9	7.0	11.0	16.4	19.7

Note: Long. = Longitudinal; Lat. = Lateral.

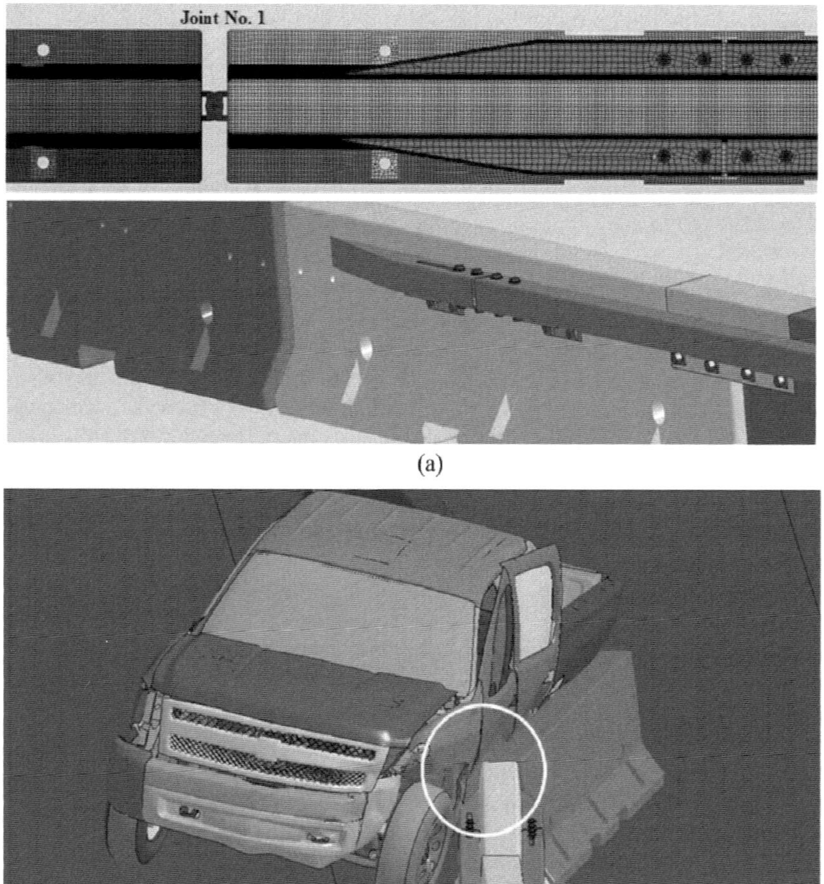

Figure 4. (*a*) Simulated tubes tapered along reduced-deflection PCB segment surface; and (*b*) simulated vehicle snag on "knee" in transition region.

Several impact points upstream from the transition joint had a predominant "knee-shape" that formed in the system and contacted the impact-side door, as shown in Figure 4*b*. This knee contact did not appear to be critically detrimental to the vehicle or the barrier performance in the simulations. However, the impact of the knee into the vehicle's door has been shown to increase vehicle instability in previous PCB tests. In one of the PCB crash tests, Test No. NELON-2, which was a modified MASH Test Designation No. 3–37 on the end of the length-of-need of a 100-ft long PCB installation, the vehicle had a roll angle in excess of 75° and failed to meet MASH safety criteria (*10*). A knee formed in the system, which extended forward laterally from the original barrier line and struck the vehicle's door. The knee contact

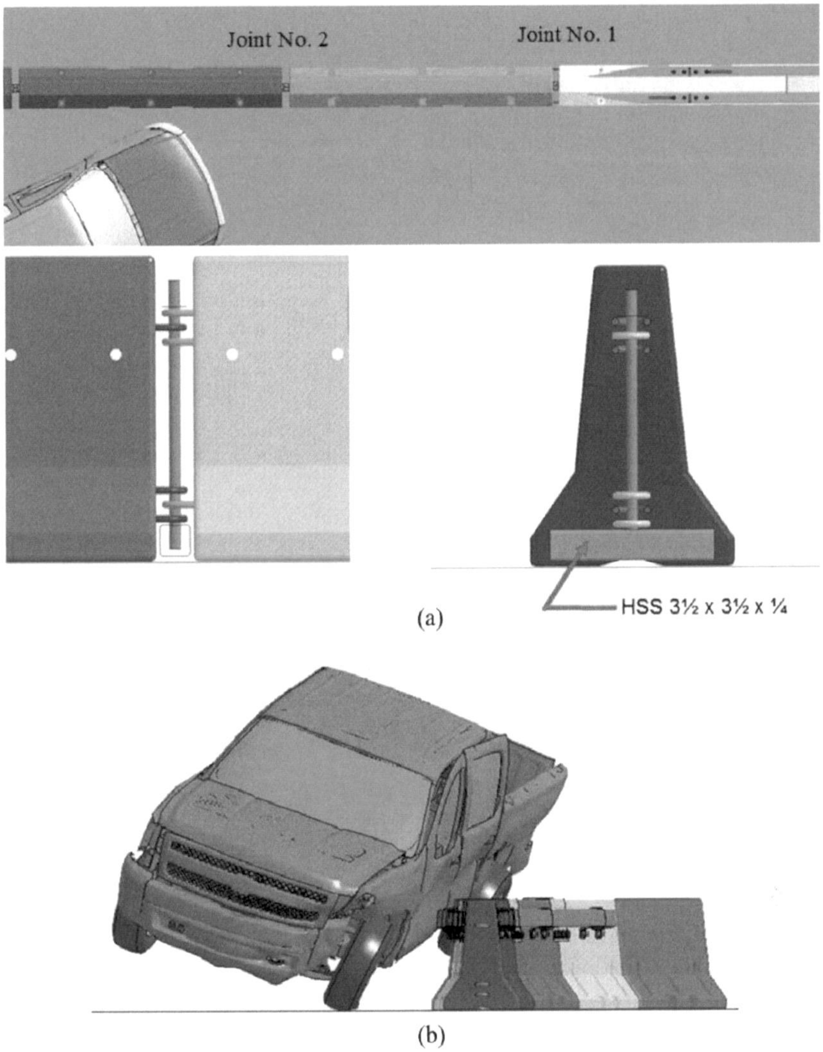

Figure 5. (*a*) Transition with single steel tube at Joint No. 1; and (*b*) knee contact on vehicle door—impact point 225.6 in. upstream from transition joint.

accentuated vehicle roll and likely contributed to the excessive roll angle. This excessive roll was not captured in computer simulations of the system. Because the same barrier model was used in this research, attention was paid to the indications of knee formation and vehicle door impact.

To mitigate the "knee" contact concern and provide a smooth and safe transition in barrier stiffness, the concept of reducing joint tolerance by adding a steel tube spacer was preferred as it was a simple option for resolving the safety concerns in the transition area. An investigation was conducted using an LS-DYNA computer simulation to evaluate the effects of adding steel tubes at the joints on the transition performance during an impact. Joint spacers with various configurations, stiffnesses, and strengths were added to an LS-DYNA model of a free-standing PCB system connected to a reduced-deflection PCB to evaluate impact performance.

Tube Spacer Added at Joint No. 1

Initially, an 18-in. long HSS3½ × 3½ × ¼ steel tube spacer was added at the base of Joint No. 1, as shown in Figure 5*a*. The nominal PCB gap distance was 4 in. ±0.5 in. Thus, an HSS3½ × 3½ × ¼ would properly fit in the gap. To pass the connection pin through the tube, a 2-in. diameter hole was created on the top surface of the tube. Simulations were conducted using the 2270P pickup truck striking the system with one tube at Joint No. 1 and with impact points from 194.1 in.

Table 2. Summary of Simulation Results—Single Steel Tube at Base of Joint Nos. 1 and 2

Impact distance from upstream joint no. 1 (in.)	Tube location	Lateral OIV (ft/s)	Longitudinal OIV (ft/s)	Lateral ORA (*g*)	Longitudinal ORA (*g*)	Roll (°)	Pitch (°)	Lateral barrier deflection (in.)	Increase in vertical bumper height (in.)
194.1	None *	−17.7	−16.1	−5.3	−5.9	−4.4	5.8	36.8	11.3
	Joint No. 1	−17.7	−16.1	−16.8	−5.5	−13.7	12.5	64.3	12.2
	Joint Nos. 1 and 2	−18.4	−16.1	−15	−5.8	−15.2	12.2	74.7	10.8
202.0	None *	−17.1	−14.8	−6.7	−4.8	−5.9	4.5	36.2	9.2
	Joint No. 1	−17.1	−15.1	−17.8	−5.5	−15.6	11.8	61.9	6.7
	Joint Nos. 1 and 2	−17.4	−14.1	−16.9	−4.8	−16.8	10.2	74.0	9.4
209.8	None	−16.1	−16.1	−19.8	−6.8	−23.2	29.1	63.8	7.0
	Joint No. 1	−16.1	−16.1	−18.0	−6.2	−20.9	11.9	65.9	7.8
	Joint Nos. 1 and 2	−16.4	−14.1	−17.6	−5.1	−17.6	9.9	75.6	5.4
217.7	None	−16.1	−13.8	−20.4	−7.9	−24.4	21.2	62.8	5.3
	Joint No. 1	−16.1	−13.5	−20.9	−6.0	−24.1	9.4	65.5	5.7
	Joint Nos. 1 and 2	−16.4	−13.1	−17.8	−5.1	−16.3	12.1	77.4	6.3
225.6	None	−16.7	−9.8	−19.0	−8.3	−23.1	9.4	65.3	6.1
	Joint No. 1	−17.4	−10.5	−20.6	−10.6	−21.9	10.9	64.1	6.2
	Joint Nos. 1 and 2	−17.7	−10.5	−17.7	−9.9	−18.7	11.9	75.5	6.3
312.2	None	−19.4	−17.1	−10.2	−4.1	−12.4	18.8	80.1	25.8
	Joint No. 1	−19.4	−16.7	−11.0	3.8	−17.1	17.6	79.2	24.9
	Joint Nos. 1 and 2	−19.0	−16.7	−11.1	−4.0	−12.8	17.6	83.1	25.4
351.6	None	−17.4	−15.1	−13.5	−6.1	−16.5	13.2	74.1	7.9
	Joint No. 1	−17.4	−15.4	−14.8	−5.9	−14.6	12.6	76.8	8.3
	Joint Nos. 1 and 2	−16.7	−15.7	−12.6	−6.3	−12.7	7.0	59.5	11.3

Note: No. = Number.
* Simulation terminated early because of numerical instabilities.

to 351.6 in. upstream from the transition joint. A summary of simulation results is shown in Table 2.

Adding a single tube at the base of Joint No. 1 improved the transition behavior slightly for knee formation and contact with the vehicle's side. However, this concern was not completely resolved. In several simulations, the knee still formed and hit the vehicle's door. The knee contact on the vehicle door and propensity for instability with the impact point at 225.6 in. upstream from the transition joint is shown in Figure 5*b*. In the cases with one tube spacer, the occupant risk values and the roll and pitch angles did not change significantly. In most of the cases with one tube, a slight increase in barrier deflection of approximately 2 in. was observed. In these simulations, the steel tube joint spacer was completely crushed. Stiffer tubes may help to prevent the knee from forming, which would provide smoother redirection.

Tube Spacers Added at Joint Nos. 1 and 2

After it was determined that adding a single HSS3½ × 3½ × ¼ tube at one joint (Joint No. 1) improved the vehicle and barrier performance, another tube (HSS3½ × 3½ × ¼) was added at the base of the next joint upstream from the previous joint with a tube spacer. Thus, a total of two joints had a tube spacer. Several simulations were conducted with the 2270P pickup truck model striking the system with one tube at Joint Nos. 1 and 2 with impact points from 194.1 in. to 351.6 in. upstream from the transition joint. A summary of simulation results is shown in Table 2. More details can be found in Asadollahi Pajouh et al. (*11*).

The simulations with one tube at two joints had lateral ORA values 10% to 20% less than the simulations with a tube at one joint. Adding tubes at two joints created a smoother redirection as more barrier segments were engaged throughout the impact. In most of the simulations, the roll angle decreased with tubes at two joints. The vehicle snag and knee hitting the vehicle's door were less pronounced in the simulations with one tube at two joints as compared with one tube at one joint. The deflection of the system increased approximately 10 in., but it was still within desired deflection range. In addition, in the simulations with one tube at two joints, the crushing of tubes was moderate.

It was believed that utilizing stiffer tubes at two joints would further decrease the potential for vehicle snag, decrease ORAs, provide smoother vehicle redirection, and mitigate knee and door contact concerns. Thus, various alternatives, including utilizing thicker tubes (HSS3½ × 3½ × ⅜ tubes), utilizing two tubes with different sizes (an HSS3½ × 3½ × ¼ tube with an HSS2½ × 2½ × ¼ tube inside), and utilizing a stack of two tubes (an HSS3½ × 1½ × ¼ tube at the base and an HSS3½ × 1½ × ¼ tube on top) at Joint Nos. 1 and 2, were investigated through simulations, as described

Table 3. Summary of Simulation Results for Impact Point 225.6 in. (5,730 mm) Upstream from Transition Joint—Single and Dual Tubes

Simulation cases	Tube location	Lateral OIV (ft/s)	Longitudinal OIV (ft/s)	Lateral ORA (g)	Longitudinal ORA (g)	Roll (°)	Pitch (°)	Lateral barrier deflection (in.)
Single ¼-in. tube	−17.7	−10.5	−17.7	−9.9	−18.7	11.9	6.3	75.5
Single ⅜-in. tube	−18.4	−12.1	−19.2	−10.4	−17.6	12.4	7.2	56.1
Dual tubes—concentric	−17.7	−10.8	−19.8	−10.5	−18.0	12.5	7.1	60.0
Dual tubes—stacked	−18.7	−13.8	−18.5	−13.4	19.1	12.5	7.5	58.8

in the following sections. The simulations included a 2270P vehicle model striking the concrete barriers 225.6 in. upstream from the transition joint to evaluate and compare the effectiveness of different stiffer tube configurations with the same impact conditions and impact location.

Single Tubes with Increased Thickness. The simplest alternative to stiffen the tube spacer was to increase the tube thickness. Computer simulations were performed with ⅜-in. thick steel tubes at Joint Nos. 1 and 2. The transition performance was significantly improved. In most of the simulations with different impact points, the concern of knee contact with the door was resolved. A comparison between these simulations with ¼-in. and ⅜-in. thick tubes with an impact point of 225.6 in. upstream from the transition joint is shown in Table 3. The occupant risk values, roll and pitch angles, and vehicle bumper climb slightly increased with the ⅜-in. thick tube. However, these values were still within the acceptable MASH limits. The dynamic deflection of the system decreased from 75.5 in. with ¼-in. thick tubes to 56.1 in. with ⅜-in. thick tubes. The tubes were partially crushed.

Dual Tubes. Another stiffened transition system utilized an HSS3½ × 3½ × ¼ tube with an HSS2½ × 2½ × ¼ tube inside, as shown in Figure 6. The other dual tube alternative with an HSS3½ × 1½ × ¼ tube at the base and another HSS3½ × 1½ × ¼ tube on top was also evaluated through simulations, as shown in Figure 6. A comparison between the cases with single and dual tubes with an impact point of 225.6 in. upstream from the transition joint is shown in Table 3. Both dual tube alternatives indicated an improved performance for knee and door contact as compared with a single HSS3½ × 3½ × ¼ tube. The knee was much less prominent than observed for the single HSS3½ × 3½ × ¼ tube simulation. The occupant risk values, roll and pitch angles, and vehicle bumper climbing were similar to the single ⅜-in. thick tube option and were within the acceptable MASH limits. The dynamic deflection of the system decreased from 75.5 in. with ¼-in. thick tubes to 60 in. and 58.8 in. with concentric dual tubes and stack of two tubes at Joint Nos. 1 and 2, respectively. The dual tubes were also partially crushed.

All three alternatives indicated a promising improvement in vehicle and barrier performance, with decreased potential for vehicle snag, decreased ORAs, smoother vehicle redirection, and mitigation of knee contact. The thicker tubes—one HSS3½ × 3½ × ⅜ at the two joints upstream from the reduced-deflection system—were recommended because of ease of fabrication and installation.

Determination of Critical Impact Points

Determination of Critical Impact Points for 2270P Pickup Truck

To identify the critical impact points for pickup truck testing with Test Designation No. 3–21, impacts on the barrier were simulated using a single HSS3½ × 3½ × ⅜ tube at the base of the two joints upstream from the reduced-deflection system. The impact points, ranging from 17.7 in. to 820.9 in. upstream from Joint No. 1, covered both regions with stiffness changes, one from the reduced-deflection PCB to the transition section and the other from the transition section to the free-standing PCB. A summary of the simulation results with thicker tubes (i.e., ⅜-in. thick tubes) with different impact points is shown in Table 4.

MASH recommends an impact point 4.3 ft upstream from the joint between PCB segments for Test Level 3 to investigate the potential for vehicle snag. At this impact point, the pickup truck fender would contact the box beam and indicate a potential for vehicle snag on the tapered box tube ends along the reduced-deflection PCB. This snag did not appear to be severe in the simulations. The tapered tube was fixed to the PCB model and could not separate from the barrier. However, the actual connection may have some flexibility and allow the truck fender to wedge between the tapered tubes and PCB face. Thus, there was potential for vehicle snag on tapered box tube ends along the reduced-deflection PCB when struck at the points upstream from the transition joint. Other measures, including occupant risk values, roll and pitch angles, and bumper climb, were acceptable.

As shown in Table 4, the impact points farther upstream from the reduced-deflection system, had larger deflections and behaved more like the free-standing PCB system. One of the design criteria was to ensure that barrier deflections range between those observed for the reduced-deflection system (40.7 in.) and the free-standing PCB system (79.6 in.). At the impact points, including 499.2 in. through 820.9 in., barrier deflections exceeded the free-standing PCB deflection (i.e., >79.6 in.). The increased deflections may be caused by the

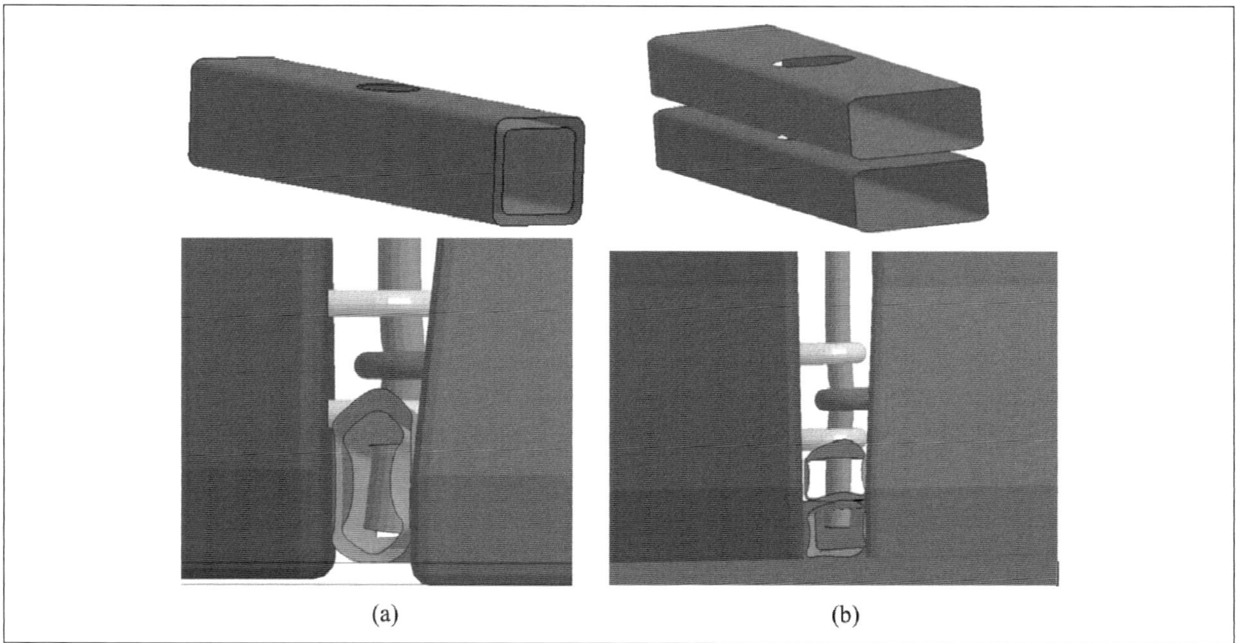

Figure 6. Dual tubes—(a) HSS3½ × 3½ × ¼ tube with HSS2½ × 2½ × ¼ tube inside added at Joint Nos. 1 and 2 (left); and (b) stack of two HSS3½ × 1½ × ¼ tubes added at Joint Nos. 1 and 2 (right).

Table 4. Summary of Simulation Results—HSS3½ × 3½ × 3/8 Single Tube at Joint Nos. 1 and 2

Simulation no.	Impact distance from upstream joint no. 1 (in.)	Tube location	Lateral OIV (ft/s)	Longitudinal OIV (ft/s)	Lateral ORA (g)	Longitudinal ORA (g)	Roll (°)	Pitch (°)	Lateral barrier deflection (in.)
1	17.7	−20.3	22.0	−11.4	−6.7	−7.6	6.6	60.8	3.5
2	33.5	−21.0	21.7	−15.0	−5.8	−11	9.8	64.3	6.6
3	36.6	−21.0	21.0	−16.7	−6.5	−11.1	10	60.3	6.2
4	51.2	−19.7	14.4	−17.8	−13	−15	11.9	53	7.6
5	128	−20.3	13.8	−15.0	−3.6	−14.9	12.4	44.5	9
6	194.4	−18	16.4	−17.5	−5.5	−13.6	12.2	66.4	9.2
7	202.8	−17.1	16.1	−16.8	−5	−12	12.4	56.2	7.3
8	205.1	−17.4	13.1	−19.3	−6.2	−15	11.3	59.4	5.7
9	217.3	−17.4	13.8	−17.9	−4.8	−16	11.6	56.3	6.9
10	225.6	−18.4	12.1	−19.2	−10.4	−17.6	12.4	56.1	7.2
11	233.5	−18.4	12.5	−18	−10.3	−18	13.6	55.4	7.5
12	272.8	−19.0	13.8	−20.3	−5.3	−17.3	13.4	66	9.6
13	281.9	−20.0	14.1	−20.2	−4.5	−17.5	12.3	69.9	10.4
14	307.1	−19.0	16.4	−12.3	−3.2	−16	14	74.7	23
15	352.0	−18.4	15.1	−17.6	−5.9	−13	11.6	80	6.5
16	359.1	−15.7	15.1	−14.6	−5.9	−14.1	10.9	80.5	5.8
17	430.3	−19.4	14.1	−18.8	−6.9	−15.5	15	77.5	9.8
18	436.0	−20.3	14.1	−20.0	5	−16.7	13.7	79.5	10.4
19	499.2	−17.4	17.4	−12	−7.8	−14.4	14	82.8	15.6
20	511.8*	−16.1	14.4	−17.5	−6.7	−17.6	5.8	57.9	9
21	590.6	−19.4	13.8	−22.5	−6.7	−18	15.3	84	10.6
22	669.3	−16.4	15.1	−15.0	−5.6	−18.5	11.4	83	6.7
23	744.1	−19.7	13.8	−22.5	6.7	−17.4	16.3	89.9	10.9
24	820.9	−18.4	15.7	−16	−9.3	−18	14.6	81.3	8.7

Note: No. = Number.

* Simulation terminated early because of numerical instability.

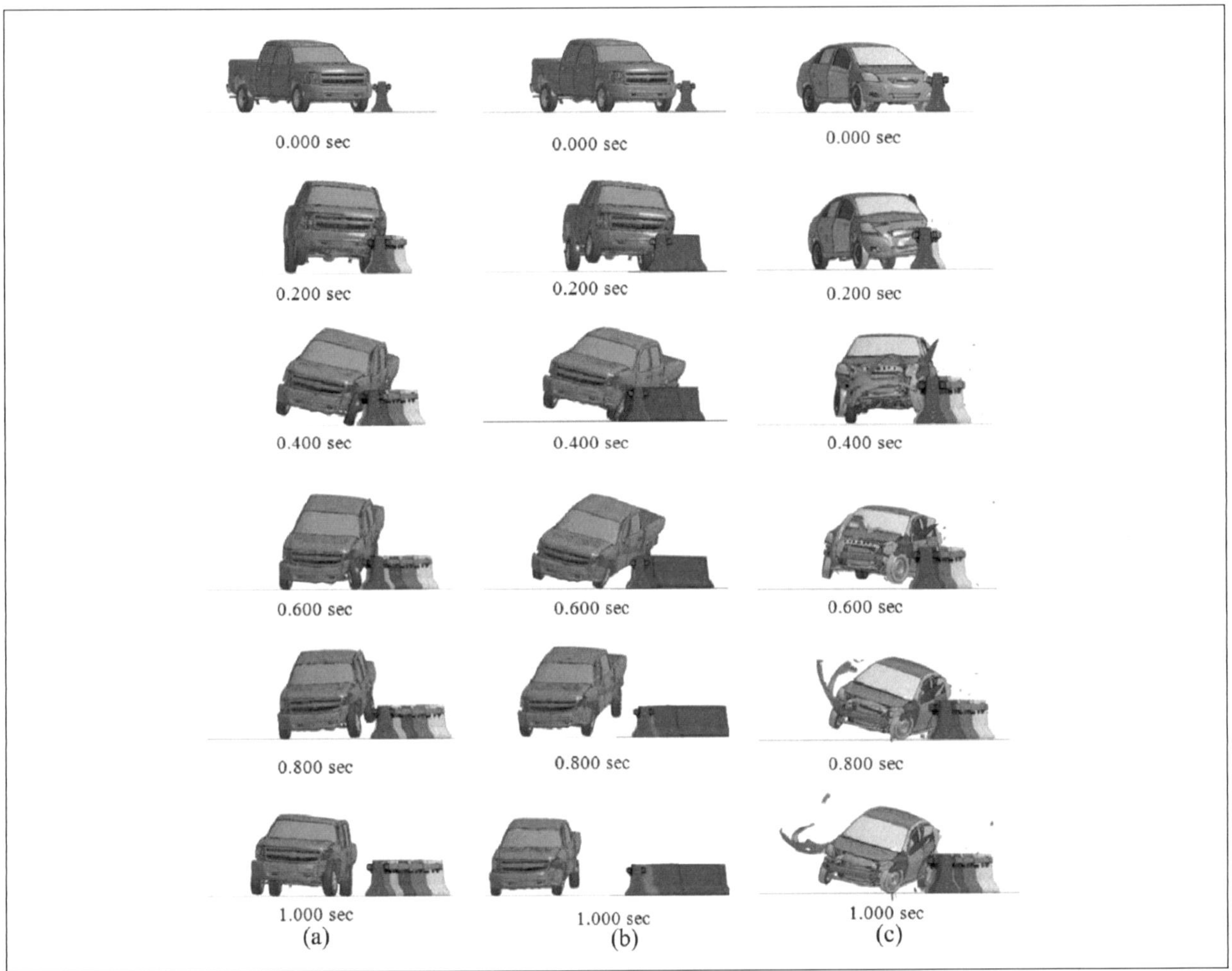

Figure 7. Sequential photographs: (*a*) impact at MASH-recommended CIP 4.3 ft upstream from transition joint—2270P; (*b*) impact at 55.8 ft upstream from transition joint—2270P; and (*c*) impact at 3.6 ft upstream from transition joint—1100C.

proximity of the impact points to the upstream end of barrier model. Because the barrier model ends were not anchored, vehicle impacts close to the upstream barrier end may cause larger deflections. In the simulation of the impact point at 820.9 in. upstream from the transition joint, the vehicle impact was 410 in. away from the end of the barrier model. As such, there was a possibility that the short distance between the impact point and the barrier end would affect the results. In the simulation of this impact point, three identical concrete segments were added to the barrier model. The results of the longer barrier with impact point at 820.9 in. are provided in Table 4.

In addition, the impact points farther upstream, including 511.8 in. through 820.9 in., generally had larger roll and pitch angles. Two impact points, 590.6 in. and 744.1 in., indicated relatively high lateral ORAs (22.5 g's). However, as previously noted, LS-DYNA simulation tends to over-predict lateral ORA. Therefore, these ORAs were not a concern. The impact points, including 669.3, 744.1, and 820.9 in. upstream from the transition joint, showed potential concern for the PCB knee contacting the vehicle's door. In these impacts, one of the PCB segments upstream from the transition joint deflected forward and hit the vehicle's door. The impact point at 669.3 in. upstream from the transition joint had a maximum forward deflection of a PCB segment of 9 in., a maximum roll angle of 18.5°, a dynamic deflection of 83 in., and the PCB knee contacted the vehicle's door. Thus, a critical impact point for 2270P vehicle crash testing test under Designation No. 3–21 was recommended to be 669.3 in. upstream from the transition joint. Sequential photographs for the simulation with an impact point at 4.3 ft and 669.3 in. (55.8 ft) upstream from transition joint are shown in Figure 7, *a* and *b*.

Determination of Critical Impact Point for 1100C Passenger Car

The numerical analysis primarily focused on the 2270P vehicle because of higher barrier deflections and loads and its higher potential for vehicle snag and climbing. However,

1100C vehicle impacts on the transition were also evaluated. Simulations were performed using single steel ⅜-in. thick tube at Joint Nos. 1 and 2. For Test No. 3–20, MASH recommends a critical impact point 3.6 ft upstream from a joint between adjacent barrier segments. This critical impact point would evaluate the potential for the small car to snag on the tapered HSS5 × 5 tube ends along the reduced-deflection system. Other impact points were not critical, because the knee contact with the vehicle's door would not be a concern as the PCB deflection in the small-car impacts was much lower than the pickup truck, and small-car instability because of knee contact seems unlikely. In addition, other safety concerns, including occupant risk values, barrier deflection, and bumper climb, were not believed to be more critical with other impact points. Thus, the 1100C vehicle striking at 3.6 ft upstream from the transition joint and tapered tubes was simulated. The simulation used a Toyota Yaris model developed by NCAC (*9*) and modified by MwRSF. Sequential photographs for the simulation with the Yaris vehicle model are shown in Figure 7*c*.

Conclusions and Recommendations

This research focused on development of a crashworthy transition between the reduced-deflection PCB and free-standing PCB through engineering analysis and LS-DYNA computer simulation. HSS5 × 5 × ¼ tubes along the sides of the reduced-deflection system were tapered down to the surface of the free-standing PCB segments at a 1:6 lateral-to-longitudinal rate to reduce the potential for vehicle snag. Several design concepts were investigated through LS-DYNA simulations to mitigate the concern of knee formation and provide a safe transition. The concept of adding an increased stiffness (HSS3½ × 3½ × ⅜) tube at the base of the two joints located upstream from the reduced-deflection PCB system was selected because of the improved transition performance in that there was less knee contact with the vehicle's door and decreased barrier deflection.

Simulations were also performed with different impact points to determine the critical impact points for full-scale crash testing of the selected design. Based on simulations, it was recommended that three full-scale crash tests be conducted, two with Test Designation No. 3–21 using a 2270P pickup truck vehicle with impact points 4.3 ft and 55.8 ft upstream from the joint between adjacent barrier segments and one with Test Designation No. 3–20 with an 1100C passenger car with an impact point 3.6 ft upstream from a joint between adjacent barrier segments. Based on the numerical study, it is believed that the concept of tapering the end section of the HSS5 × 5 × $^{3}/_{16}$ continuous tubes along the reduced-deflection PCB segments and adding an HSS3½ × 3½ × ⅜ tube at the base of the two joints upstream from the reduced-deflection PCB would have high potential to pass MASH safety criteria. Additional research is required to finalize the transition design. Crash testing is recommended to confirm the crashworthiness of the new, prototype transition system.

Acknowledgments

The authors wish to acknowledge the Wisconsin Department of Transportation for sponsoring this project. The simulation effort was completed utilizing LSTC for providing support for LS-DYNA and the Holland Computing Center of the University of Nebraska, which receives support from the Nebraska Research Initiative.

Author Contributions

The authors confirm contribution to the paper as follows: study conception, design, and simulation: Asadollahi Pajouh, Schmidt, and Bielenberg; analysis and interpretation of results: Asadollahi Pajouh, Schmidt, Bielenberg, Faller, Reid, and Emerson; draft manuscript preparation: Asadollahi Pajouh. All authors reviewed the results and approved the final version of the manuscript.

References

1. Faller, R. K., J. R. Rohde, B. T. Rosson, R. P. Smith, and K. H. Addink. *Development of a TL-3 F-Shape Temporary Concrete Median Barrier*. Report No. TRP-03-64-96, Midwest Roadside Safety Facility, University of Nebraska-Lincoln, Neb., 1996.
2. Polivka, K. A., R. K. Faller, D. L. Sicking, J. R. Rohde, B. W. Bielenberg, J. D. Reid, and B. A. Coon. *Performance Evaluation of the Free-Standing Temporary Barrier—Update to NCHRP 350 Test No. 3-11 (2214TB-1)*. Report No. TRP-03-173-06, Midwest Roadside Safety Facility, University of Nebraska-Lincoln, Neb., 2006.
3. Polivka, K. A., R. K. Faller, D. L. Sicking, J. R. Rohde, B. W. Bielenberg, J. D. Reid, and B. A. Coon. *Performance Evaluation of the Free-Standing Temporary Barrier—Update to NCHRP 350 Test No. 3-11 with 28" C.G. Height (2214TB-2)*. Report No. TRP-03-174-06, Midwest Roadside Safety Facility, University of Nebraska-Lincoln, Neb., 2006.
4. Bielenberg, R. W., T. E. Quinn, R. K. Faller, D. L. Sicking, and J. D. Reid. *Development of a Retrofit, Low-Deflection, Temporary Concrete Barrier System*. Report No. TRP-03-295-14, Midwest Roadside Safety Facility, University of Nebraska-Lincoln, Neb., 2014.
5. American Association of State Highway and Transportation Officials. *Manual for Assessing Safety Hardware (MASH)*, 2nd ed. American Association of State Highway and Transportation Officials, Washington, D.C., 2009.
6. Hallquist, J. O. *LS-DYNA Keyword User's Manual*. Livermore Software Technology Corporation, Livermore, Calif., 2007.
7. Wiebelhaus, M. J., R. J. Terpsma, K. A. Lechtenberg, J. D. Reid, R. K. Faller, R. W. Bielenberg, J. R. Rohde, and D. L. Sicking. *Development of a Temporary Concrete Barrier to Permanent Concrete Median Barrier Approach Transition*. Report No. TRP-03-208-10, Midwest Roadside Safety Facility, University of Nebraska-Lincoln, Neb., 2010.
8. Sheikh, N. M., and W. L. Menges. *Transition Design for Pinned-Down Anchored Temporary Barrier to Rigid Concrete Barrier*. TTI Research Report No. 405160-34-1, Texas Transportation Institute, Tex., 2012.

9. Vehicle Modeling. *NCAC Vehicle Model Development Update.* National Crash Analysis Center, George Washington University, Washington, D.C. http://www.ncac.gwu.edu/research/reports.html. Accessed August 20, 2014.
10. Bielenberg, R. W., D. T. Meyer, R. K. Faller, and J. D. Reid. *Length of Need and Minimum System Length for F-Shape Portable Concrete Barrier.* Report No. TRP-03-337-17, Midwest Roadside Safety Facility, University of Nebraska-Lincoln, Neb., 2017.
11. Asadollahi Pajouh, M., J. D. Schmidt, R. W. Bielenberg, and R. K. Faller. *Development of a Transition between Free-Standing and Reduced-Deflection Portable Concrete Barriers.* Report No. TRP-03-366-17, Midwest Roadside Safety Facility, University of Nebraska-Lincoln, Neb., 2017.

The Standing Committee on Roadside Safety Design (AFB20) peer-reviewed this paper (18-05516).

Article

TRR
JOURNAL OF THE TRANSPORTATION RESEARCH BOARD

Transportation Research Record
2018, Vol. 2672(39) 130–141

DOI: 10.1177/0361198118780884
journals.sagepub.com/home/trr
SAGE

Safety Impact of Weaving Distance on Freeway Facilities with Managed Lanes using Both Microscopic Traffic and Driving Simulations

Qing Cai[1], Moatz Saad[1], Mohamed Abdel-Aty[1], Jinghui Yuan[1], and Jaeyoung Lee[1]

Abstract

With the challenges of increasing traffic congestion, the concept of managed lanes (MLs) has been gaining popularity recently as a means to effectively improve traffic mobility. MLs are usually designed to be left-lane concurrent with an at-grade access/exit. Such a design forms weaving segments since it requires vehicles to change multiple general purpose lanes (GPLs) to enter or exit the ML. The weaving segments could have a negative impact on traffic safety in the GPLs. This study provides a comprehensive investigation of the safety impact of different lengths for each lane change maneuver on GPL weaving segments close to the ingress and egress of MLs through two simulation approaches: VISSIM microsimulation and driving simulator. The two simulation studies are developed based on traffic data collected from freeway I-95 in Miami, Florida. The results from the two simulation studies support each other. Based on the two simulation studies, it is recommended that 1,000 feet be used as the optimal length for per lane change at the GPLs weaving segments with MLs. The safety impact of traffic volume, variable speed limit control strategies, and drivers' gender and age characteristics are also explored. This study can provide valuable insight for evaluating the traffic performance of freeway weaving segments with the presence of concurrent GPLs and MLs in a highway safety context. It also provides guidelines for future conversion of freeways to include MLs.

Mobility and safety are two fundamental requirements of transportation services. With the increased gap between traffic demand and limited infrastructure construction, many cities are suffering from serious traffic congestion problems. Managed lanes (MLs) have emerged as an effective alternative to infrastructure expansion to help alleviate congestion problems. Based on the Managed Lanes Handbook (*1*), MLs can be accessed/exited through three types of access points: direct access/exit ramps, slip ramps, and at-grade access/exit. The at-grade access/exit is often preferred for the ML access/exit design due to the relatively lower cost compared with the direct grade-separated access. If the at-grade access is used, the most common type of ML facility is left-concurrent (i.e., the MLs are on the left side of the general purpose lanes (GPLs)). This requires drivers, especially drivers from the GPL entrance ramp or to the GPL exit ramp, to complete several lane change maneuvers on the GPLs to enter or exit the MLs. Therefore, the segment between the GPL entrance ramp and the ingress point of the ML or between the egress point of the ML and the GPL exit ramp forms a weaving area. Such weaving segments could have critically negative impacts on not only the capacity but also the safety of the GPLs. Several studies have been conducted to suggest the appropriate distance for the weaving segment from the operational perspective. However, very limited efforts have been made to investigate the safety impacts of weaving distance on the GPLs. Understanding such impacts could help enhance the safety and design of the integrated facility with the GPL and ML. Based on empirical data, this study provided a comprehensive analysis about the impacts of different weaving distances on GPL traffic safety with two simulation approaches: the VISSIM

[1]Department of Civil, Environment & Construction Engineering, University of Central Florida, Orlando, FL

Corresponding Author:
Address correspondence to Qing Cai: qingcai@knights.ucf.edu

simulation and driving simulator. These two experiments could provide valuable assessment results for the freeway facilities which are connected to the MLs in a highway safety context.

Background and Literature

The design of ML access/exit has been evaluated recently in numerous studies. Most of them recommended the distance for the weaving segment from an operational perspective. Previous studies (*1*, *2*) suggested that the weaving segment should provide between 400 feet and 1,000 feet per lane change, according to different traffic conditions and freeway capacities. Venglar et al. (*3*) analyzed the expected flow and the average speed reduction on the GPL based on a simulation study. For larger freeway facilities with five lanes or more, this study suggested the minimum and desirable distance between a freeway entrance/exit ramp and a ML ingress/egress point should be totally 2,500 feet and 5,000 feet, respectively. Liu et al. (*4*) investigated the effects of a weaving segment on GPL capacity with a function of different roadway geometric configurations (two, three, or four GPLs) and traffic conditions. Results indicated that a reduction of the cross-weaving length and a rise in cross-weave demand volumes could significantly reduce the GPL capacity. Williams et al. (*5*) recommended the distance which could maximize the capacity of the cross-weave based on a calibrated VISSIM simulation model. It was concluded that the maximum capacity could be obtained if the total weaving distance was set to be 4,000 feet for the freeway with four GPLs.

Limited studies have been conducted to investigate the safety performance at the GPL weaving segments near the ingress and egress of the MLs. Jang et al. (*6*) examined crash data from ML facilities with two different types of access: continuous and limited. The study suggested that the ML facility with a limited access offered no safety advantages over the one with continuous access. Besides, the study revealed that short access length (less than or equal to 0.25 mile) could result in higher collision rates. A recent study conducted by Abuzwidah and Abdel-Aty (*7*) examined crashes on the I-95 in Florida and confirmed that the crash risk increased on GPLs with the application of the ML system. Pulugurtha and Bhatt (*8*) investigated the relationship between crashes and weaving section characteristics and traffic variables in the Las Vegas metropolitan area. A total of 25 weaving sections between ramps with different lengths were included with only three sections shorter than 1,320 feet. It was concluded that the number of crashes would decrease with an increase in length of weaving sections. This study was conducted for regular weaving segments rather than weaving segments connected to MLs.

Microsimulation and driving simulator have been proven to be two important tools to investigate traffic safety under different conditions. For example, Yu and Abdel-Aty (*9*) investigated the feasibility of utilizing a variable speed limits (VSL) system to improve traffic safety on freeways through the microsimulation software VISSIM. Safety impacts of the VSL systems are quantified as the improved crash risk and speed homogeneity. To improve the safety of a congested expressway weaving segment, Wang et al. (*10*) conducted a VISSIM simulation study to test the effectiveness of various active traffic management strategies. This research suggested that the implementation of ramp metering and VSL strategies could significantly enhance safety without notably increasing average travel time at the weaving segment. As for the driving simulator, for example, Yan et al. (*11*) evaluated the effects of major traffic speed and driver age and gender on gap acceptance behavior at a stop-controlled intersection based on a driving simulator study. Kang and Chang (*12*) conducted a driving simulator study to evaluate the effectiveness of work zone merge control strategies. The study revealed that the VSL control strategy could effectively mitigate the speed variations. Wu et al. (*13*) proposed a hierarchical modeling framework for the driving simulator experiment to evaluate the effects of real-time warning systems on driving under fog conditions. For microsimulation and driving simulator, surrogate safety measures (e.g., time to collision (TTC), post encroachment time) were introduced to quantify the kinematics of conflicts. By using the surrogate safety measures and traffic flow characteristics to determine the safety status, both the microsimulation and driving simulator could flexibly evaluate traffic safety with different traffic conditions, different geometric designs, and different traffic control strategies. In general, the microsimulation could evaluate the safety performance from a system perspective based on all simulated vehicles while the driving simulator is able to conduct safety evaluation from an individual perspective based on a test vehicle operated by recruited participants. Hence, a comprehensive evaluation of different L_{LC} might be obtained if both microsimulation and driving simulator studies are conducted.

In this study, both VISSIM simulation and driving simulator are employed to investigate the safety impacts of different L_{LC} on GPL weaving segments close to the ingress and egress of the MLs. Several surrogate safety measures and traffic flow characteristics were used to evaluate the traffic safety status, and different geometric designs and traffic flow conditions were tested in both VISSIM simulation and driving simulator studies.

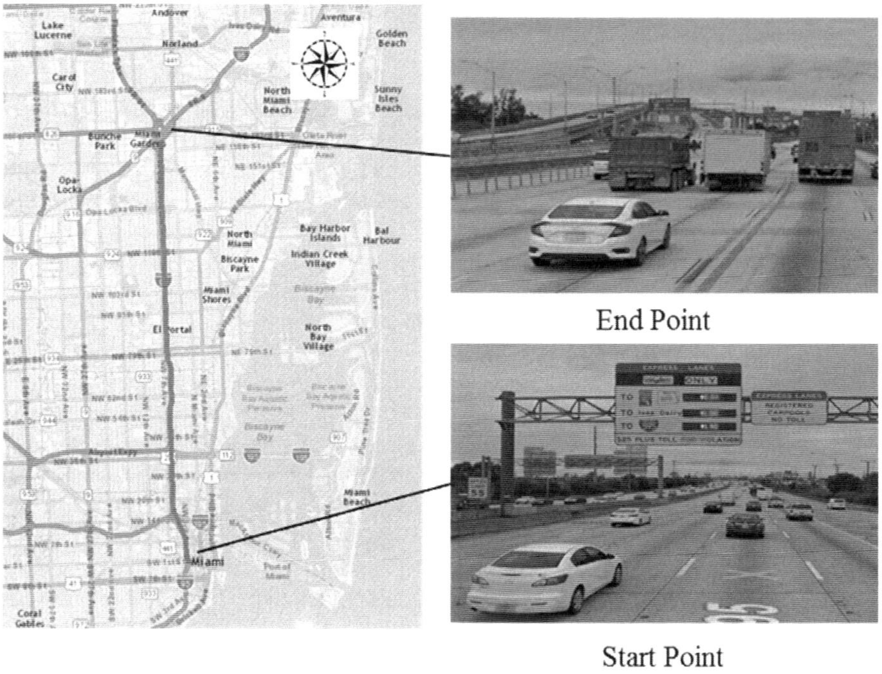

Figure 1. Illustration of the study area.

Experiment Design

Study Area

The VISSIM simulation network was developed based on a 9-mile freeway section of northbound Interstate 95 (I-95) in Miami, Florida (Figure 1). The freeway section included GPLs, MLs, and ramps. The MLs have two lanes while the GPLs have four or five lanes, which means drivers need to change three GPLs to enter or exit the MLs. It is worth noting that the current MLs have only one entrance at the start point and one exit at the end point. The detailed traffic characteristics will be provided in following section, which will be used as the input for both the VISSIM simulation and the driving simulator experiment.

VISSIM Simulation Setup

Calibration and Validation. After constructing the network, calibration and validation of the simulation model are required to guarantee a reliable VISSIM simulation study. Both the morning peak period (7:30 a.m.–8:30 a.m.) and off-peak (9:30 a.m.–10:30 a.m.) period on four Wednesdays in April 2016 were selected as the simulation periods. The simulation model was calibrated for traffic volume at 15-minute intervals and validated using 15-minute average speed in this study. The traffic volume from the simulated data and field data were used to calculate the Geoffrey E. Havers (GEH) value:

$$GEH = \sqrt{\frac{(E - V)^2}{(E + V)/2}} \quad (1)$$

where E is the simulated volume (vehicle/hour), and V is the field volume (vehicle/hour). If more than 85% of the measurement locations' GEH values are less than five, then the simulated traffic would accurately reflect the field traffic flow (*10*). As for the network validation, the absolute speed difference between simulated speeds and field speeds should be within 5 mph for more than 85% of the checkpoints (*11*). Results of ten simulation runs indicated that 91.08% of the GEH values were less than five, and 95.56% of the average speeds in the simulation were within 5 mph of field speeds. Hence, it could be concluded that the network setup in the VISSIM can accurately represent the field roadway traffic flow characteristics. It should be also noted that a logit model was used in the VISSIM software to assign the traffic on the GPLs and MLs based on the dynamic traffic conditions.

Scenario Design. Based on previous studies (*1*, *2*), three types of length per lane change (L_{LC}) were considered: 600 feet, 1,000 feet, and 1,400 feet, which could cover the range of

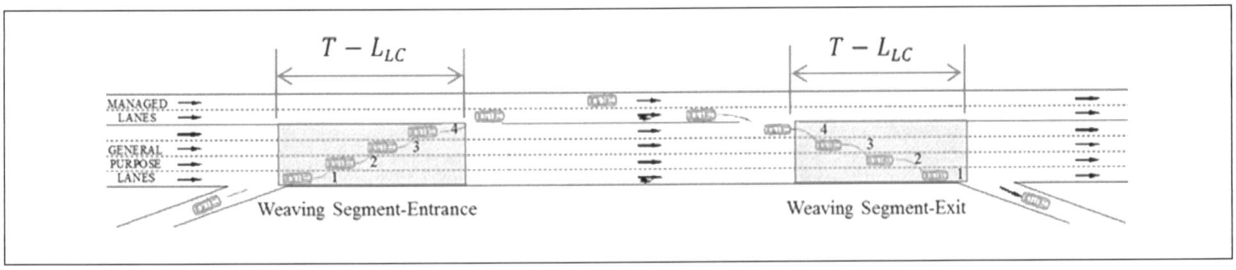

Figure 2. Schematic of lane configuration and traffic movements for the cross-weave scenario.

Figure 3. Illustration of the driving simulator: (*a*) appearance; (*b*) screenshot from a test scenario.

common weaving distances. Figure 2 illustrates a schematic of lane configuration and traffic movements for the cross-weave scenario. As shown in Figure 3, the segment includes a four-lane GPL, a two-lane ML, an entrance ramp, and an exit ramp. A vehicle in the outmost lane of the GPL needs to change three lanes (Lane 1 to Lane 4, not considering the merge from/into the acceleration/deceleration lane) on the GPL to enter the ML. Also, when exiting the ML, the vehicle needs to change three lanes (Lane 4 to Lane 1) to enter the exit ramp. Hence, the total length for vehicles to change lanes on GPL ($T - L_{LC}$, *blue box area*) should be equal to 1,800 feet, 3,000 feet, and 4,200 feet ($3 \times L_{LC}$) corresponding to the three types of L_{LC}. It should be noted that this study only focuses on the length required for drivers to make lane change maneuvers on the GPL. The merging and diverging behaviors on the acceleration and deceleration lanes between the ramps and the GPL or between the GPL and the ML should be quite different from the lane changing behaviors on the GPL. Hence, the lengths of acceleration and deceleration lanes are not included in the current study.

The three types of L_{LC} were tested under three cases with different accessibility levels in this study. Case 1 had only one ingress and one egress in the middle of the network, Case 2 had two ingresses and two egresses, and Case 3 involved three ingresses and three egresses. Compared with Case 1, Cases 2 and 3 could provide more options for drivers to enter the GPL and then enter the ML or exit the ML and then exit the GPL. At-grade accesses were adopted in this study and the ingresses and egresses were separated (Figure 3). The three cases have the same access types and all ingresses and egresses were opened at the four-lane segments (not including the acceleration and deceleration lanes).

In addition, two types of traffic volumes (off-peak and peak) were considered to test the safety impact of the L_{LC}. To sum up, there were 18 scenarios in total (3 (case) × 3 (L_{LC}) × 2 (traffic volume)) in the VISSIM simulation study.

Evaluation Measurements. Three safety measurements were collected based on the simulated traffic at each segment to evaluate the safety impact of the L_{LC}. The three measurements are explained as follows:

Speed standard deviation: the speed standard deviation of all simulated vehicles that passed the sensor deployed in each weaving segment.

Average time to collision (TTC): the time it would take for the following vehicle to hit the leading vehicle given that their current speeds are maintained. In the VISSIM simulation, all TTC values less than 1.5 s were used to calculate the average TTC. The threshold of 1.5 s has been widely used in the previous literature and could efficiently determine if a severe conflict has taken place.

Table 1. Traffic Flow Setting

	Without VSL				With VSL			
	Off-peak		Peak		Off-peak		Peak	
GPL	Average speed (mph)	Volume (veh)	Average speed (mph)	Volume (veh)	Average speed (mph)	Volume (veh)	Average speed mph)	Volume (veh)
1	60	1,385	53	1,659	60	1,511	50	1,810
2	54	1,372	47	1,667	60	1,497	50	1,819
3	63	1,384	56	1,679	60	1,510	50	1,832
4	59	1,115	52	1,524	60	1,200	50	1,624

Conflict rate: the ratio of the number of conflicts to weaving segment length and traffic volume. The surrogate safety assessment model was adopted to determine the conflict frequency with a TTC threshold of 1.5 s. The conflict rate could then be calculated as:

$$conflict\ rate = conflict\ frequency/(total\ volume{*}T - L_{LC}) \tag{2}$$

where the conflict frequency is the total number of detected conflicts in the weaving segment, total volume is the total simulated vehicles in the weaving segment, and L_{LC} is the length for each lane change maneuver. The total length of the weaving segment should be equal to three times the L_{LC} since three lane change maneuvers are required for a four-lane segment.

Scenario Design. The experiment was a 3 × 2 × 2 mixed factorial design. The between-groups independent variables was the L_{LC} (600 feet, 1,000 feet, and 1,400 feet) while the within-subjects independent variables are traffic flow (peak and off-peak) and VSL strategy (without VSL and with VSL). Each L_{LC} group has 16 participants, and each participant was required to drive four scenarios with different traffic flows and VSL strategies. The off-peak- and peak-hour traffic flow data of a four-lane GPL were collected to set the surrounding vehicles. As suggested in the previous research (*10*, *13*), the VSL strategy could effectively enhance traffic safety by improving speed homogeneity. Hence, the effect of the VSL strategy was also tested in the driving simulator experiment by setting the surrounding vehicles with more homogeneous speeds and higher traffic volumes. The traffic condition for each scenario is summarized in Table 1.

Driving Simulator Setup

Participants and Apparatus. A total of 48 licensed drivers (22 females and 26 males) participated in the experiment. Approval of the University of Central Florida Institutional Review Board was obtained. All the participants had at least one year of driving experience. The participants ranged in age from 18 to 75 years (SD = 19.4), with an average age of 38.4. The experiment lasted for about 30 minutes in total for each participant.

The driving simulator (National Advanced Driving Simulator MiniSim) used in this study is able to simulate various roadway environments and traffic conditions, as shown in Figure 2*a*. The control units of the simulator include a steering wheel, gear shift, modular accelerator, and brake pedal. The driving simulator can also provide mimetic environmental noise is equipped with vibration simulation system by three sound speakers. The simulator has a 130-degree horizontal by 24-degree vertical front view of display system. The simulator collected driving data at a rate of 60 HZ. Figure 2*b* displays a screenshot from a test scenario in the experiment.

Procedure. Upon arrival, the participants were introduced to the experiment and were asked to read and sign a consent form. All participants were advised to drive as they would in normal traffic conditions. Before the formal experiment, the participants drove the simulator to familiarize themselves with the driving simulator operations such as straight driving, lane changing, acceleration, deceleration, etc. Then, each participant performed the formal experiments with four scenarios. At the beginning of each scenario, as shown in Figure 3, drivers started from the entrance ramp and were asked to change three lanes on the GPL (Lane 1 to Lane 4) to merge into the ML. Then, after driving on the ML for about 2 miles, the drivers exited the ML and would change three lanes (Lane 4 to Lane 1) to enter the exit ramp. After finishing one scenario, participants could have at least 5 minutes to rest and they were also notified that they could quit the experiment anytime in case of motion sickness. To counterbalance the effects of time order, the four scenarios were arranged in random sequence.

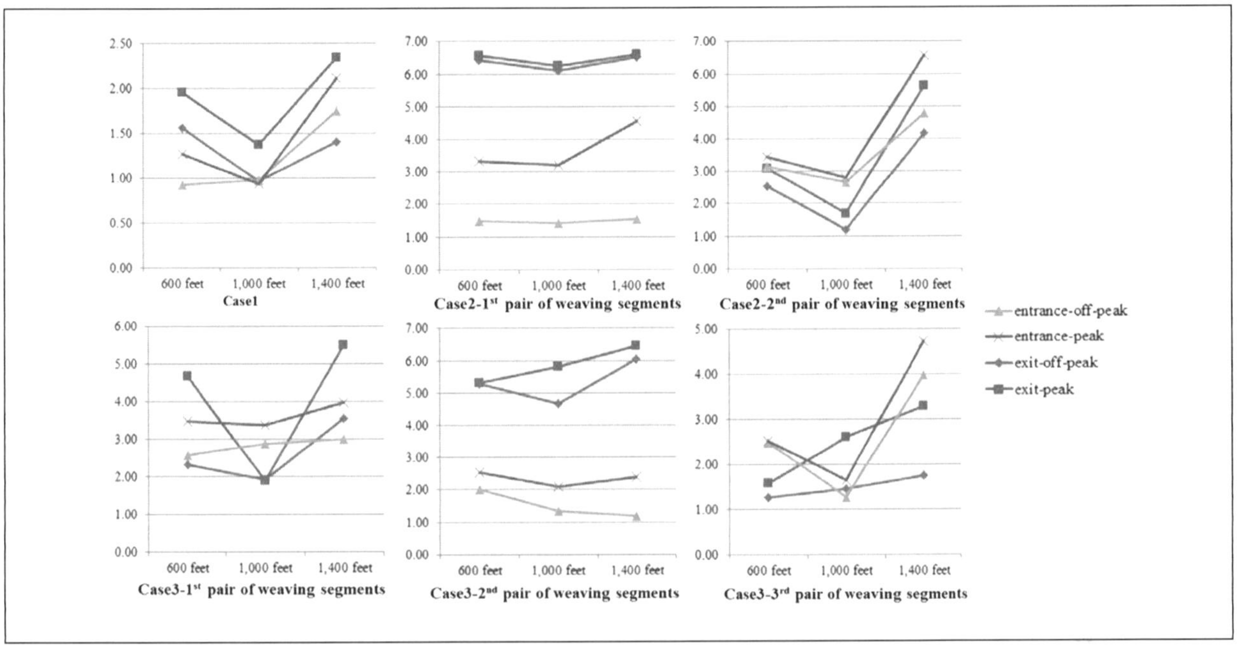

Figure 4. VISSIM simulation results: speed standard deviation (mph).

Evaluation Measurements. In the simulator experiment, the evaluation measurements included speed standard deviation, minimum TTC, and potential collision event. The measurements are explained as follows:

Speed standard deviation: the speed standard deviation of the test vehicle in each weaving segment.

Minimum TTC (MTTC): the minimum TTC between the test vehicle and the leading and lagging simulated vehicles on the target lane during each lane change process.

Potential collision event: the dangerous lane changing maneuver when the MTTC value is smaller than a specific threshold. In the driving simulator study, a potential collision event was identified with a TTC threshold of 5 s. It should be noted that a relatively larger TTC threshold was used for the driving simulator compared with the VISSIM simulation to ensure enough event cases. Although the threshold of 1.5 s for TTC has been most widely used, some previous studies extended the threshold to 2.5 s and even 5 s. It was confirmed that the long threshold could still identify relatively more dangerous conditions for the analysis.

Results

Results from the Microscopic Simulation

As discussed above, three measurements were collected in the VISSIM simulation to evaluate the traffic safety for each weaving segment. Figures 4 to 6 present the results of the measurements for each scenario. Case 1 had only one pair of weaving segments (entrance and exit) since this case opened only one ingress and one egress. Cases 2 and 3 had two and three pairs of weaving segments, respectively. For each pair of weaving segments, four measurement values could be obtained with entrance/exit weaving segment and off-peak-/peak-hour traffic volume. Hence, each L_{LC} condition could have 24 measurement values in total. Several observations can be made from the results presented in the three figures. First, it was found that all weaving segments under the off-peak-hour condition would be safer compared with the peak-hour condition based on the speed standard deviation, average TTC, and conflict rate. The result is consistent with those of previous studies that traffic volume is positively related to crash risk (*11*). Second, most of the weaving segment with the 1,000 feet L_{LC} could have smaller speed standard deviation (19 out of 24 values), larger average TTC (24 out of 24 values), and lower conflict rates (22 out of 24 values) compared with the other two counterparts (600 feet and 1,400 feet). Third, compared with the weaving segment with a 1,400 feet L_{LC}, most of the weaving segment with a 600 feet L_{LC} could have smaller speed standard deviation (22 out of 24 values) and larger average TTC (18 out of 24 values), which indicated that more potential conflicts were expected for the 1,400-feet L_{LC} condition. As for the conflict rate, it was calculated by dividing the conflict frequency by the product of volume and $T - L_{LC}$. Since the 1,400 feet $T - L_{LC}$ is much larger than the 600 feet $T - L_{LC}$, it is reasonable that most of the 600-feet L_{LC}

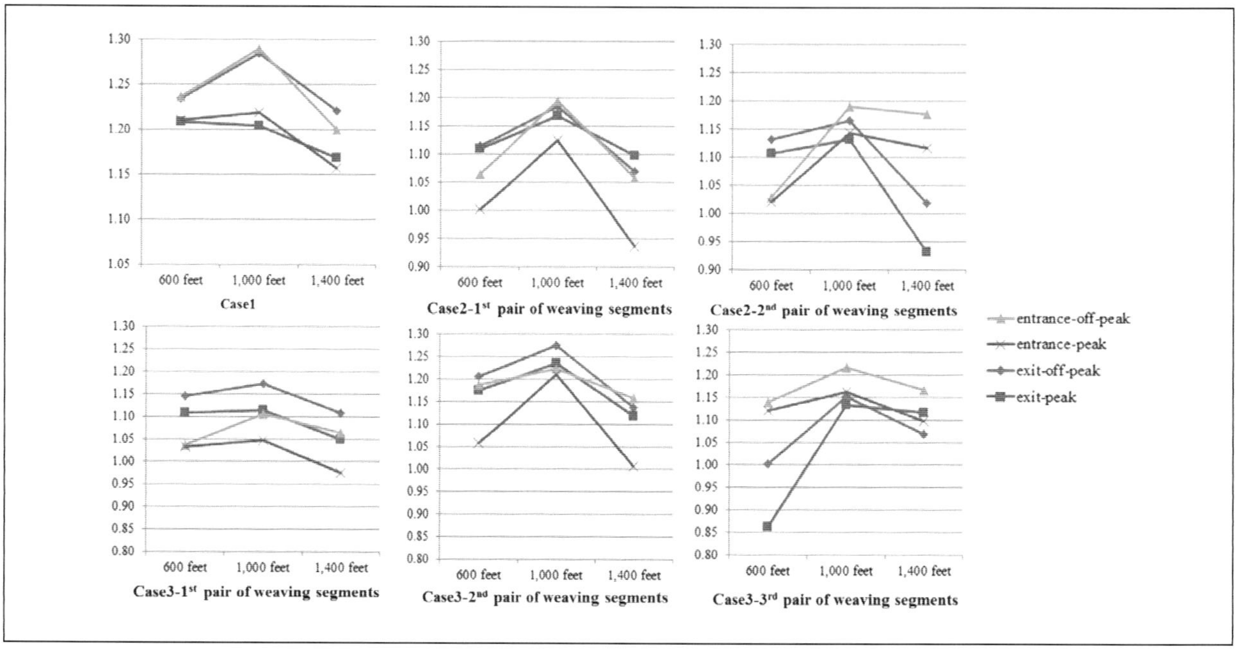

Figure 5. VISSIM simulation results: average TTC (second).

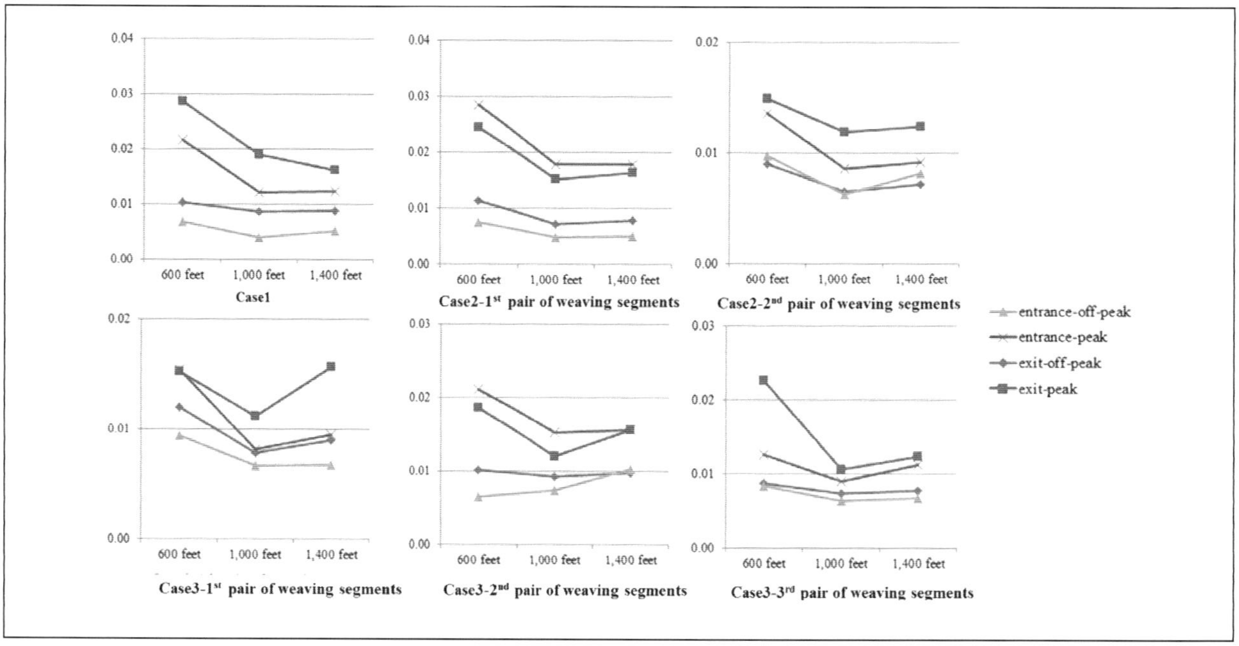

Figure 6. VISSIM simulation results: conflict rate (1/veh*mile).

condition would have higher conflict rate (22 out of 24 values) compared to the 1,400-feet L_{LC} condition.

Results from the Driving Simulator

The data of three participants were excluded since they appeared to express motion sickness during the experiment. Finally, a dataset with 180 (45 × 4) experiment results was obtained. The statistical analysis of the measurements was conducted separately for the entrance and exit weaving segments. The measurements include the speed standard deviation on the whole weaving segment; MTTC and potential conflict during each lane changing process. The measurements were analyzed

Table 2. Analysis of Variance and Descriptive Statistics for Speed Standard Deviation

Source	Entrance weaving segment			Exit weaving segment		
	F-ratio	Factors	Mean of speed SD (mph)	F-ratio	Factors	Mean of speed SD (mph)
Length per lane change (L_{LC})	1.32	600 feet	4.14	2.80*	600 feet	3.17
		1,000 feet	4.14		1,000 feet	3.14
		1,400 feet	4.65		1,400 feet	3.84
Volume	14.37**	Off-peak	3.92	1.58	Off-peak	3.28
		Peak	4.73		Peak	3.52
VSL strategy	4.25**	Without VSL	4.55	3.40*	Without VSL	3.58
		With VSL	4.1		With VSL	3.23

*Significant at the 0.1 level. **Significant at the 0.05 level.

using a 3 (L_{LC}) × 2 (traffic flow) × 2 (VSL strategy) analysis of variance.

Speed Standard Deviation. Table 2 summarizes the descriptive statistics of the speed standard deviation for the three factors and the ANOVA results based on the speed standard deviation of each driver's run. At the entrance weaving segment, the volume ($F = 14.37$, $p < 0.01$) and VSL strategy ($F = 4.25$, $P = 0.05$) significantly influenced the speed standard deviation, while there is no L_{LC} effect among the factors. The speed standard deviation was lower under off-peak conditions (M = 3.92 mph) compared with the peak condition (M = 4.73 mph). In addition, the VSL strategy (M = 4.10 mph) could lower the speed standard deviation compared with the without-VSL condition (M = 4.55 mph). On the other hand, the speed standard deviation at the exit weaving segment was significantly affected by the L_{LC} and VSL strategy, while the volume did not have significant effect. The speed standard deviation was highest if the L_{LC} is 1,400 feet (M = 3.84 mph), and there was no obvious difference between the 600 feet L_{LC} (M = 3.17 mph) and the 1,000 feet L_{LC} (M = 3.14 mph). The without-VSL condition (M = 3.58 mph) would have higher speed standard deviation than the with-VSL condition (M = 3.23 mph).

Minimum Time to Collision (MTTC). Drivers needed to complete three lane change maneuvers at the entrance (from Lane 1 to Lane 4) and exit (from Lane 4 to Lane 1) weaving segments. During each lane change process, the MTTC was collected from the driving simulator data. The repeated measures multivariate analyses of variance (MANOVA) was conducted first for the MTTC of three continuous lane change maneuvers since high correlations may exist. Repeated measures ANOVA were then conducted for the significant factors identified by the MANOVA analysis.

At the entrance of the weaving segment, the MANOVA test indicated that only volume ($F = 8.89$, $p < 0.01$) had significant effects on the MTTC and the consequent ANOVA results suggested that the volume could significantly affect the MTTC of lane change maneuvers from Lane 1 to Lane 2 and from Lane 2 to Lane 3. Figure 7*a* shows that larger MTTCs could be found under the off-peak condition. At the exit weaving segment, the MANOVA analysis revealed significant effects of the VSL strategy ($F = 8.84$, $p < 0.01$) on the MTTC. Based on the ANOVA result and Figure 7*b*, it could be concluded that drivers would have a larger MTTC when they changed lanes from Lane 3 to Lane 2 and from Lane 2 to Lane 1 with the VSL control.

Potential Conflict. For the three lane change maneuvers at the entrance and exit of the weaving segment, the potential conflict was identified with a TTC threshold of 5 s. The potential conflict intensity of the lane change between lanes is shown in Figure 8. The red color indicates the locations which experience more potential conflicts. It is indicated that, for most cases, the potential conflicts occurred when drivers changed lanes between Lane 2 and Lane 3, and between Lane 3 and Lane 4 at both the entrance and exit weaving segments.

A binary logistic model with random effects was estimated for the potential conflicts. A dummy variable indicating whether a dangerous situation occurred during each lane change maneuver was used as the dependent variable and six observations could be found for each scenario run. The regression model results are shown in Table 3. The drivers' gender and age information were also considered in the model estimation. According to the results, the 1,400 feet L_{LC} was more likely to have potential conflicts at both entrance and exit weaving segments compared with the 1,000 feet L_{LC}. A possible reason is that participants would become more relaxed when they thought the weaving distance was quite long for them to change lanes. Hence, they would be likely to concentrate less and be distracted more easily when changing lanes, resulting in higher risk of conflicts. Meanwhile, no

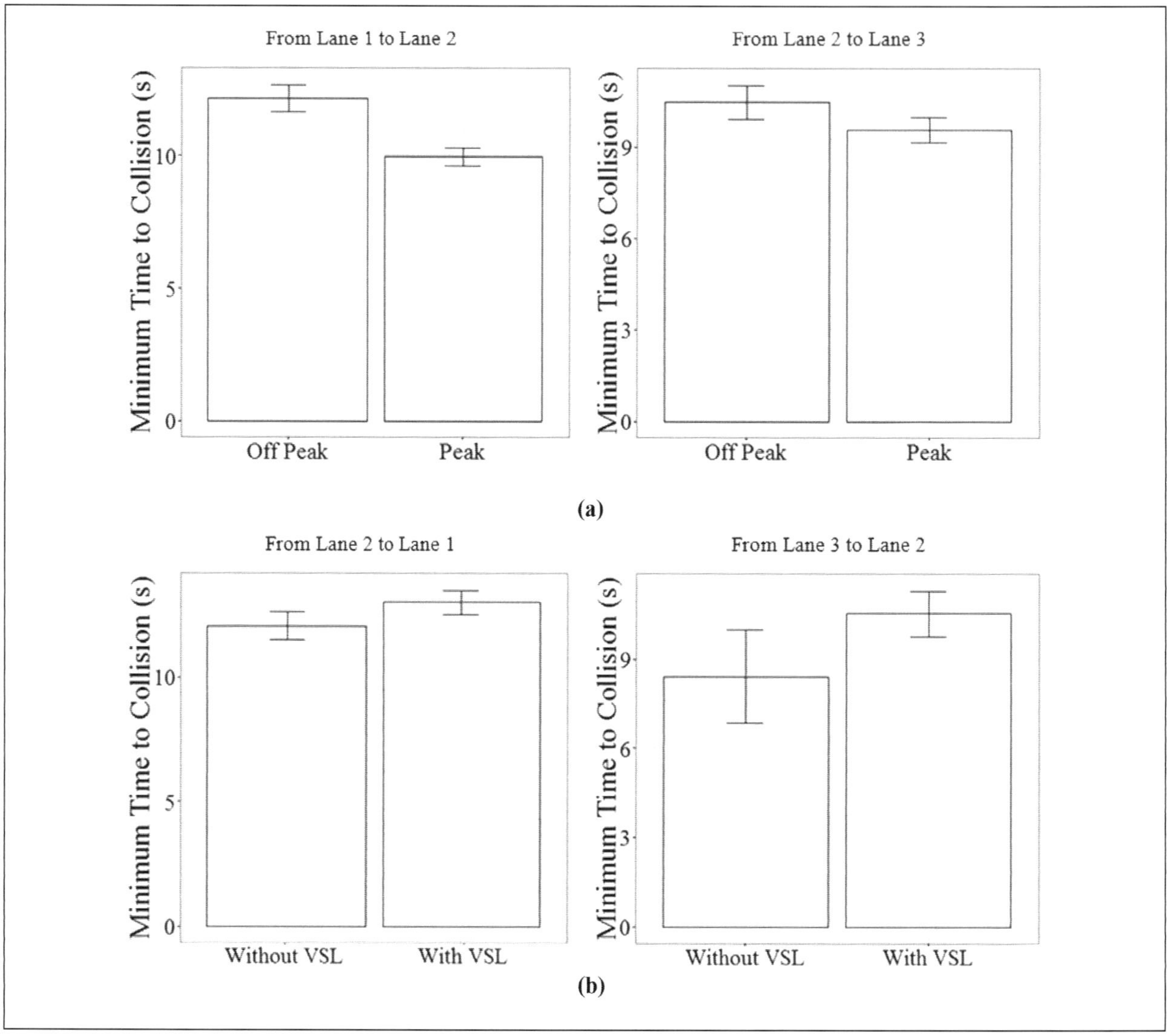

Figure 7. Minimum time to collision for the significant factors: (*a*) under different traffic conditions at the entrance weaving segment; (*b*) under different VSL strategies at the exit weaving segment.

significant difference could be found between the 1,000 feet L_{LC} and the 600 feet L_{LC}, which could be because drivers were more careful in these two L_{LC} conditions. Peak-hour traffic could increase potential conflicts at the entrance and exit weaving segments with higher traffic volume, while the VSL strategy could improve safety at the entrance weaving segment. During the three lane changing maneuvers at either entrance or exit weaving segment, the lane change between the two outer lanes (Lane 1 and Lane 2) was more dangerous than the other lane change maneuvers except the lane change process from Lane 3 to Lane 4 at the entrance of the weaving segment. As for the drivers' characteristics, it was found that older drivers (55 years old or over) tended to have fewer potential conflicts compared with the young drivers (18–24 years old) at the entrance weaving segments.

Discussion

The present study has investigated the safety impact of length per lane change on the freeway weaving segment with MLs. In order to provide a comprehensive perspective for the investigation, both VISSIM simulation and driving simulator studies were conducted on the weaving segments close to the ingress and the egress of the MLs. The experiments in the VISSIM simulation and the driving simulator were specifically designed to examine the safety performance of different L_{LC}.

Table 3. Binary Logistic Regression Results between Factors and the Potential Conflict

Variable	Reference category	Entrance weaving segment			Exit weaving segment		
		Estimate	Standard error	*P*-value	Estimate	Standard error	*P*-value
Intercept	–	−3.55**	0.6	<0.01	−6.91**	1.11	<.001
Length per lane change (L_{LC})							
600 feet	1,000 feet	0.51	0.56	0.36	0.39	0.97	0.69
1,400 feet		1.15**	0.54	0.03	1.60*	0.88	0.07
Volume							
Peak	Non-peak	0.56**	0.27	0.04	0.83**	0.39	0.03
VSL strategy							
With VSL	Without VSL	−0.49*	0.27	0.07	0.01	0.37	0.99
Lane change							
Between lane 2 and lane 3	Between lane 1 to lane 2	0.86*	0.32	0.01	1.51**	0.58	0.01
Between lane 3 and lane 4		0.01	0.37	0.99	1.35**	0.59	0.02
Gender							
Male	Female	0.10	0.43	0.81	0.52	0.71	0.46
Age							
Mid-age (25–54 years old)	Young (18–24 years old)	−0.72	0.45	0.11	0.01	0.79	0.99
Old (55 years old or over)		−1.66**	0.68	0.02	0.69	0.87	0.43
ROC	–		0.84			0.93	

*Significant at the 0.1 level. **Significant at the 0.05 level.

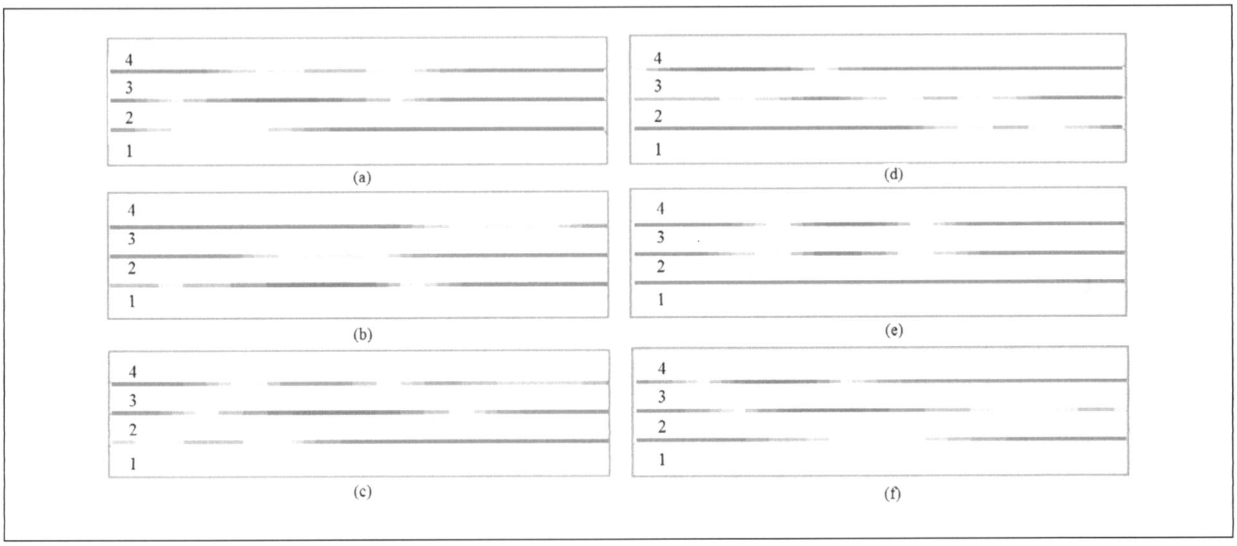

Figure 8. Spatial distribution of potential conflict locations: (a) weaving length = 600 feet (entrance); (b) weaving length = 1,000 feet (entrance); (c) weaving length = 1.400 feet (entrance); (d) weaving length = 600 feet (exit); (e) weaving length = 1,000 feet (exit); (f) weaving length = 1,400 feet (exit).

Speed standard deviation is an important measurement to evaluate traffic safety performance, and it is believed that smaller speed standard deviation indicates safer traffic conditions. As shown above, the speed standard deviations in the VISSIM simulation and driving simulator experiment are different. Specifically, the speed standard deviation in the VISSIM simulation refers to the speed standard deviation for all simulated vehicles and is calculated based on the speeds of all simulated vehicles passing detectors deployed in each weaving segment. The speed standard deviation in the driving simulator experiment is for the speed standard deviation of each tested vehicle and is computed by using the tested vehicle's speed at each time frame (1/60 s) in the weaving segment. The VISSIM simulation study suggested that the 1,000 feet L_{LC} had smaller speed standard deviation than the other two counterparts (600 feet and 1,400 feet) under most cases with different accessibility levels and traffic flows. In addition, higher speed standard deviation could be found with the 1,400 feet L_{LC} compared with the 600 feet L_{LC}. On the other hand, the driving simulator study revealed that drivers would have the highest speed standard deviation at the exit weaving segment if the L_{LC} was 1,400 feet. No significant difference could be found between the 600 feet L_{LC} and the 1,000 feet L_{LC}. In addition, smaller speed standard deviations could be found if the VSL strategy was implemented.

TTC is also an essential variable for traffic safety, and lower TTC indicates potential traffic conflicts. In the VISSIM simulation study, the largest average TTC was found when the L_{LC} was 1,000 feet under most conditions, followed by 600 feet. However, no significant effect of the L_{LC} on the minimum TTC during each lane change maneuver could be found in the driving simulator study.

Based on the TTC value, potential conflicts could be identified in both the VISSIM simulation and driving simulator experiments. In the VISSIM simulation study, the conflict rate was calculated based on the volume and total weaving length. It was revealed that the weaving segment with 1,000 feet L_{LC} had the lowest conflict rates. Interestingly, in most cases, the 1,400 feet L_{LC} could provide a relatively smaller conflict rate compared with the 600 feet L_{LC}, although the 1,400 feet L_{LC} had larger speed standard deviation and average TTC. In the driving simulator study, the logistic regression models of the potential conflicts during each lane change process indicated that drivers would be more likely to have conflicts when the L_{LC} is 1,400 feet. There was no difference between the 600 feet L_{LC} and the 1,000 feet L_{LC}. Lower probability of conflict could be found under the off-peak traffic condition. In addition, under the VSL traffic condition, drivers were less likely to have conflicts at the entrance of weaving segments. Finally, the regression models suggested that older drivers were safer when entering the ML compared with younger drivers, indicating that older drivers are sensitive to the potential risk and more careful when they change lanes (*14*).

In summary, the results from the two studies could extend and support each other. Generally, the best safety performance could be found when the L_{LC} is 1,000 feet, followed by 600 feet, and finally 1,400 feet. Previous studies (*8*, *9*, *15*) suggest the safety performance would be better as the weaving distance increases for the auxiliary lane on a freeway. The auxiliary lane should only have merging/diverging vehicles. However, the GPL in this study has not only merging/diverging vehicles but also a lot of through vehicles. Hence, more interactions between the merging/diverging vehicles and through vehicles could be expected as the L_{LC} increases. In addition, the through vehicles may have more driving behaviors (e.g., lane change, acceleration, decoration) which would have negative impacts on the traffic safety with a longer L_{LC}. Hence, it is not surprising that the safety performance initially improved and then became worse with the increase of L_{LC}.

Conclusions and Recommendations

The concept of MLs has become popular in recent decades as an effective solution to mitigate traffic congestion and produce transportation funds (in the case of high-occupancy toll [HOT] lanes). Many MLs are designed to be left-lane concurrent, where drivers need to change multiple lanes across GPLs to access or exit the MLs. Such weaving vehicles would have a negative safety impact on the weaving segments near the ingress or egress of the MLs. However, there is a lack of studies to suggest the optimal length per lane change (L_{LC}) at the weaving segments from a safety perspective. The current study contributes to providing a comprehensive evaluation of the safety impacts of different L_{LC} through two simulation approaches: VISSIM microsimulation and driving simulator. The VISSIM simulation study was developed and calibrated based on the field-collected data on a 9-mile freeway section of northbound Interstate 95 (I-95) in Miami, Florida. The driving simulator study focused on a freeway segment with four GPLs and two MLs. In the VISSIM simulation and the driving simulator studies, three L_{LC} (600 feet, 1,000 feet, and 1,400 feet) were tested under two traffic conditions (off-peak and peak). Three measurements were used for the safety evaluation in the two studies, including speed standard deviation, time to collision, and potential conflict.

The VISSIM simulation results suggested that the 1,000 feet L_{LC} could provide the best safety performance with the lowest speed standard deviation, largest average TTC, and lowest conflict rate for the whole weaving segments compared with the other two counterparts (600 feet and 1,400 feet). The exact ordering alters between the 600 feet L_{LC} and the 1,400 feet L_{LC} based on different measurements. Based on the driving simulator study, the 1,400 feet L_{LC} would increase the crash risk for each individual driver at the weaving segment based on the speed standard deviation and potential conflicts. However, no significant difference could be found between the 600 feet L_{LC} and the 1,000 feet L_{LC}. Besides, the benefit of using VSL was confirmed in the driving simulator study. In addition, better safety performance could be found under the off-peak traffic condition based on the three measurements in the two studies.

The results from the VISSIM simulation and the driving simulator experiment appear to support each other. Based on the results from the VISSIM simulation and driving simulator studies, it is recommended that 1,000 feet is the optimal length per lane change at the weaving segment near the ingress and egress of the ML. If space is limited, a length between 1,000 and 600 feet could be considered. More attention should be paid under peak-hour traffic conditions. In addition, before the weaving segment, the VSL control is highly recommended to enhance traffic safety. It could be recommended in peak-hour conditions in particular.

Although the three weaving lengths were carefully selected based on the previous literature, the recommendation from the current study could be stronger if more weaving lengths were included in both simulation studies. Also, this study could be extended to further improve the efficiency and safety of the MLs in the future. First, better VISSIM simulation results can be obtained if the drivers' lane change behavior observed in the driving simulator study could be used as input in the VISSIM simulation through the COM interface. Second, a hardware-in-the-loop system could be developed to integrate the VISSIM simulation and driving simulator to assess connected and autonomous vehicle applications with the ML facility.

Acknowledgments

This research was supported by the US Department of Transportation, Research and Innovative Technology Administration, Award No. DTRT13-G-UTC53 (SAFER-SIM).

References

1. Kuhn, B., G. Goodin, A. Ballard, M. Brewer, R. Brydia, J. Carson, S. Chrysler, T. Collier, K. Fitzpatrick, and D. Jasek. *Managed Lanes Handbook*. Texas Transportation Institute, The Texas A&M University System, College Station, 2005.
2. Fuhs, C. A. *High-Occupancy Vehicle Facilities: A Planning, Design, and Operation Manual*. Parsons Brinckerhoff Quade and Douglas, Orange, Calif., 1990.
3. Venglar, S., D. Fenno, S. Goel, and P. Schrader. *Managed Lanes: Traffic Modeling*. Report No. FHWA/TX-02/0-4160-4, Texas Transportation Institute, The Texas A&M University System, College Station, TX, 2002.
4. Liu, X., Y. Wang, B. Schroeder, and N. Rouphail. Quantifying Cross-Weave Impact on Capacity Reduction for Freeway Facilities with Managed Lanes. *Transportation Research Record: Journal of the Transportation Research Board*, 2012. 2278: 171–179.
5. Williams, J. C., S. P. Mattingly, and C. Yang. *Assessment and Validation of Managed Lanes Weaving and Access Guidelines*. FHWA/TX-09/0-5578-1. University of Texas at Arlington, 2010.
6. Jang, K., K. Chung, D. Ragland, and C.-Y. Chan. Safety Performance of High-Occupancy-Vehicle Facilities: Evaluation of HOV Lane Configurations in California. *Transportation Research Record: Journal of the Transportation Research Board*, 2009. 2099: 132–140.
7. Abuzwidah, M., and M. Abdel-Aty. Effects of Using High Occupancy Vehicle Lanes on Safety Performance of Freeways. Presented at 96th Annual Meeting of the Transportation Research Board, Washington, D.C., 2017.
8. Pulugurtha, S. S., and J. Bhatt. Evaluating the Role of Weaving Section Characteristics and Traffic on Crashes in Weaving Areas. *Traffic Injury Prevention*, Vol. 11, No. 1, 2010, pp. 104–113.
9. Yu, R., and M. Abdel-Aty. An Optimal Variable Speed Limits System to Ameliorate Traffic Safety Risk. *Transportation Research Part C: Emerging Technologies*, Vol. 46, 2014, pp. 235–246.
10. Wang, L., M. Abdel-Aty, and J. Lee. Implementation of Active Traffic Management Strategies for Safety on Congested Expressway Weaving Segments. *Transportation Research Record: Journal of the Transportation Research Board*, 2017. 2635: 28–35.
11. Yan, X., E. Radwan, and D. Guo. Effects of Major-Road Vehicle Speed and Driver Age and Gender on Left-Turn Gap Acceptance. *Accident Analysis & Prevention*, Vol. 39, No. 4, 2007, pp. 843–852.
12. Kang, K.-P., and G.-L. Chang. An Integrated Control Algorithm of the Advanced Merge and Speed Limit Strategies at Highway Work Zones. Presented at 90th Annual Meeting of the Transportation Research Board, Washington, D.C., 2011.
13. Wu, Y., M. Abdel-Aty, and J. Lee. Crash Risk Analysis during Fog Conditions Using Real-Time Traffic Data. *Accident Analysis & Prevention*, Vol. 114, 2017, pp. 4–11.
14. Cai, Q., M. Abdel-Aty, J. Lee, and N. Eluru. Comparative Analysis of Zonal Systems for Macro-level Crash Modeling. *Journal of Safety Research*, Vol. 61, 2017, pp. 157–166.
15. Wu, Y., M. Abdel-Aty, Q. Cai, J. Lee, and J. Park. Developing an Algorithm to Assess the Rear-end Collision Risk Under Fog Conditions Using Real-time Data. *Transportation Research Part C: Emerging Technologies*, Vol. 87, 2018, pp. 11–25.

The Standing Committee on Geometric Design (AFB10) peer-reviewed this paper (18-03175).

Article

TRR
JOURNAL OF THE TRANSPORTATION RESEARCH BOARD

A Framework for Prioritizing Urban Arterial/Freeway Interchange Types using Multi-Criteria Analysis

Transportation Research Record
2018, Vol. 2672(39) 142–154

DOI: 10.1177/0361198118781664
journals.sagepub.com/home/trr
SAGE

Karim Jamaleddin[1] and Isam Kaysi[2]

Abstract

An interchange is a road junction that typically uses grade separation to accommodate different volumes of traffic safely and efficiently through interconnecting roads. There are many types of interchanges, each of which has different characteristics and applications. In this paper, a prioritization framework is proposed for selecting a suitable interchange type based on a variety of performance measures, including operational performance, socio-environment, safety, and cost. The assessment is based on a multi-criteria analysis (MCA) approach. The relative importance of the criteria is obtained using the Analytic Hierarchy Process (AHP) whereby the weights of the criteria are derived using multilevel hierarchic structures. Consequently, using a linear additive model, each alternative is evaluated on a set of criteria derived from several objectives. This paper illustrates the implementation of the proposed framework in a case study of an urban interchange in the city of Riyadh. This research study considers three interchanges in the analysis to demonstrate the procedure; other interchange forms may also be appropriate at the site used in the study. This research helps in defining the criteria that play a significant role in determining the preferred design alternative, and the extent to which each of these criteria has an effect on the overall priority of a certain interchange configuration. The results of the study showed that the Diverging Diamond Interchange (DDI) outperformed its interchange counterparts and was chosen as the most preferred interchange configuration. The Single Point Urban Interchange (SPUI) ranked second while the Tight Urban Diamond (TUDI) was least preferred. Sensitivity analysis was undertaken to determine the impact of changes in criteria scoring or weights on the overall results, and revealed the robustness of the proposed model.

An interchange is a system of interconnecting roadways that provides for the movement of traffic between two or more roadways using grade separations. A wide variety of interchanges exist to accommodate different volumes of traffic safely and efficiently through interconnecting roads. The selection of the appropriate type of interchange, along with its design, is influenced by many factors such as operational performance, socio-environmental impacts, safety, and costs. This paper develops a framework for prioritizing arterial/freeway interchange types taking into account a set of different criteria that reflect the above considerations. A case study is also adopted to illustrate the typical implementation of the developed framework.

The following section gives an overview of past research related to urban arterial/freeway interchange types and summarizes the measures of effectiveness (MOEs) used to determine the performance of interchanges. The next section outlines the proposed evaluation framework, illustrates the scoring guidelines and criteria weights, and presents the online questionnaire. The subsequent section demonstrates through a case study the development of criteria scores and ranking of alternatives and presents the outcome of the prioritization analysis. The final section presents concluding remarks and states the major findings and possible extensions of this paper.

Literature Review

An interchange is a road junction that typically uses grade separation to accommodate different volumes of traffic safely and efficiently through interconnecting roads. There are many types of interchange, each of which has different characteristics and applications. Several studies have covered the topic of prioritizing different interchange alternatives to determine the preferred design. Practitioners and academics have dealt with this subject from different perspectives with many of them focusing on operational

[1]SETS International, Transport Engineer, Hamra, Beirut, Lebanon
[2]Department of Civil & Environmental Engineering, American University of Beirut, Riad El-Solh, Beirut, Lebanon

Corresponding Author:
Address correspondence to Karim Jamaleddin:
karim.jamaleddin@gmail.com

performance and safety, while few others considered cost impacts. This section summarizes the findings of the relevant literature focusing on interchange type selection and illustrates the importance of using Multi-Criteria Analysis (MCA) for prioritizing different interchange types.

Overview of Urban Arterial/Freeway Interchange Types

This paper focuses on the service interchanges connecting freeways with crossing arterial roads in an urban context. The conventional Tight Urban Diamond Interchange is compared with two other unconventional configurations of possibly comparable sizes, namely the Single-Point Urban Interchange and the Diverging Diamond Interchange.

- Tight Urban Diamond Interchange (TUDI): a prevalent form of diamnd characteterized by two signalized intersections in close proximity (typically 400 Ft or less) that are designed to operate as one system (*1*).
- Single Point Urban Interchange (SPUI): a relatively new design offering improved road capacity, safer operations, and requiring lower road right-of-way than the conventional diamond interchange (*2*). It brings the ramp terminals for the left-turning movements into one signalizd intersection with the opposing left turns proceeding simultaneously and operating to the left of each other.
- Diverging Diamond Interchange (DDI): a possible solution where the left-turning movements are high off and onto arterials. It features a reversal of the directional traffic movements on the arterial crossing road where through and left-turn vehicles are switched to the opposite sides in between the ramp terminals (*3*).

History of SPUI. In the early 1970s, a new interchange concept emerged to cater for increased traffic demand levels, and provide safer operations, known as the single point urban interchange (SPUI). Since then, several studies have been undertaken to determine the characteristics of this design and check its performance. In December 1996, the Federal Highway Administration (FHWA) in collaboration with Virginia Department of Transportation (VDOT) prepared a brief study evaluating the operational and safety characteristics of the SPUI and its conventional diamond counterpart. Guidelines were developed to identify the traffic and geometric conditions that favor one type over the other and it was concluded that the SPUI is a safe, efficient urban interchange design that can decrease delays and congestion, and uses less land area than other types (*4*). In 1997, the Department of Transportation in Michigan (MDOT) considered the SPUI as an alternative design for rehabilitating many of its freeway interchanges. To evaluate its appropriateness, site observations were made at existing SPUIs including geometric design, signal operation, and pedestrian control and researchers came to the conclusion that SPUI is a good and viable interchange and should be taken into consideration in future studies (*5*). Two years later, VDOT prepared a guideline for the selection of 10 different interchanges. The SPUI ranked first in requiring the least right-of-way and providing easier arterial coordination than conventional diamond interchanges; however, the construction cost was 10%–20% higher for SPUIs (*6*). Similarly, the Missouri Department of Transportation (MoDOT) examined the state of practice for three functional SPUIs in Missouri and offered a set of warrants for when to install a SPUI as compared with TUDI (*2*).

Other research efforts were also undertaken to examine the operational performance of the SPUI and other diamond interchanges using microscopic simulation and several measures of effectiveness were selected for performing the operational comparison, including the average speed, volume served, percent stops, and delay times (*7–9*). A 2005 study considered the performance of the SPUI and TUDI configurations from a safety perspective (*10*). A comparison of the crash history at the two interchange types was conducted using data from 27 existing TUDIs and 13 SPUIs in Washington. The SPUI was found to be safer in terms of injury and fatality frequencies whereas no significant difference was found regarding the total number of crashes.

In addition, the performance of the two diamond interchange forms was analyzed and compared in detail in the ITE Freeway and Interchange Geometric Design Handbook (*11*). To aid practitioners in selecting the appropriate interchange type at a specific location, a general qualitative comparison was presented whereby several considerations were evaluated including geometric and operational characteristics, right-of-way requirements, cost, and other pedestrian-related accomodations. The analyses make it evident that there are applications for both types of interchanges as each configuration could be preferred in different site conditions.

Evolution of DDI. In 2003, Chlewicki proposed a novel interchange design known as the Diverging Diamond Interchange to relieve congestion while accommodating different traffic patterns (*12*). The aim of this interchange is to better accommodate left-turning vehicles through switching the traffic directions in between the ramp terminals and thereby eliminating a phase in the signal cycle that could result in a better operational performance compared with the traditional design. Subsequently, other researchers analyzed the performance of the DDI with different traffic scenarios and confirmed the initial findings whereby under high traffic volumes, the DDI design outperformed the conventional counterpart (TUDI) by offering lower delays, lower stop times, a fewer number of stops, and shorter queue lengths (*13–16*).

In 2007, the (FHWA) in collaboration with the MoDOT began addressing some of the safety concerns with the DDI, such as human error due to unfamiliarity, using a driving simulator to better understand the drivers' behavior as they negotiate through the DDI and found that fewer wrong-way movements, red-light violations, and navigation errors occurred within this unconventional design as compared with the conventional counterpart (TUDI) (*17*).

Based on the DDI's potential benefits and increasing popularity in the United States, the (FHWA) opted to include this novel design in a comprehensive report that articulated its advantages and disadvantages and provided some recommendations on signal phasing, access management for nearby properties, pedestrian and bicycle accommodation, and safety requirements (*3*). This was followed by a guide that addressed a variety of conditions in the United States to achieve a most suitable DDI design for a wide array of potential users. The FHWA guide provided planning techniques and evaluation procedures for assessing safety and operational performance, in addition to other geometric design guidelines including recommended spacing between crossover intersections and evaluated pedestrian accommodation using center walkways and outside path facilities (*18*). The FHWA also published the Highway Safety Manual (1st Edition) in 2014 to help practitioners in selecting countermeasures to reduce the frequency and severity of crashes on roadways using predictive analytical tools. The manual focuses on several aspects of transportation safety including the roadway safety management process, predictive methods, and crash modification factors (*19*).

Similarly, the MoDOT released a report based on its knowledge and experience with building and operating the first DDI in the United States and provided suggestions for improving road geometrics, signing, pavement marking, medians, and pedestrian access (*20*). A year later, the (MoDOT), in collaboration with HDR Engineering, conducted an evaluation of the first DDI in the United States to validate the results of previous research work pertaining to traffic operations and to assess safety and public perceptions of the new interchange and found that a high percentage of users expressed their satisfaction as the traffic flow had improved in the DDI (*21*).

In 2012, Chlewicki prepared a study summarizing the lessons learned from six operating DDIs in the US and provided recommendations to further enhance the performance of future DDIs (*22*). Similarly, a new set of recommendations was provided pertaining to geometry, signal timing, safety and other operational considerations that could help practitioners in providing better designs for future DDIs (*23*).

SPUI versus DDI. In addition to the various publications comparing the conventional types of Diamond interchanges with other unconventional types, Stanek (*24*) and Siromaskul et al. (*25*) compared unconventional types of diamond interchanges and exposed the advantages and disadvantages of each based on traffic operation, intersection capacities, pedestrian accommodation, construction costs, and Right-of-Way costs. They concluded that SPUIs and DDIs could both be recommended whenever there are high left-turn movements at ramp terminal intersections; however, the SPUI improved signal progression and enhanced the throughput for cross-street traffic, unlike the DDI that limited the signal coordination to one direction on the cross street.

In a similar manner, Afshar et al. (*26*), Chlewicki (*27*), and Guerrieri et al. (*28*) conducted traffic performance analysis for unconventional types of interchanges and found that the SPUI provided higher throughput when the traffic split was balanced. On the other hand, the DDI outperformed the SPUI at high unbalanced splits.

Evaluation Metrics (Measure of Effectiveness)

Measures of effectiveness (MOEs) are the system performance metrics commonly used to determine the operational performance of interchanges. It was inferred from the literature that several MOEs were selected as performance criteria to ensure a fair comparison of interchanges including intersection delays, maximum queue length and average number of stops. A summary of the evaluation metrics that were used in earlier research for the different interchange types are provided in a Master's thesis prepared by Jamaleddin in 2015 (*29*).

Multi-Criteria Analysis (MCA)

Multi-criteria analysis (MCA) enables the evaluation of several strategic criteria derived from general objectives. These criteria can either be subjectively analyzed or technically quantified. MCA is based on the premise that any service, such as prioritization of interchange types, can be described by its criteria, and the extent to which the decision-makers value the interchange types depends highly on the level of importance they attach to those criteria. It can be used for identifying the single most preferred option among several alternatives; ranking and prioritizing alternatives; identifying several alternatives for a more detailed appraisal; or sorting them into acceptable and unacceptable ones (*30*).

Multi-criteria decision-making has been proposed as an approach to prioritize alternatives while addresing non-monetary variables, such as noise, accidents, air pollution and so on, in the analysis; consideration of such variables has been troublesome with other traditional approaches such as Cost-Effectiveness Analysis (CEA) and the Cost-Benefit Analysis (CBA) (*31*). MCA often involves the combination of some criteria which are valued in monetary terms, and others for which monetary valuations do not exist (*32*). Given the wide range of objectives, the diversity in criteria, and the difficulty in assigning a monetary value to all factors, the chosen approach in this research had to involve a non-monetary, multi-criteria evaluation.

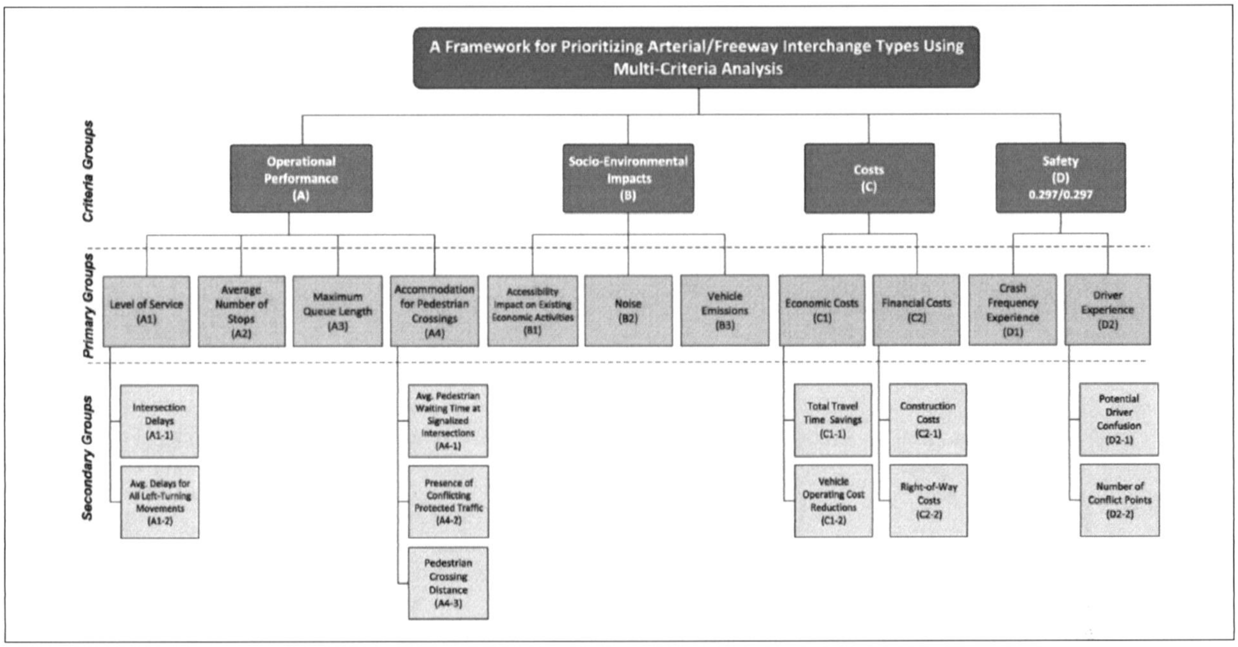

Figure 1. The proposed framework.

Proposed Evaluation Framework

In this section, the proposed evaluation framework for prioritizing arterial/freeway interchange types is introduced while establishing the goals of decision-makers and the corresponding research objectives.

Overview

In defining the evaluation framework, it is important to identify the goals and concerns of the decision-makers and the corresponding framework objectives. The overall objectives are usually established after a series of consultation processes with different decision-makers including academics, engineers, public officials and experts with the concerned domain. The proposed framework for prioritizing Urban Arterial/Freeway Interchange types consists of defining an exhaustive set of criteria to include possibly all factors that might affect the prioritization process of the different alternatives. Special care was taken to avoid redundant criteria and to ensure that each criterion was clearly defined for further assessment. Attention was also given to ensure that the criteria preference is mutually independent.

Accordingly, the objectives were further analyzed and included in the top level of the hierarchy. The selected criteria are grouped into four main groups (Criteria Groups) reflecting the broad factors that have a major impact on prioritizing and ranking the alternatives, as follows: Operational Performance (A), Socio-Environmental Impacts (B), Costs (C), and Safety (D).

Each main criterion is then divided into primary criteria; for instance, Operational Performance is seen as a key criterion, which contains sub-criteria, such as the Level of Service (A1), Average Number of Stops (A2), Maximum Queue Length (A3), and Accommodation of Pedestrian Crossings (A4). Below this primary level are additional subdivisions that describe in more detail certain aspects of the primary level. For example, Level of Service can be subdivided into two more categories, such as Intersection Delays (A1-1) and Average Delays for All Left-Turning Movements (A1-2). These criteria are represented in hierarchical form as shown in Figure 1.

Criteria Groups

Deriving Criteria. As indicated, the goals and concerns of decision-makers were instrumental in deriving the different objectives. These objectives are broken down into criteria which reflect the concerns of these decision-makers. Table 1 describes in more detail the criteria used for prioritizing Arterial/Freeway interchange types.

Scoring Guidelines. One of the advantages of the MCA approach is that criteria can be measured either quantitatively or qualitatively. In this paper, some of the criteria can be directly measured using microscopic simulation analyses with software such as VISSIM and Synchro. Measurement of other criteria can be derived through mathematical computations such as the secondary criteria under the Cost objective; others are subjectively analyzed based on expert judgment.

Table 1. Criteria Description

Criteria group	Primary group	Secondary criteria	Description
Operational performance (A)	Level of service (A1)	Intersection delays (A1-1)	The difference between the desired/free travel time and the actual travel time for the intersection as a whole, in second/vehicle on average.
		Average delays for all left-turning movements (A1-2)	The difference between the desired/free travel time and the actual travel time for the left-turning movements only, in second/vehicle on average.
	Average number of stops (A2)		A stop is counted if the speed of a vehicle reaches zero while proceeding through the interchange. The average number of stops is calculated as an average for all approaches during the peak hour.
	Maximum queue length (A3)		The maximum queue length is considered for all approaches throughout the peak hour.
	Accommodation of pedestrian crossings (A4)	Average pedestrian waiting time at signalized intersections (A4-1)	The average time needed per pedestrian waiting at signalized intersections to pass through the interchange. It depends on the signal timing plans and the number of signal-controlled intersections to be crossed.
		Presence of conflicting protected traffic (A4-2)	The hourly volume of protected left-turning traffic that conflicts with the crossing pedestrians.
		Pedestrian crossing distance (A4-3)	The distance covered by pedestrians in crossing the interchange.
Socio-environmental impacts (B)	Accessibility impact on existing economic activities (B1)		The extent of constraints imposed by the interchange on the movement of traffic trying to access local activities (e.g., continuity of through traffic along ramps).
	Noise (B2)		The level of noise generated by road traffic using the interchange. Noise exposure is related to vehicular speed, traffic volumes, vehicle mix, grades, and pavement characteristics.
	Vehicle emissions (B3)		The extent of vehicular emissions at the interchange. It is related to the traffic volumes and a set of factors including: functional road class, vehicle speed, vehicle fleet characteristics, and ambient conditions. Main air pollutants considered are CO and HO.
Costs (C)	Economic costs (C1)	Total travel-time savings (C1-1)	The savings in travel-time costs that can potentially occur at interchanges as compared with the case of at-grade signalized intersection. These costs are calculated based on the average monetary value of time (VOT).
		Vehicle operating cost reductions (C1-2)	The reductions in vehicle operating cost achieved by all vehicles travelling through the interchange as compared with the case of at-grade signalized intersection. VOC depends on vehicle usage including: fuel, tires, maintenance, repairs, and depreciation.
	Financial costs (C2)	Construction costs (C2-1)	The cost needed for constructing the interchange; particularly the concrete and steel structural works.
		Right-of-way costs (C2-2)	The cost needed for the Right-of-Way acquisition to properly accommodate the interchange.
Safety (D)	Crash frequency experience (D1)		Crash rate history/experience at similar interchanges in other locations (e.g., crashes/year). It is an indicator of safety and is used in crash data analysis.
	Driver experience (D2)	Potential driver confusion (D2-1)	Potential driver confusion while traversing the interchange. It is related to the vehicle paths, intersection area, and traffic movement directions.
		Number of conflict points (D2-2)	The number of potential vehicular conflict points including crossing, merging, and diverging points.

After measuring the performance of each alternative on each criterion, scores are assigned using a numerical scale. A score, also known as "rating," is the degree of performance of an alternative on a selected criterion. The numerical scales range from 5 to 10, where 5 is used to reflect the least preferred possible performance on each criterion and 10 for the most preferred possible performance. Any performance outside this range, above or below, would score 5 and 10, respectively. Once the end points of the scale are determined, the intermediate scores can then be established.

Basis for Scoring Various Criteria. The criteria discussed in this section can be categorized into three different types based on how their performance can be measured:

- *Direct classification using available data*, such as the "Crash Frequency Experience."
- *Direct classification using calculation of measurements*, such as "Noise" and "Vehicle Emissions" which are calculated as a function of speed and Annual Average Daily Traffic (AADT). In addition, most of the criteria under the operational performance category are computed through microscopic simulation using software like VISSIM and Synchro.
- *Indirect classification using professional judgment*, such as the "Potential Driver Confusion."

Assigning Weights. Not all criteria are equally important. The relative importance of the different criteria is reflected in the framework using weights. These weights, also known as "scale values," can be assigned using several approaches; the simplest way is to subjectively judge the weight of each criterion based on experience. However, to address the problem of subjectivity, the Analytic Hierarchy Process (AHP) is used as it allows decision-makers to move away from the open biases toward certain criteria. The advantage of this method is that it can be applied rapidly and it minimizes the subjectivity of the weight values.

The AHP involves deriving the relative weights of the criteria by organizing pairwise comparison matrices (PCM) which are converted into numbers using different scales. Saaty's scale is used in this research as it has been validated for its effectiveness in many applications against several other scales (*32*). The maximum eigen value (λ_{max}) is then determined for all PCMs and the corresponding consistency ratios (CR) are computed. Finally, the aggregate weight set is computed taking into account the PCMs of the decision-makers whose consistency ratios are acceptable, within a specified benchmark limit.

Online Questionnaire

To derive the relative weights of each criterion in each segment of the hierarchy tree, an online questionnaire was developed in which decision-makers provided their pairwise comparisons of criteria. In this research, as in similar contexts, stakeholders were considered decision-makers whose judgments are typically used to reflect the thoughts and opinions of the government and local authorities. In addition, experts in different disciplines were encouraged to participate including junior transport engineers, senior transport engineers, and professors who are familiar with this domain.

Questionnaire Structure and Results. An online survey was used to elicit the judgments of each of the stakeholders and determine the relative importance of criteria using pairwise comparisons. Before eliciting judgments and undertaking the pilot study, the survey was pre-tested on a limited number of respondents who provided significant feedback.

In this research, emails were sent to 93 potential respondents inviting them to fill out the web-based questionnaire. Survey responses were collected from September 14th, 2014 to October 30th, 2014. Of 93 potential respondents, 51 participated, giving a response rate of 55%. Over a period of seven weeks, 22 senior engineers participated in the survey, representing the highest percentage of participants among all decision-groups (around 43%). A relatively lower percentage of respondents were junior engineers (30%) and academic professors (20%), whereas only four public officials (7%) provided responses to the survey.

The online questionnaire was divided into several sections, including:

1. The purpose of the questionnaire explaining the objective of the research
2. An overview of the adopted framework that was identified for prioritization
3. A detailed description of each criterion and how interchange alternatives are expected to perform on it
4. The steps to be taken to fill out the online questionnaire
5. The preliminary ranking of the criteria to be specified by the decision-makers which would help them keep their judgments consistent during the pairwise comparisons task
6. The pairwise comparisons to be provided by the decision-makers

Derivation of the Weights

After constructing the pairwise comparison matrices using the citeria preferences extracted from the questionnaire responses, the criteria weights and consistency ratio were computed using AHP software such as Expert Choice. It was decided to use the 0.2 consistency ratio as it represents an acceptable level of randomness in the judgments (*33*). As such, the data which does not fall within the 0.2 benchmark can be purged; this resulted in the removal of only 15 matrices while keeping 97.05% valid data. Using this methodology, the weights for the 17 criteria were derived.

The results showed that the Operational Performance (A) with a weight of 0.396 was the most important criterion as perceived by the different decision-makers, Safety (D) ranked second in importance with a weight of 0.294, Costs (C) ranked third with a weight of 0.190, and Socio-Environmental Impacts (B) ranked fourth having a weight of 0.120.

Figure 2 includes the set of weights that were obtained for the criteria to be used in the determination of the priorities of

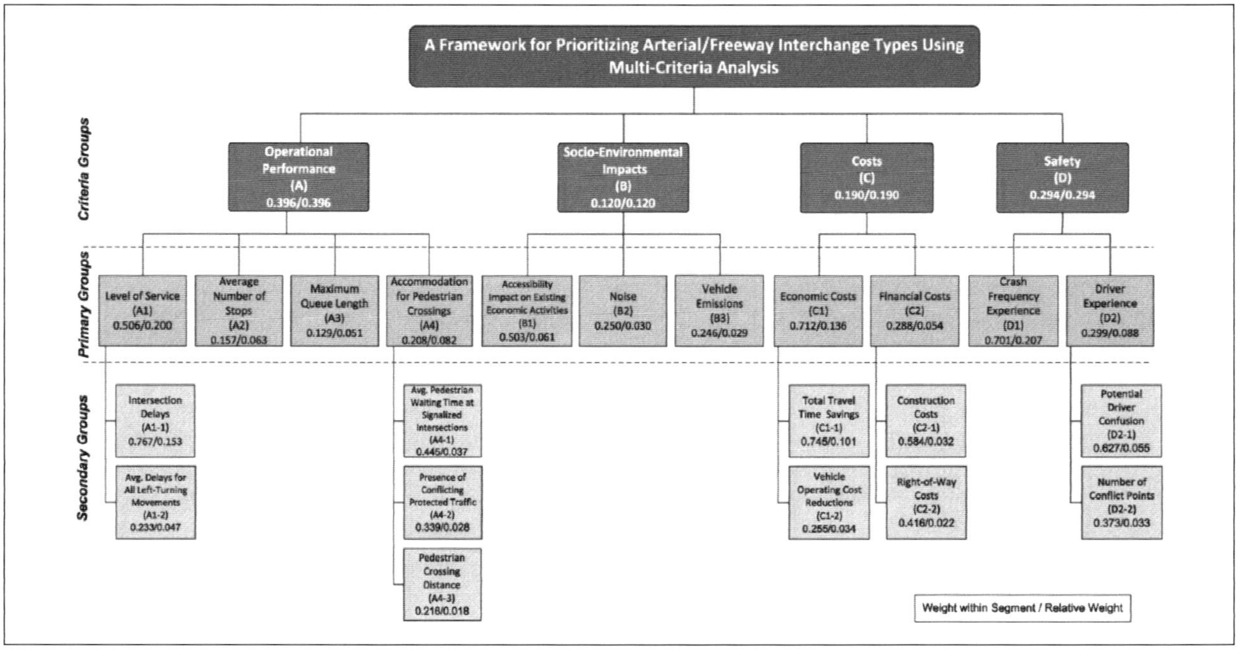

Figure 2. The proposed framework with criteria weights.

arterial/freeway interchange types. As shown, two weights are presented for each criterion at all levels (top, primary, and secondary). The weight on the left side represents the absolute weight within the segment to which the criterion belongs; the weights for criteria within the same segment/branch should add up to 1.000. The weight on the right side represents the weight within all sub-criteria to be scored; similarly, the weights for all sub-criteria to be scored should add up to 1.000.

Case Study

In this section, a site location for the case study is identified through which the prioritization framework is used to determine the preferred interchange configuration based on multi-criteria analysis. The scoring functions that are used to quantify the performance of the interchanges on the different criteria are introduced and consequently, using a linear additive model, an overall ranking of the interchanges is determined and a prioritization of the arterial/freeway interchange types is achieved. The procedure demonstrated in this study corresponds to a "Tier 3" selection process as detailed in the ITE freeway and highway geometric design handbook whereby the best alternative is selected based on conducting a quantitative rather than qualitative evaluation for the different alternative schemes (*11*).

Interchange Selection

Riyadh, the capital of the Kingdom of Saudi Arabia (KSA), is one of the fastest growing cities in the Middle East. Over the past couple of decades, the city of Riyadh has experienced tremendous growth in all aspects of urban life. According to a recent study, the city's population is expected to increase by 40% from 5.8 million in 2012 to 8 million in addition to a 60% rise in travel demand by the year 2030 (*34*).

One of the road corridors which will play an important role in supporting Riyadh's expansion is the "Sheikh Jaber Road" corridor. This road, being a major arterial road in Riyadh, has a significant strategic role in the city's road network, as it falls on the eastern part of the Second Ring Road as shown in Figure 3. With the ongoing plans to upgrade this road to cater for the increased traffic demand, it is proposed that this road act as a freeway to take the pressure off the congested nearby arterial roads. The grade-separated junction at the intersection between Sheikh Jaber road (proposed freeway) and Prince Bandar Bin Abdul Aziz road (existing arterial road) is selected as a case study for prioritizing different interchange types using multi-criteria analysis. The selection of the case study was made due to the availability of the traffic data in Riyadh in addition to the available geometric design plans on Sheikh Jaber road.

In order to assess the traffic conditions along Sheikh Jaber road, we contacted the engineering consulting firm that was commissioned to undertake the highway upgrade studies. It was noted that the (Emme/3) traffic model was used to assign the traffic onto the road network of Riyadh. This model was developed for the target year 2030 covering the whole city of Riyadh. Therefore, the forecasted traffic conditions at the case study site location were provided as shown in Figure 3.

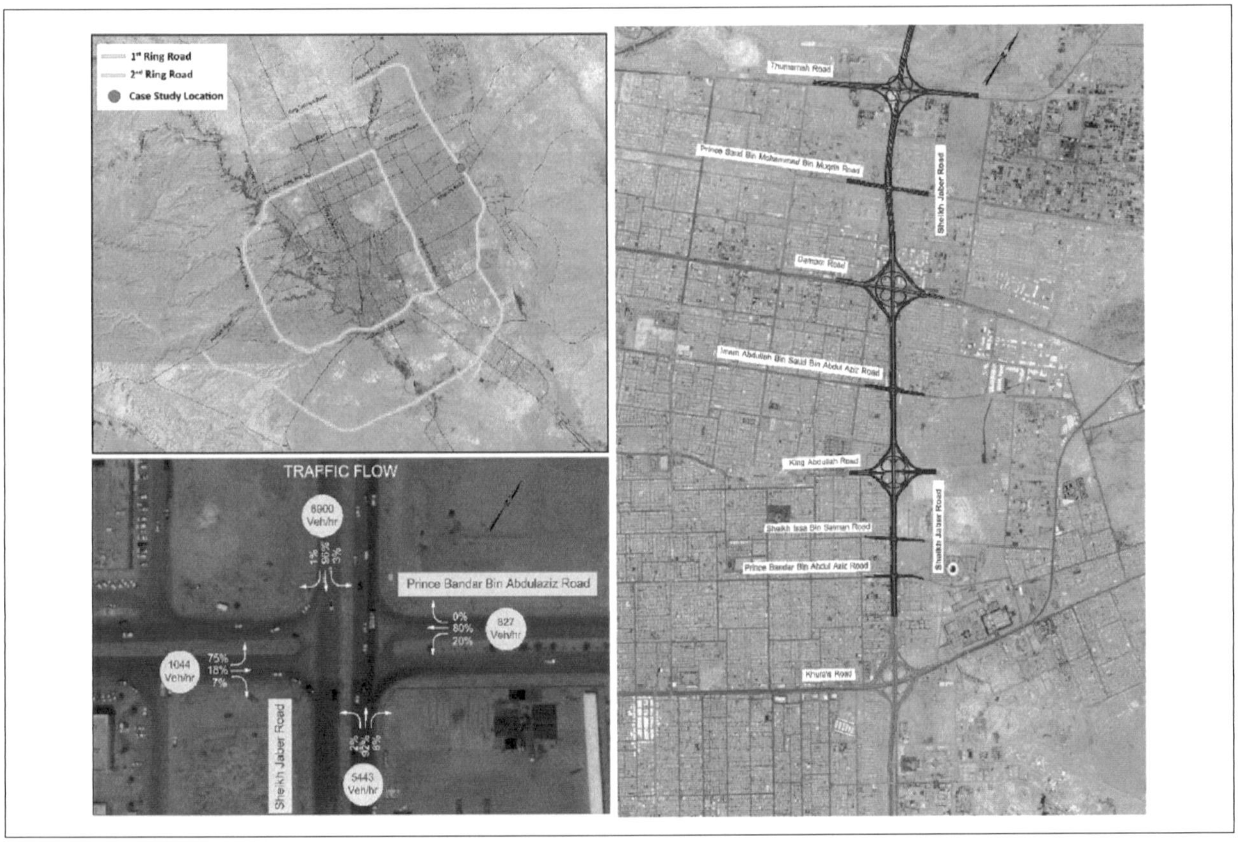

Figure 3. Aerial image of the case study site location.

Models Used in Scoring

Out of the 17 criteria that were used to rank and prioritize the arterial/freeway interchange types, three main modeling tools were used to generate the data needed for determining the scores. The tools were: VISSIM Traffic Model, Traffic Noise Model (TNM), and Vehicle Emissions Model.

VISSIM is one of the most important models used in this paper as it was useful in determining the operational performance of interchanges, evaluating their environmental impact, and computing their corresponding economic costs.

In addition, many models and applications have been suggested to assess the impact of road traffic on the environment. In this research, TNM version 2.5 was used for computing the noise levels and evaluating the acoustic impact at interchanges as it combines the noise emission levels with internal speed computations (*35, 36*).

As for the impact of vehicular emissions generated at the interchanges, several studies and standards were reviewed and considered. According to the EPA standards and to previous studies (*37*), greenhouse gas emission rates were found as a function of several attributes such as vehicle speed, vehicle fleet characteristics, functional road class, type of fuel used, and weather conditions. It was decided to use these functions to estimate the emission levels on the different types of interchanges as they provide a good indicator of the impact that changes in vehicular volumes and speeds have on emission rates.

Scoring the Criteria

This section provides a detailed description of the scoring functions and the corresponding criteria scores for six of the 17 criteria, due to paper-length limitations. The remaining descriptions and further details are provided in the thesis report prepared by Jamaleddin (*29*). The selection of these six criteria was based on the main criteria groups while showing a good representation of how the interchanges perform on the different criteria. Each interchange alternative receives a score ranging from "5" to "10" on each of those criteria. Some criteria were scored according to piecewise linear functions whereas others were scored based on discrete thresholds.

Intersection Delays (A1-1). Criterion A1-1 measures the performance of the interchanges according to the projected intersection delays. The scoring functions shown in Table 3 below are based on the level of service (LOS) criteria that are provided in the Highway Capacity Manual for signalized intersections. In general, the lower the intersection delays that are observed, the better the LOS an interchange achieves and consequently the higher it scores on this criterion.

Using the VISSIM traffic model, the operational performance for each of the alternative interchange configurations was determined. These results are conforming to previous research whereby the DDI outperformed the other interchange types on this criterion particularly due to its requiring only 2 signal phases (40 sec cycle length) whereas the TUDI performed the worst due to its 4 signal phases (140 sec cycle length).

Average Pedestrian Waiting Time at Signalized Intersections (A4-1). Pedestrian movements are considered in the study since the analysis relates to urban interchanges. Criterion A4-1 measures the performance of the interchanges based on pedestrian waiting times at signalized intersections. The scoring functions are taken from the informational guide of signalized intersections that is prepared by the Federal Highway Administration (*38*). Using Synchro model, the signal timing plans were optimized and the waiting time needed for the pedestrians to cross was determined. It was found that the DDI outperformed its interchange counterparts.

Noise (B2). Criterion B2 evaluates the performance of the interchanges based on the level of noise generated. According to the Highway Traffic Noise Abatement Guidelines, the noise attenuation requirements set by the FHWA indicate a maximum hourly sound level of 72 dBA for highway projects adjacent to developed lands and properties (*39*). Accordingly, the scoring functions were defined in relation to the areas having sound levels exceeding the specified noise level. Using the Traffic Noise Model, the TUDI ranked first on this criterion followed by the SPUI and DDI respectively.

Total Travel Time Savings (C1-1). Criterion C1-1 measures the performance of each of the alternatives according to the savings in travel time costs that these interchanges can potentially result in. Using the VISSIM model, the reductions in travel times were computed as compared with the at-grade signalized intersection. Both DDI and SPUI resulted in high travel time savings whereas the TUDI had lower savings and was thus assigned a lower score.

Construction Costs (C2-1). Criterion C2-1 measures the performance of the interchanges according to their construction costs. The higher the cost needed to construct the interchanges, the lower the score the interchange receives on this criterion. The DDI resulted in the lowest construction costs unlike the SPUI that scored the least due to its wide central intersection and large bridge structure.

Number of Conflict Points (D2-2). Criterion D2-2 rates the performance of the interchanges based on the number of potential vehicular conflict points such as crossing, merging, and diverging points. There are no clear guides/references that indicate the ranges of potential conflict points; therefore the scoring functions were proposed based on professional judgment. The DDI was found to have only 14 possible conflict points and consequently scored highest while the TUDI that had 26 conflict points and thus ranked the lowest.

Table 2 below summarizes the scoring functions used for the six criteria described earlier and illustrates how the interchages performed and scored on each of these criteria.

Ranking the Alternatives

According to the methodology adopted, Table 3 below displays the computed weights and introduces the scores of each interchange alternative for each of 17 criteria. Using the linear additive model, the weighted scores were then derived as shown in the far right side of Table 3 and a rank is then determined for each alternative. It was found that the DDI ranked first and was chosen as the most preferred interchange configuration for the case study. The SPUI ranked second while the TUDI was the least preferred. Due to the free flow movement for the left-turning traffic and high level of service achieved, the DDI ranked first under the majority of operational performance criteria, unlike the TUDI that experienced high delays and consequently ranked least. In addition, due to the geometric configuration of the different interchanges, the DDI ranked first under the "Financial Cost" primary criteria unlike the SPUI that ranked least due to the high construction cost needed to span the intersection with a large bridge structure.

Sensitivity Analysis

Uncertainty in the evaluation and ranking results may stem from the imprecision of the model inputs or the disagreement between the different decision-makers on the relative importance of different criteria. It can exist in different areas of the model such as the measurements of the performance of the alternatives on each criterion and the weights. Sensitivity tests explore the effects of changes in the criteria scores or weights on the overall results. If the results are relatively stable in the face of such changes, then the model is considered to be robust.

Sensitivity with Respect to Changing Volumes. In addition to using the forecast traffic volumes throughout a case study, it was important to study the sensitivity with respect to traffic volumes on a more continuous scale to determine the robustness of the model. The impact of increasing the traffic volumes by an additional 25%, 50%, 75%, and 100% on the performance of alternatives and the corresponding scores on the related criteria was analyzed thoroughly. It was found that the DDI remained the most preferred alternative for all scenarios tested. Similarly, SPUI ranked second in all scenarios whereas the TUDI ranked last.

Table 2. Criteria Scoring Functions and Corresponding Criteria Scores

Criterion	Scoring function		Criteria scoring			
A1-1	Delay range (sec/veh)	Function used		TUDI	SPUI	DDI
	[80, ∞)	A1-1 score = 5	Intersection delay (sec/veh)	57.31	11.56	10.30
	[55, 80)	A1-1 score = 6 − (1/25) (Delay − 55)	Score	5.91	8.84	8.97
	[35, 55)	A1-1 score = 7 − (1/20) (Delay − 35)				
	[20, 35)	A1-1 score = 8 − (1/15) (Delay − 20)				
	[10, 20)	A1-1 score = 9 − (1/10) (Delay − 10)				
	[0, 10)	A1-1 score = 10 − (Delay/10)				
A4-1	Waiting time range (sec)	Function used		TUDI	SPUI	DDI
	[60, ∞)	A4-1 score = 5	Waiting time (sec)	50.00	36.50	20.00
	[40, 60)	A4-1 score = 6 − (1/20) (waiting time − 40)	Score	5.50	6.35	8.00
	[30, 40)	A4-1 score = 7 − (1/10) (waiting time − 30)				
	[20, 30)	A4-1 score = 8 − (1/10) (waiting time − 20)				
	[10, 20)	A4-1 score = 9 − (1/10) (waiting time − 10)				
	[0, 10)	A4-1 score = 10 − (waiting time/10)				
B2	% Area range (noise level >72 dBA)	Function used		TUDI	SPUI	DDI
	[80, 100)	B2 score = 6 − (1/20) (%Area − 80)	% Area (noise level >72 dBA)	31%	48%	50%
	[60, 80)	B2 score = 7 − (1/20) (%Area − 60)	Score	8.47	7.60	7.49
	[40, 60)	B2 score = 8 − (1/20) (%Area − 40)				
	[20, 40)	B2 score = 9 − (1/20) (%Area − 20)				
	[0, 20)	B2 score = 10 − (%Area/20)				
C1-1	% Travel time savings range	Function used		TUDI	SPUI	DDI
	[90,100]	C1-1 score = 10 − (1/10) (100 − %Savings)	% Travel time savings	66.18%	74.01%	74.16%
	[80, 90)	C1-1 score = 9 − (1/10) (90 − %Savings)	Score	6.62	7.40	7.42
	[70, 80)	C1-1 score = 8 − (1/10) (80 − %Savings)				
	[60, 70)	C1-1 score = 7 − (1/10) (70 − %Savings)				
	[50, 60)	C1-1 score = 6 − (1/10) (60 − %Savings)				
	[0, 50)	C1-1 score = 5				
C2-1	Cost range (10^6 $)	Function used		TUDI	SPUI	DDI
	[20, ∞)	C2-1 score = 5	Cost (10^6 $)	10.24	16.78	6.70
	[16, 20)	C2-1 score = 6 − (1/4,000,0000) (Cost − 16,000,000)	Score	7.44	5.80	8.32
	[16, 12)	C2-1 score = 7 − (1/4,000,0000) (Cost − 12,000,000)				
	[8, 12)	C2-1 score = 8 − (1/4,000,0000) (Cost − 8,000,000)				
	[4, 8)	C2-1 score = 9 − (1/4,000,0000) (Cost − 4,000,000)				
	[0, 4)	C2-1 score = 10 − (Cost/4,000,000)				
D2-2	No. of conflict points range	Function used		TUDI	SPUI	DDI
	[35, ∞)	D2-2 score = 5	Potential number of conflict points	26	22	14
	[28, 35)	D2-2 score = 6 − (1/7) (Number − 28)	Score	6.29	6.86	8.00
	[21, 28)	D2-2 score = 7 − (1/7) (Number − 21)				
	[14, 21)	D2-2 score = 8 − (1/7) (Number − 14)				
	[7, 14)	D2-2 score = 9 − (1/7) (Number − 7)				
	[0, 7)	D2-2 score = 10 − (Number/7)				

Sensitivity with Respect to Criteria Weights. In addition to testing the sensitivity of changing the traffic volumes, additional tests were needed to explore the effects of changes in the criteria weights on the overall results.

Figure 4 below displays the sensitivity of rankings with respect to the weights of the main criteria groups. Using the computed weight (0.396) of the aggregate group for "Operational Performance (A)" criteria group, the DDI

Table 3. Weighted Score Table

	Weight					Score			Weighted score		
Criterion	Top	Primary	Secondary	Overall	Rank	TUDI	SPUI	DDI	TUDI	SPUI	DDI
A2	0.396	0.157	1.000	0.063	4	7.40	9.32	9.12	0.460	0.580	0.567
A3	0.396	0.129	1.000	0.051	7	7.42	8.96	9.06	0.379	0.457	0.463
B1	0.120	0.503	1.000	0.061	5	9.11	5.25	7.64	0.549	0.316	0.460
B2	0.120	0.250	1.000	0.030	13	8.47	7.60	7.49	0.254	0.228	0.225
B3	0.120	0.246	1.000	0.029	14	7.92	8.66	8.67	0.234	0.255	0.256
D1	0.294	0.701	1.000	0.207	1	5.00	5.00	5.00	1.033	1.035	1.033
A1-1	0.396	0.506	0.767	0.153	2	5.91	8.84	8.97	0.906	1.356	1.376
A1-2	0.396	0.506	0.233	0.047	8	5.20	8.46	8.93	0.243	0.395	0.417
A4-1	0.396	0.208	0.445	0.037	9	5.50	6.35	8.00	0.202	0.233	0.293
A4-2	0.396	0.208	0.339	0.028	15	8.81	8.82	8.64	0.246	0.246	0.241
A4-3	0.396	0.208	0.216	0.018	17	7.64	7.40	7.28	0.136	0.131	0.129
C1-1	0.190	0.712	0.745	0.101	3	6.62	7.40	7.42	0.667	0.746	0.748
C1-2	0.190	0.712	0.255	0.034	10	7.73	7.77	7.60	0.266	0.268	0.262
C2-1	0.190	0.288	0.584	0.032	12	7.44	5.80	8.32	0.238	0.186	0.266
C2-2	0.190	0.288	0.416	0.022	16	5.77	5.53	5.88	0.131	0.126	0.134
D2-1	0.294	0.299	0.627	0.055	6	10.00	8.50	7.00	0.553	0.470	0.387
D2-2	0.294	0.299	0.373	0.033	11	6.29	6.86	8.00	0.206	0.225	0.263
							Weighted score		6.702	7.254	7.518
							Rank		3	2	1

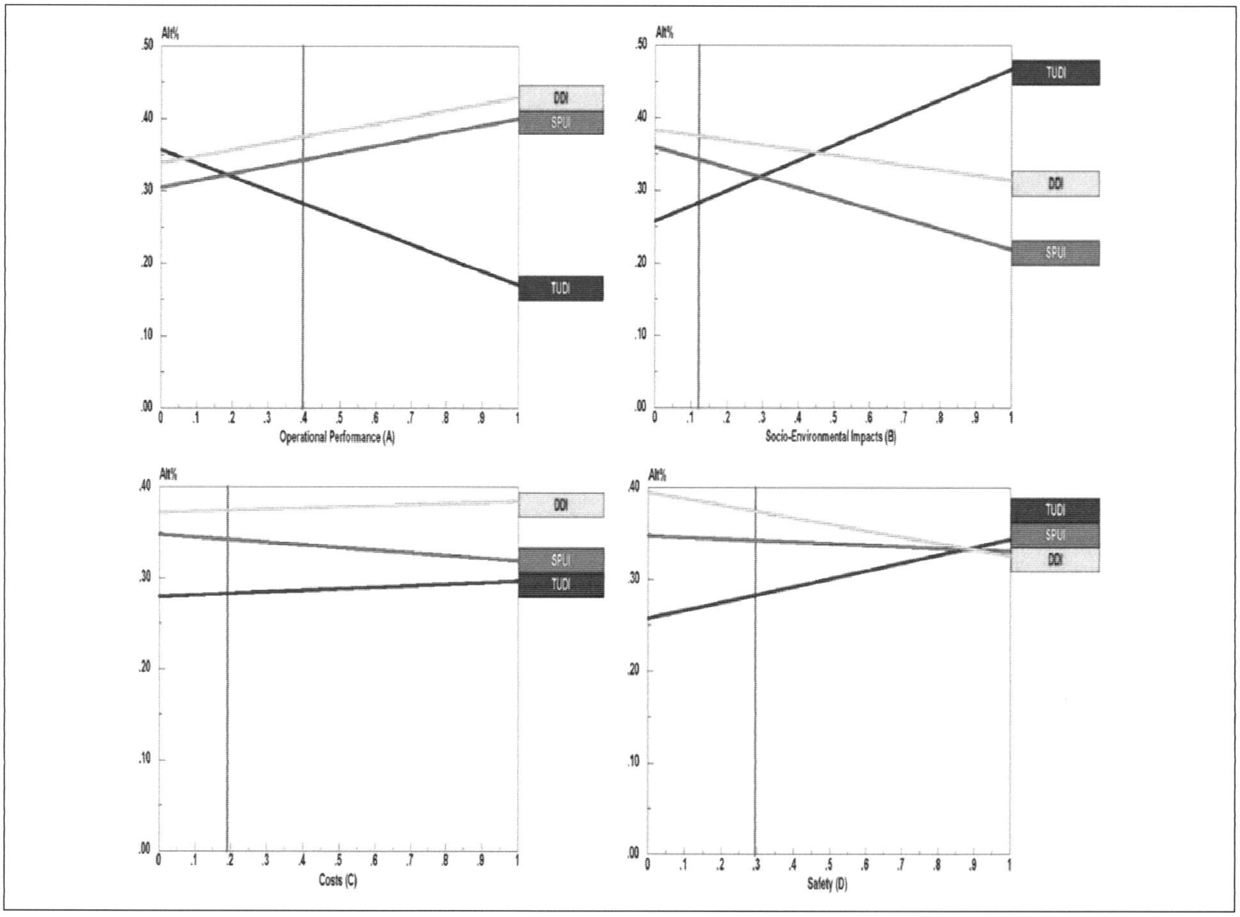

Figure 4. Impact of main criteria group weights on the prioritization of interchanges.

ranked first followed by the SPUI and TUDI, respectively. Based on the figure, the DDI retains the first rank whenever the weight of criterion (A) is above 10%, the SPUI ranks second for weights above 20%, whereas the TUDI could rank first for an assigned weight less than 10%. Similarly, the DDI retains the first rank whenever the weight of criterion (B) remains below 43%. On the other hand, the TUDI overtakes the ranking of its interchange counterparts when the weight assigned to this criterion exceeds 43%. As for the SPUI, its rank remains second for all weights below than 28% on this criterion; beyond this weight the SPUI becomes the least preferred interchange type. The same analysis is applied for Costs (C) and Safety (D).

Conclusions

This paper aimed to develop a framework for prioritizing urban arterial/freeway interchange types using multi-criteria analysis. The proposed framework involves defining a set of criteria to include possibly all factors that might affect the prioritization process of the different alternatives including operational performance, socio-environment, safety, and cost. Through an online questionnaire, the judgments of the decision-makers were elicited to determine the relative importance of criteria and the corresponding weights were assigned for each of the 17 criteria presented in the framework. As for computing the scores, some of the criteria were directly measured using microscopic simulation analyses, others were derived through mathematical computation while the remaining were subjectively analyzed based on expert judgment.

Through a case study implemented at the intersection of Sheikh Jaber road and Prince Bandar Bin Abdul Aziz road in the city of Riyadh, three different interchange types were prioritized using multi-criteria analysis. Using a linear additive model, the weighted scores of each alternative were computed. The results of the study showed that the DDI outperformed its interchange counterparts and was chosen as the most preferred interchange configuration. The SPUI ranked second while the TUDI was least preferred. In addition, this paper demonstrated the robustness of the model results through considering sensitivity tests on different criteria weight scenarios.

Compared with previous studies, this research makes several contributions. A comprehensive review of the literature pertaining to the operational performance, socio-environmental impacts, costs, and safety of three arterial/freeway interchanges was conducted and a summary of the applied evaluation metrics was introduced. As for prioritizing the urban arterial/freeway interchange types, a multi-criteria analysis was adopted rather than a conventional assessment approach (such as cost-benefit analysis or cost-effectiveness analysis) that does not address non-monetary factors. Furthermore, several scoring functions were developed to compute the scores of interchanges on the different criteria while the relative importance of different criteria was determined using an online questionnaire.

Although this research undertook a comprehensive analysis of three different urban arterial/freeway interchange types and examined their performance based on an actual case study, several areas could be considered in further research such as examining other similar in-use interchange configurations, such as roundabout diamond interchanges. In addition, the prioritization framework could be expanded to include all interchange classifications, such as arterial/arterial interchanges and freeway/freeway interchanges, and the like. Moreover, driver behavior at interchanges could be considered more closely in the evaluation by utilizing driving simulator experiments to provide some of the factor scores such as driver confusion.

Author Contributions

The authors confirm contribution to the paper as follows: study conception and design: Jamaleddin, Kaysi; data collection: Jamaleddin, Kaysi; analysis and interpretation of results: Jamaleddin, Kaysi; draft manuscript preparation: Jamaleddin, Kaysi. All authors reviewed the results and approved the final version of the manuscript.

References

1. Chaudhary, N. A., and C.-L. Chu. *Guidelines for Timing and Coordinating Diamond Interchanges with Adjacent Traffic Signals*. Report TX-00/4913-2. Texas Transportation Institute, 2000.
2. Qureshi, M., G. Spring, R. Lasod, and N. Sugathan. *Design of Single Point Urban Interchanges*. No. RDT 04-011. MODOT, Jefferson City, 2004.
3. Hughes, W., R. Jagannathan, D. Sengupta, and J. E. Hummer. *Alternative Intersections/Interchanges: Informational Report (AIIR)*. No. FHWA-HR-09-060. USDOT, McLean, VA, 2010.
4. Garber, N. J., and M. J. Smith. *Comparison of the Operational and Safety Characteristics of the Single Point Urban and Diamond Interchanges*. VTRC 97-R6. Virginia Transportation Research Center, VA, 1996.
5. Dorothy, P. W., T. L. Maleck, and L. Aylsworth-Bonzelet. Field Analysis of Operation and Design of Single-Point Urban Interchanges. Presented at 76th Annual Meeting of the Transportation Research Board, Washington, D.C., 1997.
6. Garber, N. J., and M. Fontaine. *Final Report-Guidelines for Preliminary Selection of the Optimum Interchange Type for a Specific Location*. No. VTRC 99-R15. VDOT: Virginia Transportation Research Council, 1999.
7. Sharp, W. H., and M. J. Selinger. Comparison of SPUI and TUDI Interchange Alternatives with Computer Simulation Modeling. Presented at the ITE Annual Meeting Compendium of Technical Papers, Washington, D.C, 2000.
8. Jones, E. G., and M. J. Selinger. A Comparison of Operations of Single-Point and Tight Urban Diamond Interchanges. *Transportation Research Record: Journal of the Transportation Research Board*, 2003. 1847: 29–35.
9. Bared, J. G., A. Powell, and E. Kaisar. Traffic Planning Models for Single-Point and Tight Diamond Interchanges.

Transportation Research Record: Journal of the Transportation Research Board, 2003. 1847: 52–57.
10. Bared, J. G., A. Powell, E. Kaisar, and R. Jagannathan. Crash Comparison of Single Point and Tight Diamond Interchanges. *Journal of Transportation Engineering*, Vol. 131, No. 5, 2005, pp. 379–381.
11. Leisch, J. P., and J. M. Mason. *Freeway and Interchange Geometric Design Handbook*. ITE, Washington, D.C., 2016.
12. Chlewicki, G. New Interchange and Intersection Designs: The Synchronized Split-Phasing Intersection and the Diverging Diamond Interchange. *Proc., Prepared for 2nd Urban Street Symposium*, Anaheim, CA, 2003.
13. Bared, J. G., P. K. Edara, and R. Jagannathan. Design and Operational Performance of Double Crossover Intersection and Diverging Diamond Interchange. *Transportation Research Record: Journal of the Transportation Research Board,* 2005. 1912: 31–38.
14. Sharma, S., and I. Chatterjee. Performance Evaluation of the Diverging Diamond Interchange in Comparison with the Conventional Diamond Interchange. Presented at Transportation Scholars Conference, Iowa State University, Ames, 2007.
15. Bared, J. G. *Double Crossover Diamond Interchange*. No. FHWA-HRT-09-054. McLean, VA, 2009.
16. Salley, J., and A. Christo. Easing Congestion in a New Way: The First Diverging Diamond Interchange in Chicagoland. *Proc., 2012 ITE Midwestern District Conference and TRB 4th Urban Street Symposium*, Chicago, IL, 2012.
17. Bared, J., T. Granda, and A. Zineddin. *Drivers' Evaluation of the Diverging Diamond Interchange*. FHWA-HRT-07-048. FHWA, U.S. Department of Transportation, 2007.
18. Schroeder, B., C. Cunningham, B. Ray, A. Daleiden, P. Jenior, and J. Knudsen. *Diverging Diamond Interchange: Informational Guide*. No. FHWA-SA-14-0672014, Federal Highway Administration, Washington, D.C., 2014.
19. American Association of State Highway and Transportation Officials. *Highway Safety Manual*, 1st ed. American Association of State Highway and Transportation, Washington, D.C., 2010.
20. MoDOT. *Missouri's Experience with a Diverging Diamond Interchange: Lessons Learned*. No. OR 10-021. Jefferson City, MO, 2010.
21. Chilukuri, V., S. Siromaskul, M. Trueblood, and T. Ryan. *Diverging Diamond Interchange Performance Evaluation (I-44 and Route 13)*. No. OR11.012. HDR Engineering for Missouri Department of Transportation, Chicago, IL, 2011.
22. Chlewicki, G. Learning from Six (Plus) Operational DDIs in the US. *Proc., 2012 ITE Midwestern District Conference and TRB 4th Urban Street Symposium*, 2012.
23. Vaughan, C., C. Cunningham, B. Schroeder, and J. E. Hummer. Early Findings on the Operational Impacts of Double Crossover Diamond Interchanges. Presented at the ITE 2012 Annual Meeting & Exhibit, Atlanta, GA, 2012.
24. Stanek, D. Innovative Diamond Interchange Designs: How to Increase Capacity and Minimize Cost. Proc., Institute of Transportation Engineers Annual Meeting, Pittsburgh, PA, 2007.
25. Siromaskul, S., and S. B. Speth. A Comparative Analysis of Diverging Diamond Interchange Operations. Presented at the ITE 2008 Annual Meeting and Exhibit, Anaheim, CA, 2008.
26. Afshar, A. M., J. G. Bared, S. Wolf, and P. K. Edara. Traffic Operational Comparison of Single-Point and Diverging Diamond Interchanges. Presented at 88th Annual Meeting of the Transportation Research Board, Washington, D.C., 2009.
27. Chlewicki, G. Should the Diverging Diamond Interchange Always Be Considered a Diamond Interchange Form? *Transportation Research Record: Journal of the Transportation Research Board*, 2011. 2223: 88–95.
28. Guerrieri, M., and F. Corriere. An International Review on One and Two Level Innovative Unconventional Intersection and Interchange. *ARPN Journal of Engineering and Applied Sciences*, Vol. 8, No. 12, 2013, pp. 1059–1070.
29. Jamaleddin, K. *A Framework for Prioritizing Arterial/Freeway Interchange Types Using Multi-Criteria Analysis*. Masters, 2015.
30. Kaysi, I., H. Al-Hathlool, F. Darwish, and M. Sayour. Prioritization of National Road Projects in Saudi Arabia: Framework and Implementation. *Proc., 17th International Road Federation (IRF) World Meeting and Exhibition*, Riyadh, Saudi Arabia, 2013.
31. Department of Communities and Local Governments. *Multi-Criteria Analysis: A Manual*, Communities and Local Government Publications, London, UK, 2009.
32. Tudela, A., N. Akiki, and R. Cisternas. Comparing the Output of Cost Benefit and Multi-Criteria Analysis: An Application to Urban Transport Investments. *Transportation Research Part A: Policy and Practice*, Vol. 40, No. 5, 2006, pp. 414–423.
33. Saaty, T. L., and K. P. Kearns. *Analytical Planning: The Organization of Systems*. Pergamon Press, Oxford, 1985.
34. Arab News. Metro Work Will Disrupt Traffic, Governor Tells Riyadh Residents. *Arab News*, 2004. http://www.arabnews.com/news/619571.
35. Hankard, M., J. Cerjan, and J. Leasure. *Evaluation of the FHWA Traffic Noise Model, TNM, for Highway Traffic Noise Prediction in the State of Colorado*. No. CDOT-2005-21. Colorado Department of Transportation, Research Branch, 2006.
36. Menge, C., C. Rossano, G. Anderson, and C. Bajdek. FHWA Traffic Noise Model: Technical Manual. FHWA Traffic Noise Model, Version 1: Technical Manual. Federal Highway Administration, US Department of Transportation, 1998.
37. Sbayti, H., M. El-Fadel, and I. Kaysi. Effect of Roadway Network Aggregation Levels on Modeling of Traffic-Induced Emission Inventories in Beirut. *Transportation Research Part D: Transport and Environment*, Vol. 7, No. 3, 2002, pp. 163–173.
38. Rodegerdts, L. A., B. Nevers, B. Robinson, J. Ringert, P. Koonce, J. Bansen, T. Nguyen, J. McGill, D. Stewart, J. Suggett, and T. Neuman. *Signalized Intersections: Informational Guide*. No. FHWA-HRT-04-091, Federal Highway Administration, Washington, D.C., 2004.
39. FHWA. *Highway Traffic Noise Abatement*. Project Development Procedures Manual, 2009.

The Standing Committee on Geometric Design (AFB10) peer-reviewed this paper (18-01181).

Article

TRR
JOURNAL OF THE TRANSPORTATION RESEARCH BOARD

Transportation Research Record
2018, Vol. 2672(39) 155–165

DOI: 10.1177/0361198118793521
journals.sagepub.com/home/trr
SAGE

Enhanced MASH TL-3 Short-Radius System: Promising Design to Accommodate Flat Terrain and Ditches

Akram Y. Abu-Odeh[1], Roger P. Bligh[1], Christopher Lindsey[2], and Wade Odell[3]

Abstract

A challenging guardrail installation situation presents itself when two roadways intersect. Combining the guardrails from intersecting roadway results in what is commonly known as a short radius or T-intersection. It is difficult if not physically impossible to provide the required tensile capacity to the geometrical constraints of the curved section. Researchers and practitioners in the roadside safety area have been investigating the short-radius issue for many years. Investigators conducted numerous crash tests for different short-radius guardrail designs, yet none of those designs passed the National Cooperative Highway Research Program (NCHRP) Report 350 Test Level 3 (TL-3) criteria. In 2009, the crash testing guidelines were updated in the *Manual for Assessing Safety Hardware* (MASH). MASH guidelines increased the impact severity for TL-3 tests over those in NCHRP 350. This paper presents a MASH TL-3 short-radius design that was successfully crash tested for both a flat terrain and a 3H:1V sloped terrain behind the installation. The impact conditions adopted from the MASH terminal/crash cushion matrix were MASH 3-33, 3-32, 3-31, and 3-35 for the flat terrain. Additionally, a slightly modified design that was installed in front of a 3H:1V slope was successfully evaluated using MASH 3-33 and 3-32 test conditions. These tests used a 25° impact angle since it was shown to be more critical for installation during simulation of the system.

It is difficult if not physically impossible to provide the required barrier length along a highway if it intersects private driveways and county roads. The presence of site constraints such as a bridge rail or a culvert along the highway may not allow the placement of a properly designed guardrail. In these cases, the alternatives are to relocate the site constraint if possible, shorten the guardrail length, or provide a curved guardrail design. The curved guardrail design is known as a short radius or T-intersection depending on specific constraints.

Researchers and practitioners in the roadside safety area have been investigating the short-radius issue for many years. Although numerous tests were conducted for different short-radius guardrail designs, none of them passed National Cooperative Highway Research Program (NCHRP) Report 350 Test Level 3 (TL-3) (*1*) requirements. The crash testing guidelines were updated in the Manual for Assessing Safety Hardware (MASH) (*2*). MASH increased the impact severity for TL-3 tests. Satisfying that impact severity is now even more challenging for any short-radius system.

This paper presents a MASH TL-3 short-radius design that was successfully crash tested for both flat terrain and sloped terrain applications. This short-radius design utilizes a thrie-beam application through its perimeter as the basic element of containment. This thrie beam is augmented with energy dissipation sand barrels and a tension-carrying cable over the nose section. The number of sand barrels is different for the sloped terrain to accommodate the increased capture capacity required by the presence of the 3H:1V slope.

Background

One of the earliest known series of tests of a short-radius guardrail system was conducted by the Southwest Research Institute (SwRI) per NCHRP Report 230

[1]Roadside Safety and Physical Security Division, Texas A&M Transportation Institute, College Station, TX
[2]Design Division (Roadway Design Section), Texas Department of Transportation, Austin, TX
[3]Research & Technology Implementation Office, Texas Department of Transportation, Austin, TX

Corresponding Author:
Address correspondence to Akram Y. Abu-Odeh: abu-odeh@tamu.edu

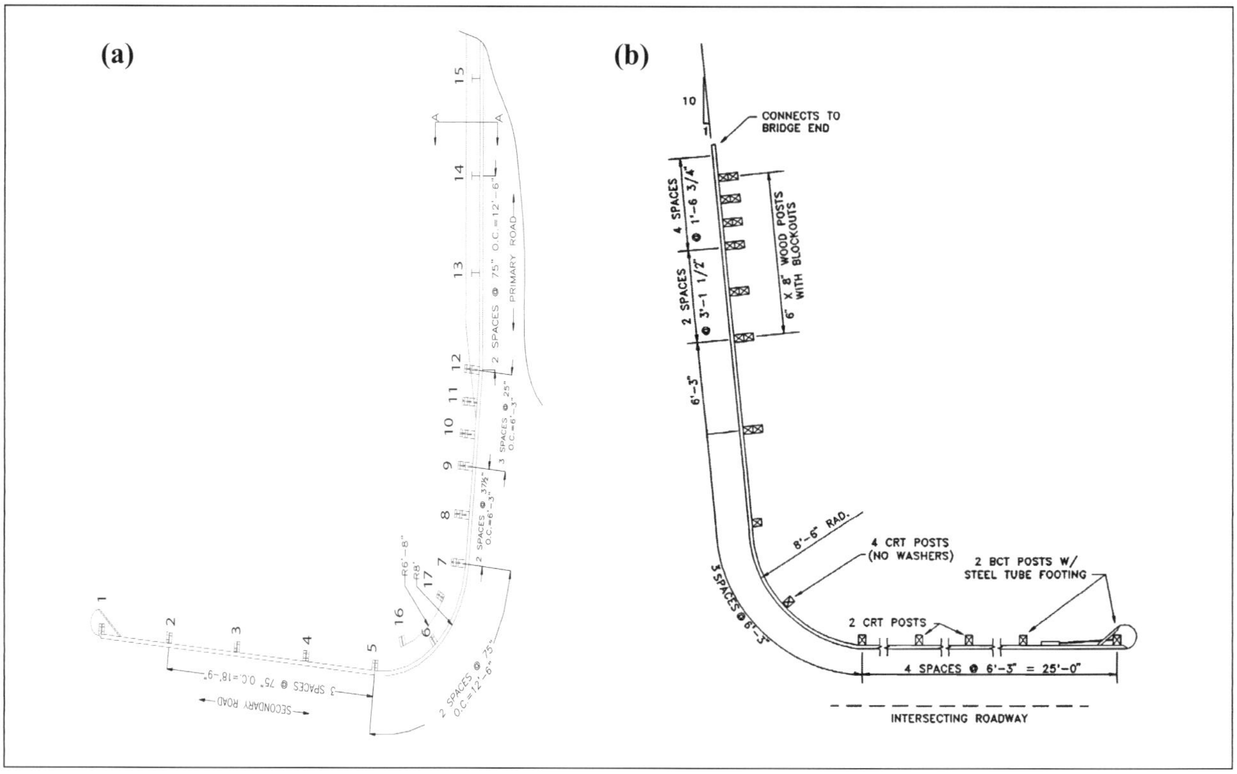

Figure 1. Short-radius installation conducted by SwRI in (*a*) 1981 and (*b*) 1989.

guidelines in 1981 (*3*). The short-radius test installation is shown in Figure 1*a*. As a result, a system was developed to contain both a 4,400 lb (2,000 kg) sedan and an 1,800 lb (820 kg) small car, although the performance was marginal. During the course of the testing, SwRI found that a tangent guardrail length of 25 ft (7.620 m) or more was needed to develop sufficient tensile capacity in the rail when impacted in the radius region.

Later, SwRI conducted a series of tests for the Yuma County, Arizona, Public Works Department, and the tests were evaluated per NCHRP Report 230 criteria (*4*). The system was developed with an 8 ft (2.44 m) radius and designed for low-volume roads, as shown in Figure 1*b*. The system was successfully tested at this lower service level. The Federal Highway Administration (FHWA) developed the guidance for curved W-beam guardrail installations at minor roadway intersections issued in 1992 (*5*). The recommended designs were developed based on the research results of SwRI.

In 2010, the Texas A&M Transportation Institute (TTI) conducted a study to investigate the performance of previously tested T-intersection guardrail systems to determine if some of these systems would meet NCHRP Report 350 TL-2 criteria (*6*). The evaluations performed in that study indicate that the Yuma County T-intersection guardrail design meets NCHRP Report 350 TL-2 criteria.

Another series of tests was conducted by TTI per NCHRP Report 230 and NCHRP Report 350 guidelines under two different research projects in the early 1990s (*7, 8*). In those tests, 7 in. (178 mm) diameter round wood posts were used in the system. The posts used in the curved section had 3.5 in. (89 mm) holes that were similar to the holes in the controlled released terminal (CRT) wood post. There were no freestanding posts in these tests. The weakened round wood posts broke readily once impacted by the test vehicles, similar to the CRT wood posts. The system was judged to meet NCHRP Report 230 but not NCHRP Report 350 guidelines. The short-radius test installation with the thrie beam is shown in Figure 2*a*.

More recently, the Midwest Roadside Safety Facility (MwRSF) embarked on testing a T-intersection curved rail design per NCHRP Report 350 and subsequent MASH TL-3 impact conditions (*9–12*). A short-radius system consists of a curved and slotted thrie-beam nose section that is supported by breakaway posts, as shown in Figure 2*b*. One side of the system was attached to a TL-3 steel post approach transition, while the other was attached to a TL-2 end terminal. However, efforts to

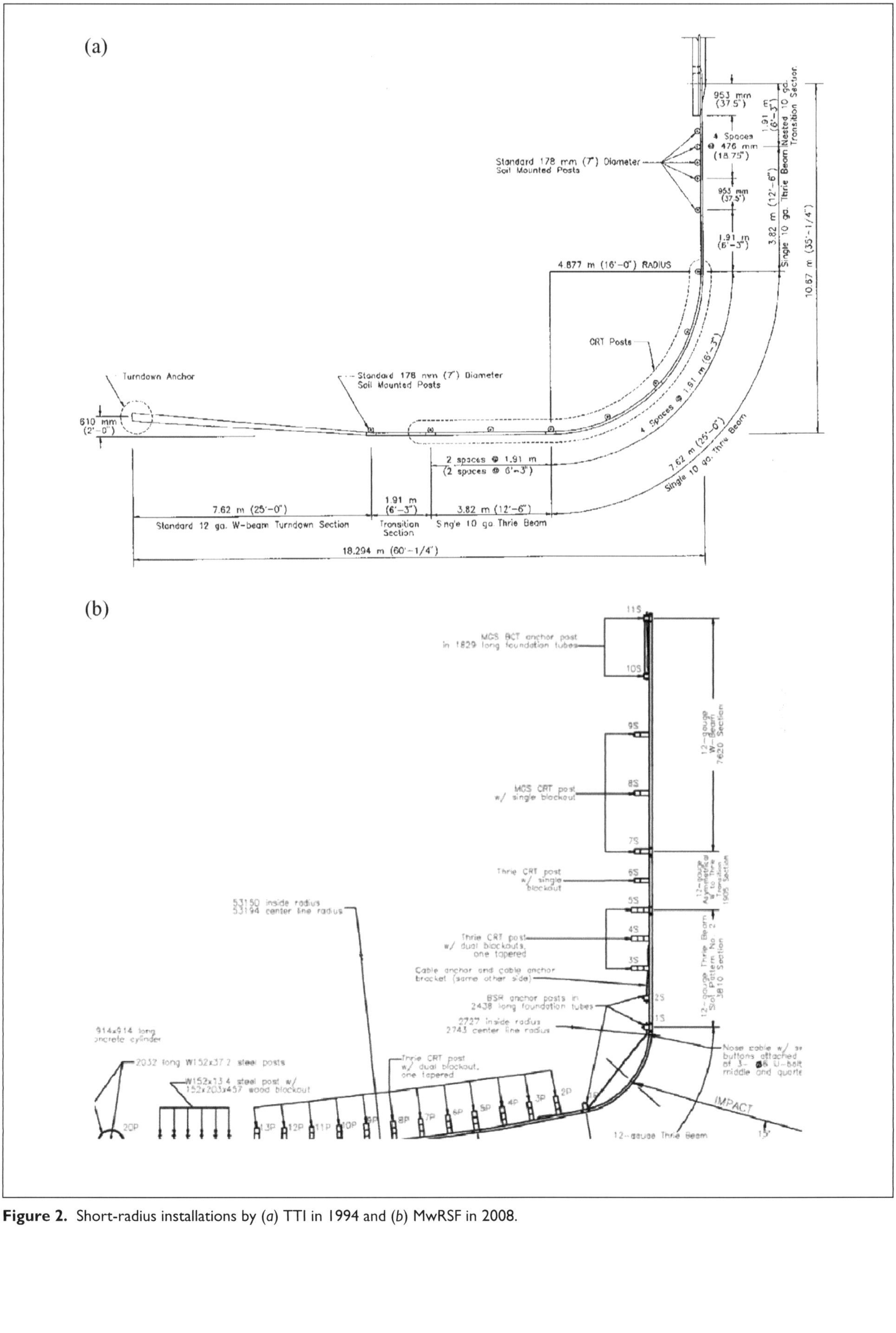

Figure 2. Short-radius installations by (a) TTI in 1994 and (b) MwRSF in 2008.

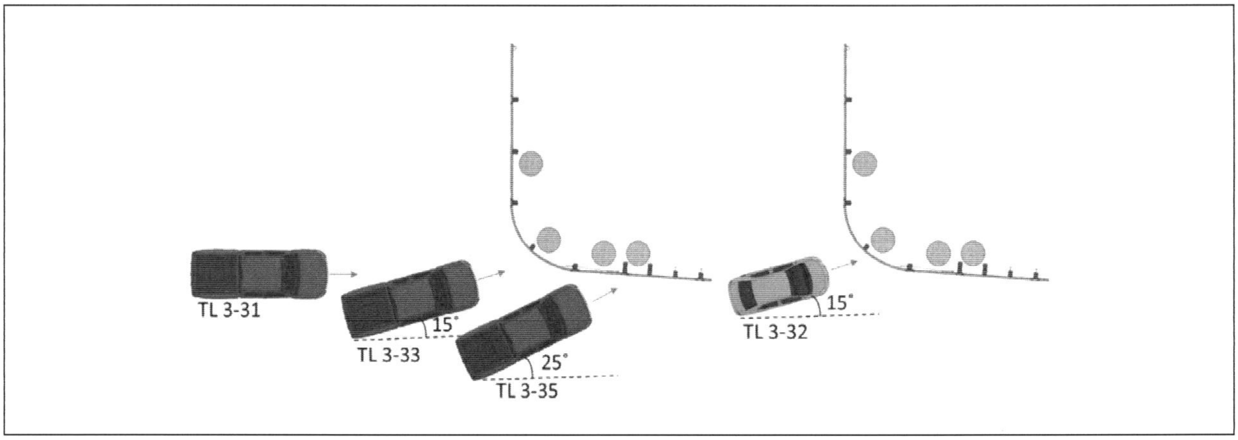

Figure 3. MASH TL-3 recommended test matrix for short radius on flat terrain.

date have not been successful at achieving a TL-3 compliant short-radius design.

Research Approach

Investigation of the geometry of a short-radius system suggests that the critical test conditions are 3-33, 3-32, 3-31, and 3-35 from the MASH crash cushions test matrix. The recommended tests are shown in Figure 3. Other tests in the matrix (3-36, 3-37, and 3-38) are not applicable to the short radius since they focus on interaction with the rigid hazard of the crash cushion. This fact should be considered in the design of the transition to the bridge parapet. The redirecting test (3-39) is not considered critical since its design is softer at the nose and flared out 4° for the reverse impact condition. Impact at the nose seems to be the most challenging test due to lack of tension in a curved section, which constitutes the nose portion of the system.

Several concepts were developed through this research based on the realization that the short-radius railing components were not sufficient by themselves to capture an impacting vehicle unless additional contribution was added from another feature. These features included specially designed posts with mass containers, cable components, and inertial mass. In summary, the simulation effort narrowed the potential design to focus on augmenting the short-radius railing with inertial contribution from frangible sand barrels. This narrowed the problem to designing the optimum layout of sand barrels to capture a heavy vehicle within the desired minimum stopping distance while stopping a small vehicle without violating the occupant risk factor. Refined modeling and subsequent simulations helped the researchers develop an optimum short-radius design that satisfied both conditions.

The short-radius installation on a sloped terrain was tested with MASH 3-33 and 3-32 but using a 25° impact angle since the simulation indicated that 25° was more critical than 15°.

Detailed Description of the Short-Radius Installation

Each flat terrain test installation consisted of a 31 in. tall, 58 ft 10 in. long, thrie-beam short-radius guardrail system constructed with a 27 ft 7 ¼ in. long primary road leg (as measured along the guardrail) that transitioned to a section of bridge parapet, and an 18 ft 9 in. long secondary road leg that terminated with a rounded thrie-beam end section. The curved 12 ft 6 in. (post-to-post) arc length thrie-beam section was rolled to an 8 ft 4 ½ in. radius. The primary roadside thrie beam of the system was flared to the field side 4° from the tangent line of the parapet face, and the secondary roadside thrie beam was perpendicular to the parapet face tangent. Four sand barrels were strategically placed on the inboard, field side of the installation. The end anchor on the secondary roadway and the modified breakaway cable terminal (BCT) foundation tube were analyzed and designed to withstand expected loads due to vehicular impact. Figure 4 depicts the short-radius drawing for the short radius on flat terrain. This short-radius system had a rail height of 31 in. (787 mm) utilizing a gauge 12 thrie-beam element to make up all the railing components in the system.

For the sloped field terrain installation, two additional sand barrels were used to accommodate the demands of having a sloped fill behind the installation. Figures 5 and 6 show the short-radius drawings for the sloped terrain application.

The installations of both the flat terrain short radius and the sloped terrain short radius are shown in Figure 7. The slope break was placed 5 ft away from the primary and the secondary rail in the sloped terrain installation.

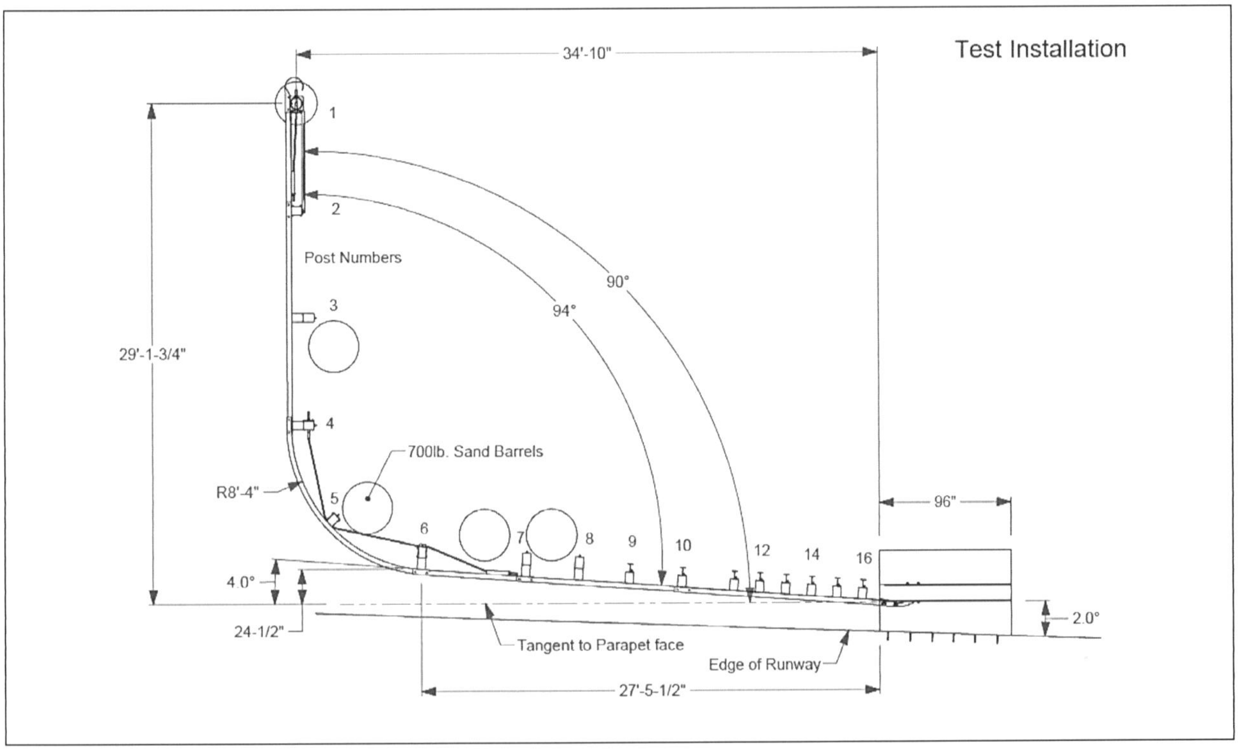

Figure 4. An overall view of the short-radius installation for MASH 3-33, 3-31, and 3-32 tests, and the first MASH 3-35 test for flat terrain.

Full-Scale Crash Tests for Short Radius on Flat Terrain

MASH Test 3-33

MASH Test 3-33 involved a 2270P vehicle weighing 5,000 lb (2,270 kg) traveling at 62 mph (100 km/h) and impacting the nose section of the short radius at 15° relative to the main roadway. As shown in Figure 8, the system was able to capture the vehicle and bring it to a controlled stop.

MASH Test 3-32

MASH Test 3-32 examined the behavior of the short-radius system during an oblique impact on the nose of the system. Occupant risk and vehicle trajectory were the main concerns with regard to this test. The 1100C vehicle impacted the center of the radius of the system at a 15° angle and 62 mph (100 km/h). As shown in Figure 9, the system was able to capture the vehicle and bring it to a controlled stop.

MASH Test 3-31

MASH Test 3-31 is designed to show whether a system is capable of safely and stably decelerating a 2270P vehicle to a stop. For this system, the 2270P vehicle needed to be redirected or safely captured when the impact was parallel to one of the sides of the system. As shown in Figure 10, the system was able to capture the vehicle and bring it to a controlled stop.

MASH Test 3-35 (First Test)

MASH Test 3-35 was conducted to determine the rail's capacity for containing or redirecting a truck. The impact location is defined in MASH as the beginning of length of need at a 25° angle. The truck impacted the system at a 25° angle in the first section of guardrail along the primary roadway after the radius. Although the system was able to redirect the pickup truck, the vehicle began to roll clockwise once it started to redirect, as shown in Figure 11.

MASH Test 3-35 (Second Test with System Modifications)

Modifications to the short-radius design were made to overcome the challenges imposed by Test 3-35. The modifications were made only outside the maximum deflected region of MASH Test 3-33 in order to not affect the performance of the earlier successful tests.

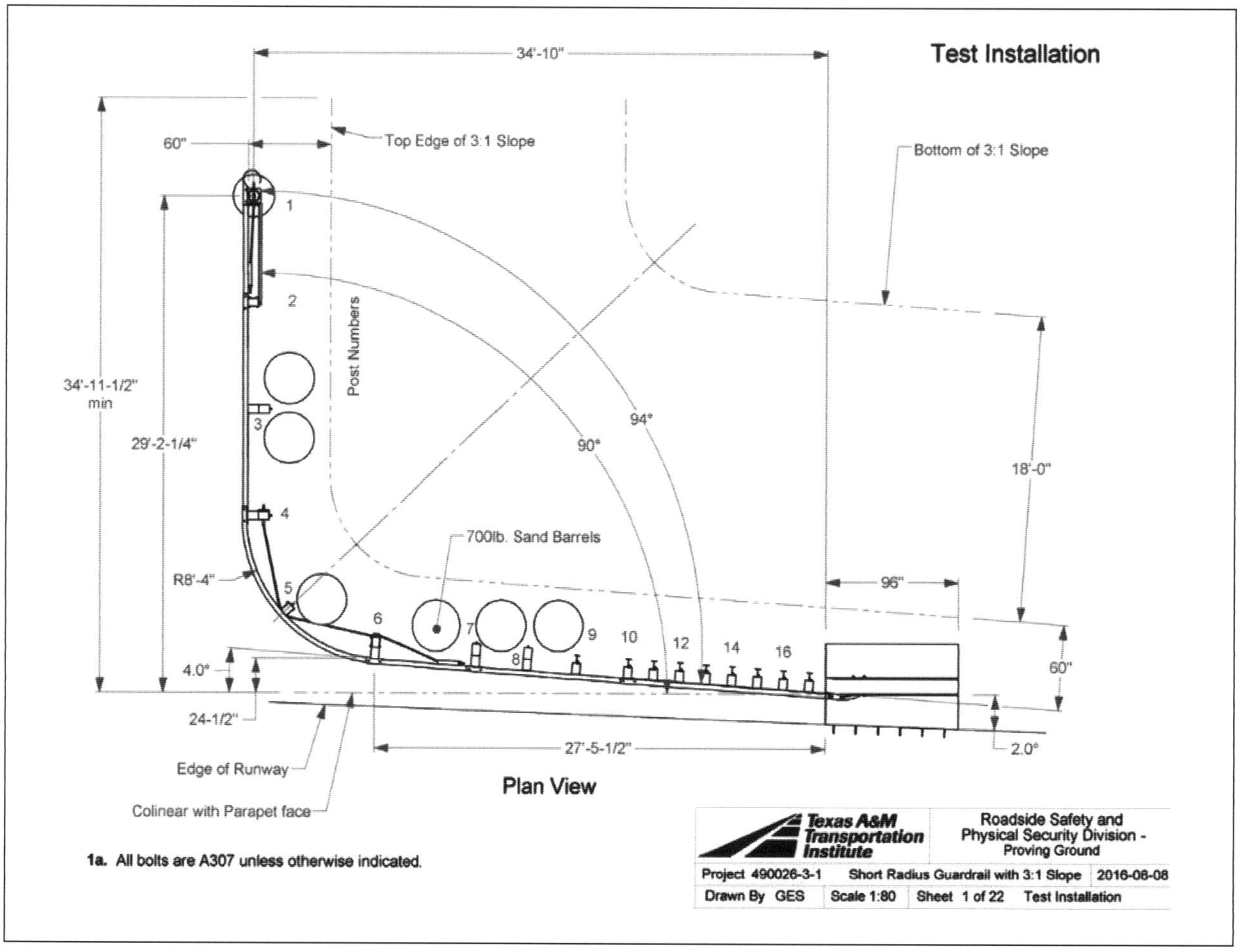

Figure 5. An overall view of the short-radius installation for MASH 3-33 and MASH 3-32 tests for sloped terrain.

The test installation for the second MASH 3-35 test had an extra steel post in the thrie-beam section between post 10 and the parapet. This design resulted in a total of 17 posts for the installation. Furthermore, a shorter blockout (14 in. instead of 18 in.) was used at posts 9, 10, and 11.

The guardrail was not bolted to posts 12 and 13. At post 10, the upstream single thrie-beam section was sandwiched between the nested double thrie beams. Figure 12 illustrates the modified installation.

With this new configuration, the system was able to capture the vehicle and bring it to a controlled stop, as shown in Figure 13.

Full-Scale Crash Tests for Short Radius on Sloped Terrain

Simulations indicated that a critical impact angle for MASH 3-33 and MASH 3-32 was 25° for the same short-radius system but with a 3H:1V sloped terrain behind it. The system configuration was based on the second MASH 3-35 test installation shown in Figure 12. However, simulation studies indicated that two extra sand barrels were needed; this system configuration for the short radius with sloped terrain is shown in the lower image in Figure 7.

MASH Test 3-33

MASH Test 3-33 involved a 2270P vehicle weighing 5,000 lb (2,270 kg) traveling at 62 mph (100 km/h) and impacting the nose section of the short radius at 25° relative to the main roadway. Although the right rear wheel of the pickup truck straddled the thrie beam, as shown in Figure 14, the system was able to capture the vehicle and bring it to a controlled stop.

MASH Test 3-32

MASH Test 3-32 examined the behavior of the short-radius system during an oblique impact on the nose of

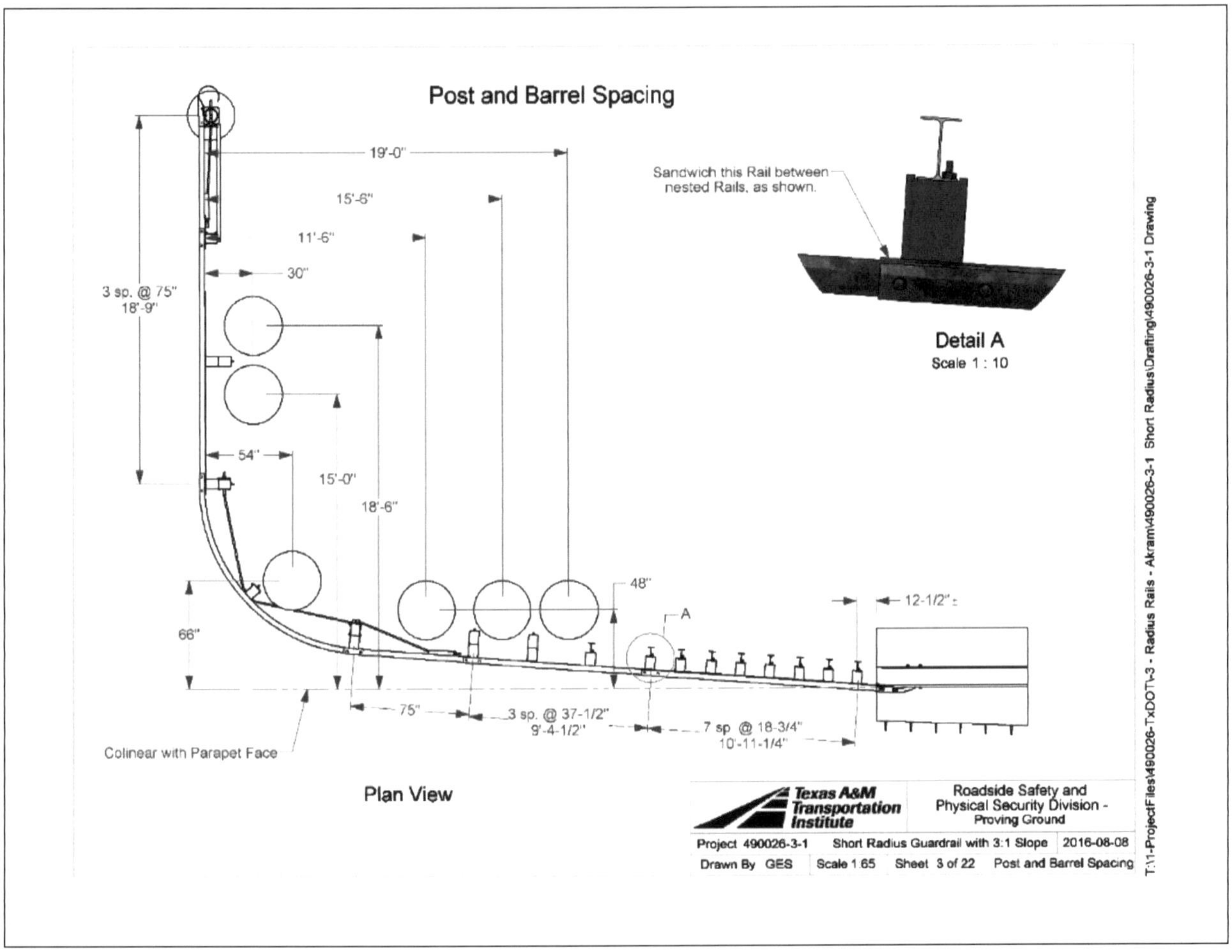

Figure 6. An overall view of the short-radius installation tests for sloped terrain showing barrel locations.

the system. Occupant risk and vehicle trajectory were the main concerns with regard to this test. The 1100C vehicle impacted the center of the radius of the system at a 25° angle and 62 mph (100 km/h). The system was able to capture the vehicle and bring it to a controlled stop, as shown in Figure 15.

Conclusions and Implementation

A MASH TL-3 compliant short-radius system was developed and tested by TTI for both a flat terrain and a 3H:1V sloped terrain configuration. High-fidelity simulations were conducted that accurately predicted the performance of the subsequent full-scale crash tests (*13*). The final short-radius system that was simulated and crash tested consisted of a thrie beam that was 18 ft 9 in. long placed along the secondary roadway. The radius itself was 8 ft 4 in. and connected to the thrie beam on the primary roadway, which was 27 ft 5 in. long. The primary road rail section included a transition section to connect the rail to the concrete parapet design. A combination of BCT and CRT wood posts was utilized to provide quick post releases for the capture impacts. A tension cable began on the primary roadway, ran across the nose section along the ground, and anchored on the secondary roadway. This tension cable helped to maintain the tension in the rail for impacts such as those tested in MASH 3-35 and MASH 3-31. Frangible sand barrels were spaced behind the rail to help slow the vehicle down without violating occupant impact velocity and ride-down acceleration thresholds while maintaining a desired stopping distance behind the rail. Four sand barrels were needed for the short-radius system with flat terrain, and six sand barrels were needed for the system with the 3H:1V sloped terrain.

Figure 7. Short-radius installation for both flat terrain (above) and sloped terrain (below).

Figure 8. MASH Test 3-33 (flat terrain) at impact and final vehicle position.

This new short-radius system requires a placement footprint of 34 ft 10 in. along the primary road and 29 ft 3 in. along the secondary road. This 32 in. tall short-radius system is MASH TL-3 compliant; thus, it can be used on primary roads where TL-3 (or lower) safety features are recommended.

The system with four 700 lb (318 kg) sand barrels requires a graded flat ground behind it at a slope of 10H:1V or flatter. However, the system with six 700 lb (318 kg) sand barrels can be installed with a 3H:1V sloped terrain behind it.

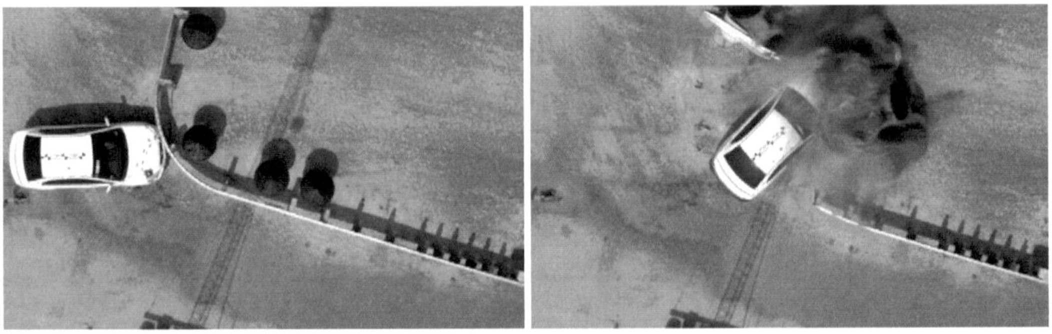

Figure 9. MASH Test 3-32 (flat terrain) at impact and final vehicle position.

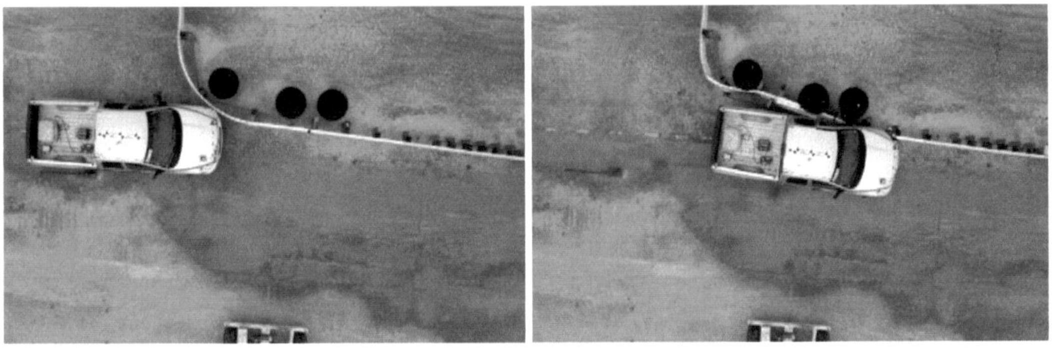

Figure 10. MASH Test 3-31 (flat terrain) at impact and final vehicle position.

Figure 11. MASH Test 3-35 (flat terrain, first test) at impact and vehicle rolling over as it exits the system.

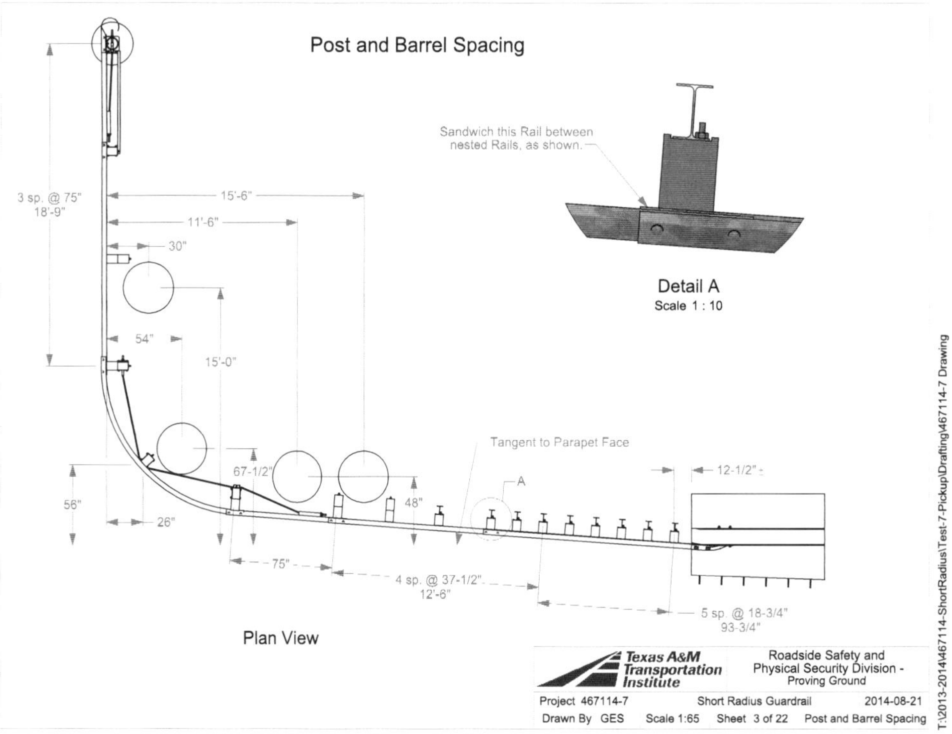

Figure 12. An overall view of the short-radius installation for the second MASH 3-35 test for flat terrain.

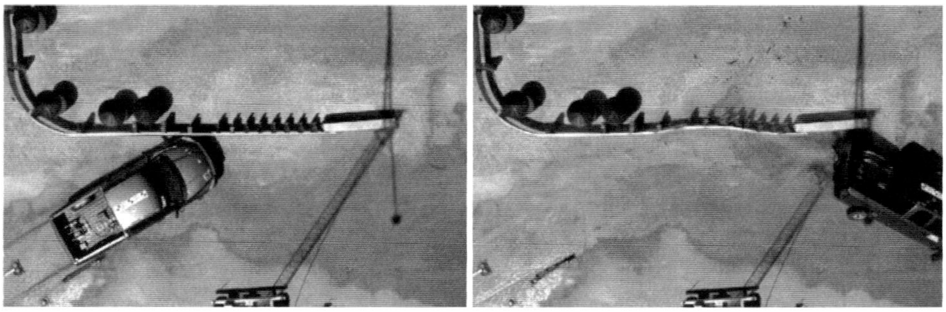

Figure 13. MASH Test 3-35 (flat terrain, second test) at impact and vehicle exiting the system without rolling over.

Figure 14. MASH Test 3-33 (sloped terrain) at impact and final vehicle position.

Figure 15. MASH Test 3-32 (sloped terrain) at impact and final vehicle position.

Acknowledgments

This research project was conducted under a cooperative program between the Texas A&M Transportation Institute, the Texas Department of Transportation (TxDOT), and the Federal Highway Administration. The authors acknowledge and appreciate TxDOT guidance and assistance throughout the performance period of the project.

Author Contributions

The authors confirm contribution to the paper as follows: study conception and design: AA-O, RB, CL, WO; data collection: AA-O, RB; analysis and interpretation of results: AA-O, RB, CL, WO; draft manuscript preparation: AA-O. All authors reviewed the results and approved the final version of the manuscript.

References

1. Ross, H. E., D. L. Sicking, R. A. Zimmer, and J. D. Michie. *NCHRP Report 350: Recommended Procedures for the Evaluation of Highway Features*. HRB, National Research Council, Washington, D.C., 1993.
2. American Association of State Highway and Transportation Officials. *Manual for Assessing Safety Hardware*. AASHTO, Washington, D.C., 2009.
3. Michie, J. D. *NCHRP Report 230: Recommended Procedures for the Safety Performance Evaluation of Highway Appurtenances*. HRB, National Research Council, Washington, D.C., 1981.
4. Mayer, J. B. *Full-Scale Crash Testing of Approach Guardrail for Yuma County Public Works Department*. Southwest Research Institute, San Antonio, TX, 1989.
5. Federal Highway Administration. *Curved W-Beam Guardrail Installations at Minor Roadway Intersections (T5040.32)*. April 13, 1992. http://safety.fhwa.dot.gov/roadway_dept/policy_guide/road_hardware/barriers/techadvs/t504032.cfm. Accessed February 2011.
6. Abu-Odeh, A. Y., K. Kim, and D. C. Alberson. *Evaluation of Existing T-Intersection Guardrail Systems for Equivalency with NCHRP Report 350 TL-2 Test Conditions*. Texas Transportation Institute, College Station, TX, 2010.
7. Ross, H. E., R. P. Bligh, and C. B. Parnell. *Bridge Railing End Treatments at Intersecting Streets and Drives*. Texas Transportation Institute, College Station, TX, 1992.
8. Bligh, R. P., H. E. Ross, and D. C. Alberson. *Short-Radius Thrie Beam Treatment for Intersecting Streets and Drives*. Texas Transportation Institute, College Station, TX, 1994.
9. Bielenberg, R. W., J. D. Reid, R. K. Faller, J. R. Rohde, D. L. Sicking, and E. A. Keller. *Concept Development of a Short-Radius Guardrail System for Intersecting Roadways*. Midwest States Regional Pooled Fund Program, Lincoln, NE, 2000.
10. Bielenberg, R. W., R. K. Faller, J. C. Holloway, J. D. Reid, J. R. Rohde, and D. L. Sicking. *Phase II Development of a Short-Radius Guardrail for Intersecting Roadways*. Midwest Roadside Safety Facility, Lincoln, NE, 2003.
11. Stolle, C. S., K. A. Polivka, J. D. Reid, J. R. Rohde, R. W. Bielenberg, R. K. Faller, and D. L. Sicking. *Phase IV Development of a Short-Radius Guardrail for Intersecting Roadways*. Midwest Roadside Safety Facility, Lincoln, NE, 2008.
12. Stolle, C. S., D. L. Sicking, J. D. Reid, J. R. Rohde, K. A. Polivka, R. K. Faller, and R. W. Bielenberg. *Phase III Development of a Short-Radius Guardrail for Intersecting Roadways*. Midwest Roadside Safety Facility, Lincoln, NE, 2007.
13. Abu-Odeh, A. Y., K. M. McCaskey, R. P. Bligh, W. L. Menges, and D. L. Kuhn. *Crash Test and MASH TL-3 Evaluation of the TxDOT Short Radius Guardrail*. Texas A&M Transportation Institute, College Station, TX, 2015.

The Design Section (AFB00) peer-reviewed this paper (18-06773).

Article

TRR
JOURNAL OF THE TRANSPORTATION RESEARCH BOARD

Transportation Research Record
2018, Vol. 2672(39) 166–176

DOI: 10.1177/0361198118777379
journals.sagepub.com/home/trr
SAGE

American Complete Streets and Australian SmartRoads: What Can We Learn from Each Other?

Alexa Delbosc[1], James Reynolds[2], Wesley Marshall[3], and Andrew Wall[4]

Abstract

Road management in both Australasia and America has historically focused on facilitating vehicle movement and reducing congestion. More recently, however, there has been a shift to acknowledge the wider role that roads play in society. Road safety, equity impacts, considerations of "place" and the needs of different road users (including transit, pedestrians, and cyclists) are all gaining prominence. Two relatively new approaches to road design and management—Complete Streets in the United States and SmartRoads network operations planning in Australia—embody the spirit of this change. This paper summarizes the development of the Complete Streets movement in America and introduces the SmartRoads management framework, which was developed in the state of Victoria, Australia. In the SmartRoads process, roads within a network are classified in a multimodal Road User Hierarchy, network issues are identified using multimodal level of service measures, and possible operational or design solutions are compared using decision-making Network Fit Assessment software. We compare the scope, emphasis, and approach of the two frameworks; although they were developed at around the same time, the two approaches differ in significant ways. Yet the two approaches can learn from each other in order to significantly improve the management and design of roads in both Australia and the United States.

Road network management has historically focused on vehicle movement, congestion relief, and other traffic engineering issues. More recently, however, road authorities and cities have developed a greater appreciation of the role of streets as places for activity, pedestrians, bicycles, transit, and as part of the public realm and overall urban environment. This is evident in the growing use of Complete Streets frameworks in the United States and network operations planning (NOP) in Australasia. The VicRoads SmartRoads framework (*1*), the National Association of City Transportation Officials (NACTO) Urban Street Design Guide (*2*) and the UK Department for Transport's *Manual for Streets* (*3, 4*) provide examples of systems that have developed out of these new concepts.

To a certain extent, Australasian NOP and American Complete Streets approaches share a common ancestry. They have developed at a similar point in time and in response to the difficulty of capturing the diversity of street types and uses within a traditional movement vs. access road hierarchy. Authors of these new systems and researchers active in this area are likely to have been influenced by the *woonerf* (*5*) concept, the ideas of Hans Monderman (*6*) and other European views on shared space and road management, despite being firmly rooted in the vehicle-focused approaches typical in Australiasia and America. However, because the Complete Streets and NOP frameworks have developed separately, each has its own distinctive traits. It is not yet known what features they have in common and how the two approaches might learn from each other.

This research paper describes in detail a leading Australian approach to NOP, the VicRoads SmartRoads framework, which has gradually influenced NOP across many states of Australia and New Zealand. We use a qualitative synthesis approach to compare the key concepts and attributes of SmartRoads with the American

[1]Institute of Transport Studies, Department of Civil Engineering, Monash University, Clayton, Victoria, Australia
[2]Public Transport Research Group, Institute of Transport Studies, Department of Civil Engineering, Monash University, Clayton, Victoria, Australia
[3]College of Engineering and Applied Science, Department of Civil Engineering, University of Colorado, Denver, CO
[4]Integration Services, VicRoads, Kew, Victoria, Australia

Corresponding Author:
Address correspondence to Alexa Delbosc: alexa.delbosc@monash.edu

Complete Streets framework and identify aspects that might be missing from each. Although different countries will continue to have their own guides, frameworks, and approaches that reflect local conditions and directions, we will identify where the approaches to road management in North America might be improved by considering or integrating concepts from Australia, and vice versa. Many of the differences highlighted are likely the result, in part, of different cultural meanings and approaches to road use and management. This paper will take a fairly descriptive approach, in the hope that further work may delve deeper into the underlying reasons for these differences.

This paper is structured as follows: the next section describes the shared underpinnings of the Complete Streets and SmartRoads frameworks. This is followed by a summary of both approaches and a qualitative synthesis of how the approaches differ. The paper concludes with a discussion of opportunities for improvement to current practices, possible future research areas and a summary of the findings.

Complete Streets and Smartroads: A Shared Philosophy

Both Australia and the United States recently had a significant shift in approaches to road planning and traffic management. Previously, road design and traffic management focused on moving motorized vehicles (primarily cars and trucks) as efficiently as possible. In many—if not most—locations, this car focus was provided at the expense of on-road public transport, walking, or cycling infrastructure.

The mid-2000s saw the beginning of a shift in how road space was conceived and managed. In the United States in 2005, the National Complete Streets Coalition was launched under the premise that streets should accommodate all users, regardless of age, physical ability, or mode used. Meanwhile, in Australia in 2007 the Victorian state road authority (VicRoads) launched the SmartRoads NOP framework (*7*), which uses multimodal level of service (LOS) to manage the competing demands of different road uses across the network.

Although these approaches are by no means universally adopted across the United States or Australia, their use has expanded significantly in the last decade. Complete Streets policies have expanded from 31 agencies in 2005 to 1,232 agencies across the local, regional, or state level in 2016 (*8*). The SmartRoads framework was first applied in the Australian state of Victoria, and the underpinning principles of NOP have since been adopted into Australian national guidance (*9*). Within Australia, this framework is currently being adopted for use in the states of Queensland, South Australia, and Western Australia, although not all of these are based specifically on the SmartRoads frameworks. Additionally, NOP has been used in the cities of Auckland, Hamilton, Tauranga, Christchurch, and Wellington in New Zealand.

Although these two philosophies are inspired by a shared multimodal perspective, they differ in their scope, emphasis, and approach. The next two sections discuss the scope, emphasis and approach of Complete Streets and SmartRoads, in turn, in order to highlight similarities and differences.

Complete Streets

The components that underpin the Complete Streets concept (such as traffic calming, bike and pedestrian plans, and context-sensitive design) are not new; various components have been advocated and applied across the United States for decades. In 2003 the term "Complete Streets" was coined by the America Bikes organization to package together these principles under a single concept (*10*). Soon thereafter in 2005, the National Complete Streets Coalition was established. They define Complete Streets as "those designed and operated to enable safe access and travel for all users" (*11*). A wide range of design guides and resources have been released that support the adoption and implementation of Complete Street policies by governmental organizations in the USA. For example in 2007, both Louisville, KY, and Charlotte, NC, adopted street design manuals based on Complete Streets thinking (*10*). In 2010, the Institute of Transportation Engineers, in conjunction with the Congress for the New Urbanism, released an ITE-recommended practice titled *Designing Walkable Urban Thoroughfares*, which begins to integrate Complete Streets mindset with conventional design practice. Vega-Barachowitz et al. (*12*) also highlight the release of the NACTO *Urban Bikeway Design Guide* in 2011 and the development of its *Urban Street Design Guide* from 2012 as examples of how the Complete Streets approach has been incorporated into practice and reconciled with existing policies such as the AASHTO *Policy on Geometric Design of Highways and Streets* (the "Green Book").

The following subsections briefly describe the scope, emphasis, and approach of Complete Streets, drawing on a range of literature.

The Scope and Emphasis

The guiding philosophy of Complete Streets is to provide safe roadways for all users, regardless of age or physical ability (*11*). It grew out of decades of underinvestment in infrastructure for active modes, which contributes to thousands of deaths every year on America's roads (*13*).

These two components—*improving road safety* and considering *all road users*—form the core scope of Complete Streets policies.

Complete Streets policies also tend to have a strong equity component. Most vision statements mention that Complete Streets should be designed for people of all ages and abilities. A report by the Complete Streets Coalition emphasized that people of color, older adults, and low-income areas are overrepresented among pedestrian deaths, highlighting the need for change in how streets are managed to address systematic bias.

Although the wording of Complete Streets policies mentions "all road users," there is a tendency to focus on walking and cycling. To even consider a policy to be "Complete Streets," the national policy scorecard requires "walking and bicycling"; public transportation, freight, or other road users are not necessary but do count for additional points (*8*). For instance, a review of policies officially designated as Complete Streets in 2014 found that 23 out of 185 policies did not mention transit at all (*14*). Many provided no guidance for designing stops or traveled ways, and very few provided any transit-specific performance measures.

Some plans may include environmental impacts or placemaking, but these are not central to the Complete Streets philosophy. Although such components are encouraged, the Complete Streets Coalition does not include these elements in their scorecard (*8*).

The Approach

Complete Streets is largely a policy framework that helps transport organizations change their vision and emphasis. Although a commitment to Complete Streets is only considered a "first step," 33% of policies reviewed in 2016 did not include any implementation steps, and 83% did not include performance measures (*8*). There is little formal guidance on how best to implement Complete Streets policies. The Coalition suggests that organizations follow four key steps, including 1) revising procedures, 2) adopting more appropriate design guidelines, 3) education and training in Complete Streets, and 4) developing some measure of performance (*8*). Although including "at least one performance measure" is considered best practice, the type of measure is left to organizations to decide.

Although there is the common misconception that "completing the street" simply means adding street design components such as sidewalks and bike lanes, the core of the Complete Streets movement is not intended to be a prescriptive set of directives but rather a "frame of mind" (*8*, p. 4). Yet, the emphasis in Complete Streets policies remains on appropriate street design rather than operation. Adopting new design policies and guidelines is listed as an important step in implementation (*8*). There is a great deal of discussion about using the best and most recent design guidance, particularly for bicycle and pedestrian facilities, and using the Americans with Disabilities Act (ADA) standards for accessible design. Examples given of "typical elements" are street design characteristics such as sidewalks, bicycle lanes, bus lanes, pedestrian crossings, curb extensions, median islands, and accessible bus stops (*11*). Following this, examples of performance measures used to judge the success of Complete Streets policies tend to track changes to design, such as measuring the length of new and repaired sidewalks, new bike lanes, new and repaired curb ramps, crosswalk improvements, new transit stops, and the percentage of accessible transit stops (*8*).

SmartRoads

SmartRoads is the framework developed by VicRoads for managing current and future operations of Victoria's road network (*15*). It provides tools to manage competing road use priorities, manage congestion, assess level of service (LOS), and compare the performance of road improvement proposals. SmartRoads is a special case of Australian NOP, which has now been adopted in the Australian national guidelines (*9*).

SmartRoads was developed to respond to the significant challenges associated with the rapid and continual growth of Melbourne's population. Melbourne is projected to grow from 4.5 million in 2017 to almost 8 million by 2050, with the goal to provide an integrated transport system to connect jobs and housing (*16*). Yet Melbourne is also conscious of maintaining its listing among the world's most livable cities with a desire to improve environmental sustainability and support a strong sense of "place."

The Scope and Emphasis

SmartRoads draws its scope and emphasis from the Transport Integration Act 2010, the state of Victoria's principal transport statute. The Act sets the following six objectives for Victoria's transport system.

1. Social and economic inclusion;
2. Economic prosperity;
3. Environmental sustainability;
4. Integration of transport and land use;
5. Efficiency, coordination, and reliability; and
6. Safety, health, and wellbeing.

The primary emphasis of SmartRoads is to provide decision-making tools to improve the efficiency of the road network for all road users, balancing the needs of general traffic, freight, transit, cycling, and walking. The key component of this definition is that it is *multimodal* with a *network* perspective.

SmartRoads is largely a procedural tool used to:

- classify roads based on a multimodal Road User Hierarchy,
- prioritize the needs of different modes,
- develop options for road network improvements, and
- assess those options.

The software tools used to support SmartRoads allow decision makers to set out their desired network operation. Using multimodal LOS measures, they can experiment with alternative road designs and operational schemes to estimate the impact on all road users. In theory, this allows the road authority to select options that provide the greatest benefits to the network as a whole. Although SmartRoads can be used to assess different road designs, it is often used to assess operational changes such as changes to signal timing or transit signal priority.

This brief introduction provides the first key difference between SmartRoads and Complete Streets. Although both have a multimodal perspective, SmartRoads has more of a focus on improving road network operations. Road safety does not have an explicit consideration in the SmartRoads toolkit, although there is a perception that the use of SmartRoads implicitly improves safety outcomes by encouraging different modes to use different routes.

The Approach

SmartRoads provides an operational framework to tailor the operation of the road network to the needs and challenges of different precincts or corridors. The SmartRoads process is conducted using simple software called "Network Fit Assessment," which has been made freely available for any road authority to use upon request.

Figure 1 outlines the steps undertaken during the SmartRoads process, which is generally applied to a section of a road network centered around a road corridor, precinct, or activity center. Although the process in Victoria is overseen by the state government road authority, consultation with stakeholders is a key component, particularly in developing the Road User Hierarchy and brainstorming alternative treatment options. Stakeholders may include local government, transit operators, and special interest groups. The steps in the process are described in detail in the following sections.

Set Network Objectives. The first step of the SmartRoads process is to set network objectives. These are drawn from the six objectives for the Victorian transport system

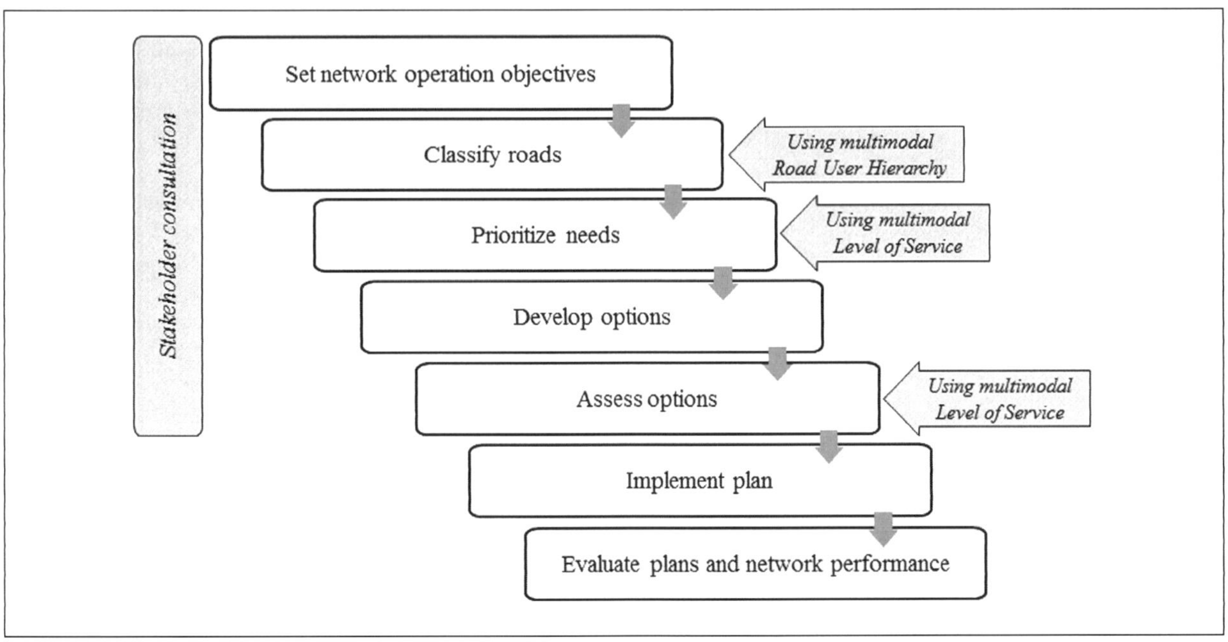

Figure 1. The SmartRoads operational framework. *Source*: Adapted from *Guide to Traffic Management Part 4: Network Management*, 2016 (9).

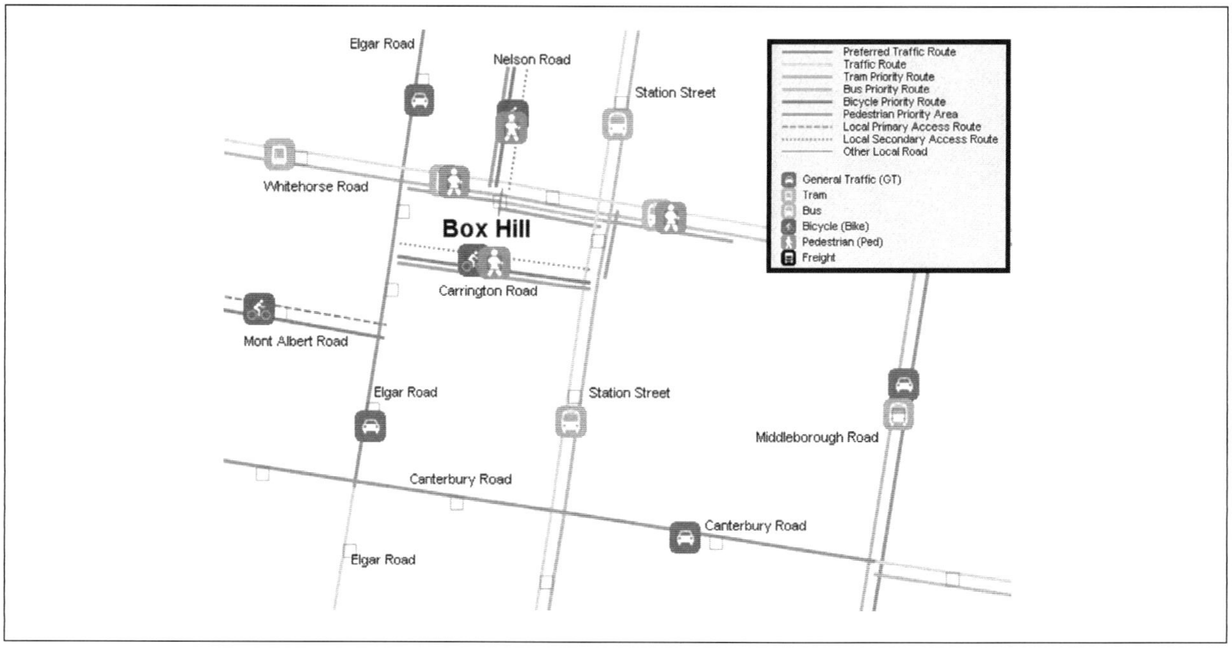

Figure 2. Example Road User Hierarchy for Box Hill activity center.

described above, but may be informed by local conditions and stakeholder needs within the area under consideration.

Classify Roads. After setting the general operational objectives for the network, the second step of the SmartRoads process is to classify the road network using a multimodal "Road User Hierarchy." In a Road User Hierarchy, different modes are emphasized in different roads within the network. Although SmartRoads uses the term "priority route" to describe routes for tram, bus, bicycle, or pedestrians, this does not necessarily mean that specific priority treatments, for example bus or bicycle lanes, are to be provided on that road segment. Rather, it implies that the needs of a given mode should be emphasized along that route.

Figure 2 provides an example of a simplified road network around an activity center in Melbourne (Box Hill). Bus (orange) and tram (green) are given primary emphasis to feed into the activity center, with cycling (purple) emphasized on separate routes and pedestrian activity (pink) emphasized within the activity center. Through-traffic (blue) is encouraged to pass around the center rather than through it.

As far as practicable, different modes should be emphasized on different roads to reduce conflict between modes. The Road User Hierarchy can vary by time of day; for example, through-traffic could be emphasized during peak hours, with local pedestrian traffic emphasized in off-peak times.

Prioritize Needs. Once roads have been classified using a Road User Hierarchy, needs can be identified based on the current performance of the road network. This performance is assessed using a multimodal LOS framework, summarized in Figure 3.

This LOS framework differs from the American *Highway Capacity Manual 2010* (*17*). It is a largely movement-based LOS framework, emphasizing the importance of reducing delay and travel time variability particularly for transit, freight, and car traffic. Pedestrian LOS is based on minimizing delay for crossing movement (rather than sidewalk design). Cycling LOS is the exception to this, as it is primarily assessed based on the safety of the infrastructure provided.

Using the Network Fit Assessment software, each segment of the Road User Hierarchy is assigned a LOS rating, which is weighted by the volume of each road user group. The outcome of the scoring is a numerical result that describes the weighted shortfall of the road with respect to the ideal and desired operational levels. The size of these operational gaps is represented graphically in the software program using pie charts (see Figure 4). Larger pies represent more significant delays, and the size of the segments within the pies indicates the road users who are experiencing the greatest shortfalls.

In the case shown in Figure 4, for example, the greatest operational shortfalls are experienced at the intersection of Station Street and Whitehorse Road, particularly at the south and west approaches. The majority of the delay is experienced by buses and pedestrians. Note that although general traffic and freight experiences LOS D,

Level of Service	Tram / Bus	Pedestrian	Bicycle	Freight	Car
A	No route delay. Always runs to timetable.	Opportunities to cross within 50m. Minimal crossing delay.	High degree of separation. Minimal delay.	No delay. No variability.	No delay. No variability.
B					
C	Stop at every set of signals. Within 5 min of timetable.	Crossing within 200m. Average crossing delay is 45 sec.	On-road bicycle lane.	Stop at every set of signals.	Stop at every set of signals.
D					
E	Takes at least 3 signal cycles to clear intersection.	Crossing within 400m. Average crossing delay is 90 sec.	Bicycles share traffic lanes.	Takes at least 3 signal cycles to clear intersection.	Takes at least 3 signal cycles to clear intersection.
F					

reducing performance

Figure 3. SmartRoads multimodal level of service framework.

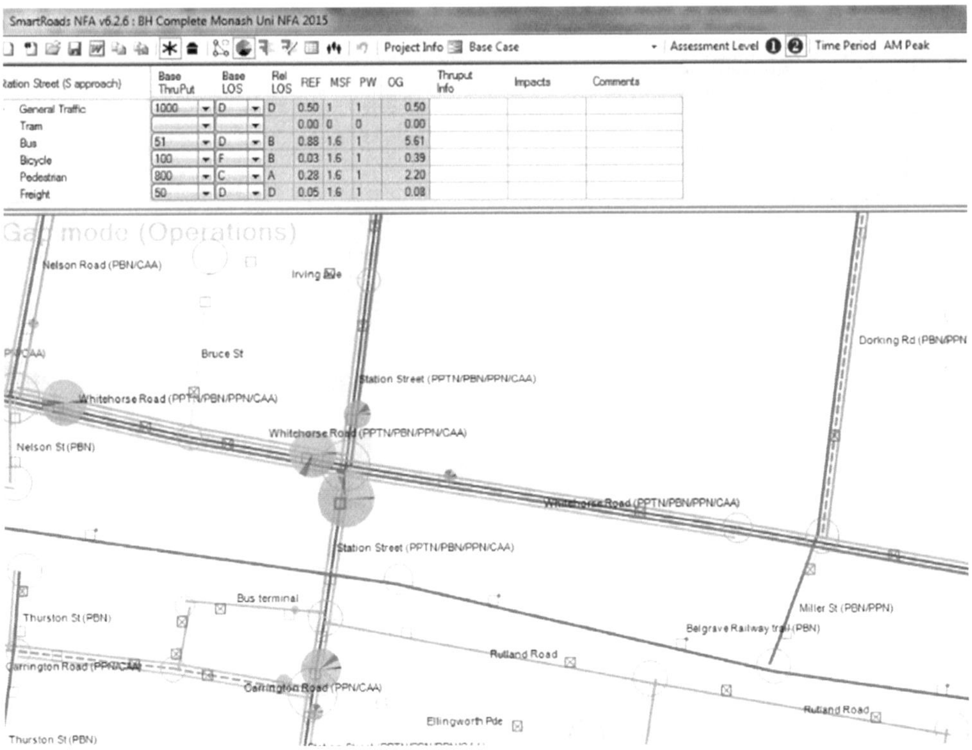

Figure 4. SmartRoads Network Fit Assessment software showing operational shortfalls for Box Hill.

as they are not assigned priority in the Road User Hierarchy for this intersection, the delay they experience is given little weight.

Develop Options. The operational gaps highlighted in the previous step help inform which options should be considered to improve the network performance. In the case shown in Figure 4, the operational gaps suggest that buses should be given greater priority on the south approach of Station Street. This could be achieved through operational solutions (e.g., shorter traffic signal timings or active signal priority) or road design solutions (a dedicated bus lane).

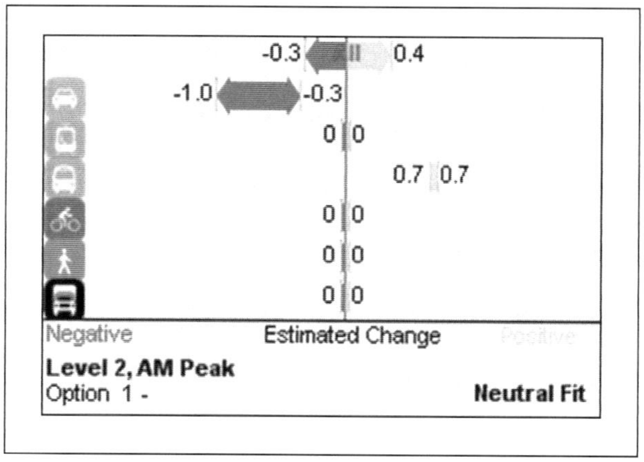

Figure 5. Example SmartRoads fit assessment.

While alternative options should be developed in consultation with relevant stakeholders and take into account the operation of the entire network and secondary impacts on other modes, the SmartRoads approach provides highly specific recommendations to network planning.

Assess Options. Once alternative proposals have been developed, their impact on network operation is then estimated for each link in the road network. The process involves identifying the change in LOS and throughput volumes for each mode on each link due to the proposed works. These estimates can be calculated using microsimulation software or professional judgment, depending on the time and resources available to decision makers. These LOS changes are input into the Network Fit Assessment software along with the level of confidence in these estimates.

The software produces an overall assessment of the impacts of the proposed works; an example is given in Figure 5. In this example, the proposed works will worsen the operational conditions for general traffic but improve the conditions for buses. The width of the bands indicates the degree of confidence in the estimates. In this case, the overall network fit is neutral with an overall score in the range of −0.3 to 0.4.

For smaller projects (such as adjusting signal timing), this process can quickly identify the overall effects of changes. For larger projects (such as major road redesigns), decision makers can use this process to shortlist network improvements for further investigation and cost–benefit analysis.

Case Studies of SmartRoads in Melbourne

SmartRoads can be applied to a range of road networks and projects, from minor signal changes to major investments such as new roads. It can also help clarify the overall priorities for different modes in different locations. The next section describes two examples of SmartRoads applications in Melbourne to illustrate the range of projects.

The Melbourne Central Business District

Melbourne's city center is serviced by a dense network of trams and two bus corridors, as well as a radial train network with stations that ring the city. The central city has seen incredible growth in job and housing density in recent decades. The resident population more than tripled from 11,800 in 2001 to over 37,000 in 2016 (*18*, *19*). The number of jobs in the central business district increased from 146,200 in 2001 to 220,800 in 2016 (*19*, *20*). This concentration of activity puts enormous pressure on the road network, including extreme pedestrian densities. It is not unusual to see sidewalks with more than 3,500 pedestrians an hour (*21*). In contrast, car access to the city is not prioritized; less than a quarter of workers use a car to access the city.

The Road User Hierarchy developed for Melbourne's city center reflected the priority provided to pedestrian and tram movements with some priority for cyclists and buses (see Figure 6). No roads provide major emphasis for through-traffic, as the high-level strategy for the city center is to encourage access by public transport and discourage through-traffic.

This exercise illustrated that the greatest issues in Melbourne's city center do not involve cars but trams and pedestrians. This helped to justify a range of design and operational changes, such as justifying protected bicycle lanes on Latrobe Street. It also meant signal timings support walking as first priority, trams second and then general traffic. This generally means the traffic signals have a maximum cycle time of 90 seconds even where this means much less efficient movement of car traffic.

The Dingley Bypass

The Dingley bypass provides a very different application of the SmartRoads framework. The bypass was one section of the Dingley Arterial Project, a long-term project to link major highway and freeway routes in Melbourne's south-eastern suburbs. It is located in a corridor with a number of major traffic routes, two important bus corridors and a limited cycling network. The proposed alignment of the segment connecting Springvale Road to Perry Road included four at-grade intersections (see Figure 7).

The Dingley Bypass was a traditional traffic engineering project with an approved scope and budget of AU$80

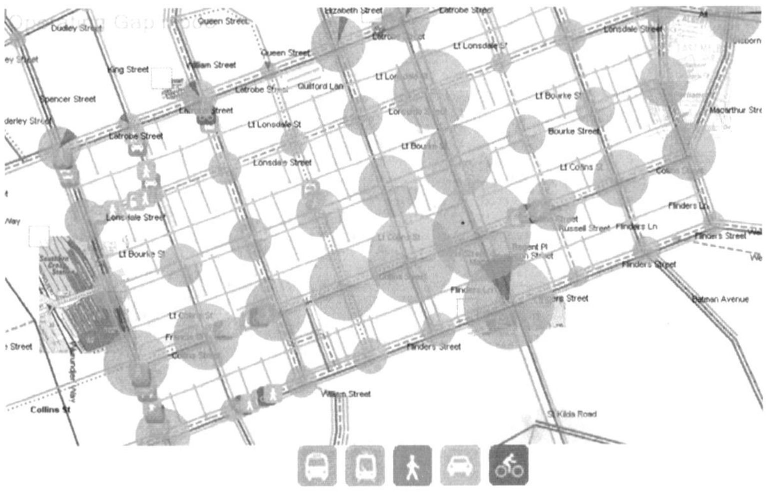

Figure 6. Road User Hierarchy and network operation gaps for Melbourne City Center.

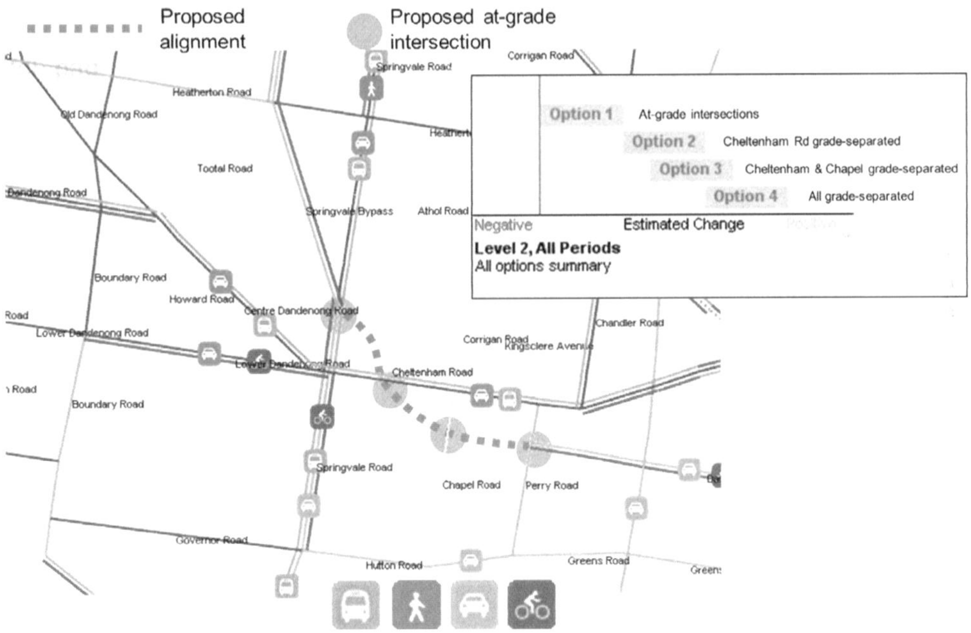

Figure 7. The Dingley corridor Road User Hierarchy and bypass proposal.

million. SmartRoads was used as a tool to enable the project team to challenge the wisdom of the traditional traffic analysis and develop a quite innovative solution. A Network Fit Assessment showed that the new intersection on Cheltenham Road would introduce additional delays along a bus priority corridor (see Figure 7). The SmartRoads analysis revealed significant risks that the original scope may not deliver any net transport benefit.

The same analysis demonstrated that a revised proposal, which grade-separated the intersection with Cheltenham Road (Option 2), would deliver both better transport outcomes but also more certainty around those outcomes. However, there was only a small additional improvement from also grade separating Chapel Road and other intersections (options 3 and 4), despite the significant additional costs of associated with these options.

Table 1. Synthesis Table Comparing Complete Streets and SmartRoads

	Complete streets	SmartRoads
Introduced	2004	2007
Adoption	Top-down implementation. Adopted in over 1,200 jurisdictions in United States	Bottom-up implementation. Adopted throughout Victoria. Adapted into Australian national guidance; currently being adapted in the states of South Australia and Queensland. Similar forms applied in parts of New Zealand
Core themes	Safety, equity, vulnerable road users	Multimodal movement, congestion management
Scope, emphasis, and objectives	Improving road safety, Considering all road users	Reducing conflict between modes, reducing congestion
Modes considered	Walking and cycling core modes. Most consider transit to some extent. Few consider other modes	Walking, cycling, public transport, freight, private vehicles
Safety considered?	Core consideration; often measured	Indirectly; not measured
Equity considered?	Directly or indirectly considered	Not considered
Place considered?	Depends on jurisdiction	Recently adopted through Link and place framework
Format	Policy framework	Operational management and decision-making tool
Approach	Revising procedures, adopting more appropriate design guidelines, education and training in Complete Streets, developing some measure of performance	Setting network operation objectives, classifying roads, prioritizing needs, developing options, assessing options, implementing plan, evaluating plans and network performance
Role of road design	Core component of framework	Considered alongside network operation
Role of road operation	Rarely mentioned	Core component of framework
Network considerations	Not always considered	Central to Road User Hierarchies
Performance indicators	Usually based on measuring improvements to road design; possibility of using multimodal LOS	Based on multimodal LOS assessment that is used to calculate a score describing the shortfall from ideal conditions

For this reason, Option 2 was ultimately selected and constructed on time and within the original fixed budget.

Comparing SmartRoads with Complete Streets

The previous sections have described the Complete Streets and SmartRoads approaches and provided case studies of how SmartRoads has been used in practice. This section compares and contrasts the two approaches directly. Table 1 shows a synthesis of the key characteristics of Complete Streets and SmartRoads.

Despite their almost concurrent periods of development and similar influences, there are many differences between the SmartRoads and Complete Streets approaches. Chief among these is the emphasis of SmartRoads as a tool for network operations assessment and improvement, while Complete Streets is more focused on the policy framework guiding street design.

Complete Streets had its formal origins at national level, through the establishment of the National Complete Streets Coalition, and has spread from the top-down to state and local jurisdictions. In contrast, SmartRoads was developed at the state level as a special case of the national NOP guidelines and is now having a bottom-up influence on the national guidelines and NOP approaches in other states.

The core themes of the two approaches significantly differ. The focus of Complete Streets is safety, equity, and vulnerable road users (with a strong emphasis on walking and cycling). In contrast, SmartRoads has a greater focus on congestion management and all modes, not just pedestrians and cyclists. Road safety considerations are not explicitly measured in SmartRoads, although they are considered implicitly. This may be due to the historical stand-alone road safety programs in Australia and New Zealand, such as the National Black Spot Program (*22*), that continue to be an important part of road management in Australasia. Equity is a growing consideration for Complete Streets policies but not an explicit consideration in SmartRoads. Consideration of the concept of place and placemaking is not yet a core component of either approach.

There are also significant differences between Complete Streets and SmartRoads in their format and approach. SmartRoads' origin as a tool to guide and assist road management is evident in the development of the structured Network Fit Assessment software. Because this software tool is freely provided on request, a jurisdiction anywhere in the world would be able to load its road network into the tool and start undertaking SmartRoads assessments with only minor modifications needed. In contrast, Complete Streets is a policy framework that provides top-down direction for jurisdictions, which must provide their own tools and methods. The emphasis is on revising procedures and design guidelines, education and training, and the development of jurisdiction-specific performance measures.

Complete Streets and SmartRoads differ considerably on the role of road design and operations. Street design is a key focus of the Complete Streets approach, and network operations are rarely considered in Complete Street policies. In contrast, the focus of SmartRoads is on road network operations, but the SmartRoads approach can also be used to guide road designs, particularly at the conceptual level.

SmartRoads has a key focus on impacts on the road network as a whole, not just on individual road sections or intersections. This is evidenced by the use of a network approach when assigning Road User Hierarchies, as well as the network-wide Network Fit Assessment process. The Complete Streets approach, in contrast, does not necessarily have such a wide emphasis, and network effects are not always considered.

Performance indicators in the Complete Streets approach vary between jurisdictions but tend to focus on design-related aspects. There is some evidence that American jurisdictions are adopting the latest *Highway Capacity Manual* multimodal LOS indicators, but this is by no means universal (*23*). In contrast, the SmartRoads approach is based entirely on using multimodal LOS to assess service shortfalls.

What Can Complete Streets and SmartRoads Learn from Each Other?

Road management in both Australasia and America has historically focused on facilitating vehicle movement, reducing congestion, and providing sufficient road capacity to meet demand. However, more recently there has been a wider appreciation that this is not enough to meet the needs of road users, modern cities, and society at large. The SmartRoads and Complete Streets approaches have developed concurrently in Victoria and the United States in response to the challenge of how to better manage road networks.

Having developed at similar times and from a similar need to move beyond vehicle-focused management and design of streets, it is remarkable how different the SmartRoads and Complete Streets approaches turned out to be. While many of these differences might be attributed to the different institutional, historical, and cultural settings, there are enough similarities between road management in Australasia and the United States that this is a significant opportunity to improve practices in both places.

Aspects of Complete Streets that might be adopted to improve SmartRoads include:

- the inclusion of road safety as an integral part of the approach,
- a greater emphasis on equity considerations, and
- a greater emphasis on providing a framework for jurisdictions to adopt SmartRoads, such as support to train staff and adopt updated design guidelines.

Aspects of SmartRoads that might be adopted to improve Complete Streets include:

- more of a focus on the road network, rather than just individual road sections,
- the inclusion of all modes, in particular public transport, as an integral part of policies,
- a more structured method for measuring the outcomes of Complete Streets policies, such as a multimodal LOS indicator, and
- the incorporation of both operational management and concept design assessment into the same process.

These recommendations echo the findings from recent studies and advocates. For example Babb and Watkins (*14*) argue that transit should be more actively considered in Complete Streets policies, and Hui et al. (*24*) point out that most Complete Streets frameworks do not provide a sufficient method for measuring the success of a scheme. Some have proposed multi-criteria outcome indicators [e.g., (*25*)] but it appears that in practice this approach is not commonly applied.

SmartRoads continues to be improved by VicRoads and a major update is currently under development. This is expected to more fully incorporate road safety as a core component of the approach, actively consider the "place" status of roads and introduce design guidance into the approach. Whether and how these new elements are adopted by other jurisdictions in Australasia remains to be seen.

Both of these approaches are in great need of further research. Although these frameworks have been used for over 10 years, there is remarkably little research to show whether they successfully deliver on their intended outcomes. Do Complete Streets policies actually improve road safety for pedestrians and cyclists? Do SmartRoads projects deliver better outcomes than traditional traffic engineering projects? Without this information, we cannot move beyond adopting what sounds good about each framework to adopting what actually works from each framework. Yet objectively measuring the effectiveness of these policies is a very challenging task (*24, 25*).

Practitioners in Australasia, America, and much of the world are all facing similar challenges transitioning from the previous vehicle-movement focus of traffic engineering to new ways of thinking about streets as existing for people and movement, and as places in their own right. Fortunately, Australasia and America have taken those initial steps in slightly different directions, and so can learn from each other's experiences and approaches. Hopefully comparisons between these different approaches, such as the one presented in this paper, can help practitioners learn from each other and thereby manage and develop street and road environments more successfully than might otherwise be the case.

Author Contributions

The authors confirm contribution to the paper as follows: study conception and design: A.D.; data collection: A.D., J.R., W.M.; analysis and interpretation of results: A.D., J.R., W.M.; draft manuscript preparation: A.D., J.R., W.M. All authors reviewed the results and approved the final version of the manuscript.

References

1. VicRoads. *SmartRoads*. December 2, 2016. https://www.vicroads.vic.gov.au/traffic-and-road-use/traffic-management/smartroads. Accessed May 3, 2017.
2. National Association of City Transportation Officials. *Urban Street Design Guide*. New York, NY, 2013.
3. Department for Transport. *Manual for Streets*. London, 2007.
4. Department for Transport. *Manual for Streets 2; Wider Application of the Principles*. Chartered Institution of Highways and Transportation, London, 2010.
5. Collarte, N. *The American Woonerf: Creating Livable and Attractive Shared Streets* (C. Cousineau, ed.), PhD dissertation. ProQuest Dissertations Publishing, 2014.
6. Vanderbilt, T. The Traffic Guru. *The Wilson Quarterly*, Vol. 32, No. 3, 2008, pp. 26–32.
7. Wall, A. Network Operation Planning: A New Approach to Managing Congestion. In *30th Australiasian Transport Research Forum (ATRF)*, Melbourne, Victoria, Australia, 2007.
8. National Complete Streets Association. *The Best Complete Streets Policies of 2016*. Smart Growth America, 2017.
9. Austroads. *Guide to Traffic Management Part 4: Network Management*, 2016.
10. Zehngebot, C., and R. Peiser. Complete Streets Come of Age: Learning from Boston and Other Innovators. *Planning Magazine*, May 2014.
11. Smith, R., S. Reed, and S. Baker. *Street Design: Part 1—Complete Streets*. Federal Highway Administration, 2010.
12. Vega-Barachowitz, D., P. Koonce, and M. Flynn. Changing the DNA of City Streets: NACTO's Urban Street Design Guide and the New City Street Design Paradigm. *ITE Journal*, Vol. 83, No. 12, 2013, pp. 36–40.
13. National Complete Streets Association. *Dangerous by Design 2016*. Smart Growth America, 2017.
14. Babb, A., and K. E. Watkins. Complete Streets Policies and Public Transit. *Transportation Research Record: Journal of the Transportation Research Board*, 2016. 2543: 14–24.
15. VicRoads. *SmartRoads Guidelines Version 1.17*. VicRoads, 2011, p. 65.
16. Victoria State Government. *Plan Melbourne 2017–2050*, 2017.
17. *Highway Capacity Manual 2010*. Transportation Research Board of the National Academies, Washington, D.C., 2010.
18. Australian Bureau of Statistics. *Census Community Profiles*, 2016.
19. Australian Bureau of Statistics. *Census Community Profiles*, 2001.
20. City of Melbourne. *City of Melbourne Census of Land Use and Employment Profile 2016*, 2016.
21. City of Melbourne. *Pedestrian Counting System*, 2016.
22. Australian Government Department of Infrastructure and Regional Development. *Black Spot Programme*. April 12, 2016. http://investment.infrastructure.gov.au/funding/blackspots/.
23. Carter, P., F. Martin, M. Núñez, S. Peters, L. Raykin, J. Salinas, and R. Milam. Complete Enough for Complete Streets? Sensitivity Testing of Multimodal Level of Service in the Highway Capacity Manual. *Transportation Research Record: Journal of the Transportation Research Board*, 2013. 2395: 31–40.
24. Hui, N., S. Saxe, M. Roorda, P. Hess, and E. J. Miller. Measuring the Completeness of Complete Streets. *Transport Reviews*, Vol. 38, No. 1, 2017, pp. 1–23.
25. Kingsbury, K., M. Lowry, and M. Dixon. What Makes a "Complete Street" Complete? A Robust Definition, Given Context and Public Input. *Transportation Research Record: Journal of the Transportation Research Board*, 2011. 2245: 103–110.

The Standing Committee on International Cooperation (A0010) peer-reviewed this paper (18-03662).

Sage Publications USA
sagepub.com

ISBN 978-1-5108-8247-8

CURRAN ASSOCIATES INC.
proceedings
.com

Travel Demand Forecasting 2018

Transportation Research Record: Journal of the Transportation Research Board Volume 2672, Issue 49

SAGE Publishing